BIOREMEDIATION AND PHYTOREMEDIATION TECHNOLOGIES IN SUSTAINABLE SOIL MANAGEMENT

Volume 2

Microbial Approaches and Recent Trends

Bioremediation and Phytoremediation Technologies in Sustainable Soil Management, 4 volume set:
ISBN: 978-1-77463-720-3 (hbk)
ISBN: 978-1-77491-100-6 (pbk)
ISBN: 978-1-00328-133-7 (ebk)

Volume 1: Fundamental Aspects and Contaminated Sites
ISBN: 978-1-77463-718-0 (hbk)
ISBN: 978-1-77463-958-0 (pbk)
ISBN: 978-1-00328-065-1 (ebk)

Volume 2: Microbial Approaches and Recent Trends
ISBN: 978-1-77463-719-7 (hbk)
ISBN: 978-1-77463-892-7 (pbk)
ISBN: 978-1-00328-068-2 (ebk)

Volume 3: Inventive Techniques, Research Methods, and Case Studies
ISBN: 978-1-77463-987-0 (hbk)
ISBN: 978-1-77463-988-7 (pbk)
ISBN: 978-1-00328-117-7 (ebk)

Volume 4: Degradation of Pesticides and Polychlorinated Biphenyls
ISBN: 978-1-77491-038-2 (hbk)
ISBN: 978-1-77491-039-9 (pbk)
ISBN: 978-1-00328-120-7 (ebk)

Innovations in Agricultural and Biological Engineering

BIOREMEDIATION AND PHYTOREMEDIATION TECHNOLOGIES IN SUSTAINABLE SOIL MANAGEMENT

Volume 2

Microbial Approaches and Recent Trends

Edited by

Junaid Ahmad Malik, PhD

Megh R. Goyal, PhD, PE

AAP APPLE ACADEMIC PRESS

First edition published 2023

Apple Academic Press Inc.
1265 Goldenrod Circle, NE,
Palm Bay, FL 32905 USA
4164 Lakeshore Road, Burlington,
ON, L7L 1A4 Canada

CRC Press
6000 Broken Sound Parkway NW,
Suite 300, Boca Raton, FL 33487-2742 USA
4 Park Square, Milton Park,
Abingdon, Oxon, OX14 4RN UK

Library and Archives Canada Cataloguing in Publication

Title: Bioremediation and phytoremediation technologies in sustainable soil management / edited by Junaid Ahmad Malik, PhD, Megh R. Goyal, PhD, PE, Khursheed Ahmad Wani, PhD.
Names: Malik, Junaid Ahmad, 1987- editor. | Goyal, Megh R., editor. | Wani, Khursheed Ahmad, editor.
Series: Innovations in agricultural and biological engineering.
Description: First edition. | Series statement: Innovations in agricultural and biological engineering | Includes bibliographical references and indexes. | Content: Volume 2: Microbial Approaches and Recent Trends.
Identifiers: Canadiana (print) 20220155194 | Canadiana (ebook) 20220155224 | ISBN 9781774637197 (v. 2 ; hardcover) | ISBN 9781774638927 (v. 2 ; softcover) | ISBN 9781003280682 (v. 2 ; ebook)
Subjects: LCSH: Soil remediation. | LCSH: Phytoremediation. | LCSH: Bioremediation.
Classification: LCC TD878.48 .B56 2022 | DDC 628.5/5—dc23

Library of Congress Cataloging-in-Publication Data

CIP data on file with US Library of Congress

ISBN: 978-1-77463-719-7 (hbk)
ISBN: 978-1-77463-892-7 (pbk)
ISBN: 978-1-00328-068-2 (ebk)

ABOUT THE BOOK SERIES: INNOVATIONS IN AGRICULTURAL AND BIOLOGICAL ENGINEERING

Under this book series, Apple Academic Press Inc. is publishing book volumes over a span of 8–10 years in the specialty areas defined by the American Society of Agricultural and Biological Engineers (www.asabe.org). Apple Academic Press Inc. aims to be a principal source of books in agricultural and biological engineering. We welcome book proposals from readers in areas of their expertise.

The mission of this series is to provide knowledge and techniques for agricultural and biological engineers (ABEs). The book series offers high-quality reference and academic content on agricultural and biological engineering (ABE) that is accessible to academicians, researchers, scientists, university faculty and university-level students, and professionals around the world.

Agricultural and biological engineers ensure that the world has the necessities of life, including safe and plentiful food, clean air and water, renewable fuel and energy, safe working conditions, and a healthy environment by employing knowledge and expertise of the sciences, both pure and applied, and engineering principles. Biological engineering applies engineering practices to problems and opportunities presented by living things and the natural environment in agriculture.

ABE embraces a variety of the following specialty areas (www.asabe.org): aquaculture engineering, biological engineering, energy, farm machinery and power engineering, food, and process engineering, forest engineering, information, and electrical technologies, soil, and water conservation engineering, natural resources engineering, nursery, and greenhouse engineering, safety, and health, and structures and environment.

For this book series, we welcome chapters on the following specialty areas (but not limited to):

1. Academia to industry to end-user loop in agricultural engineering.
2. Agricultural mechanization.
3. Aquaculture engineering.
4. Biological engineering in agriculture.

5. Biotechnology applications in agricultural engineering.
6. Energy source engineering.
7. Farm to fork technologies in agriculture.
8. Food and bioprocess engineering.
9. Forest engineering.
10. GPS and remote sensing potential in agricultural engineering.
11. Hill land agriculture.
12. Human factors in engineering.
13. Impact of global warming and climatic change on agriculture economy.
14. Information and electrical technologies.
15. Irrigation and drainage engineering.
16. Micro-irrigation engineering.
17. Milk Engineering.
18. Nanotechnology applications in agricultural engineering.
19. Natural resources engineering.
20. Nursery and greenhouse engineering.
21. Potential of phytochemicals from agricultural and wild plants for human health.
22. Power systems and machinery design.
23. Robot engineering and drones in agriculture.
24. Rural electrification.
25. Sanitary engineering.
26. Simulation and computer modeling.
27. Smart engineering applications in agriculture.
28. Soil and water engineering.
29. Structures and environment engineering.
30. Waste management and recycling.
31. Any other focus areas.

Books in the Innovations in Agricultural & Biological Engineering Series

- Advances in Food Process Engineering: Novel Processing, Preservation and Decontamination of Foods
- Advances in Green and Sustainable Nanomaterials: Applications in Energy, Biomedicine, Agriculture, and Environmental Science
- Biological and Chemical Hazards in Food and Food Products: Prevention, Practices, and Management

- Bioremediation and Phytoremediation Technologies in Sustainable Soil Management
    - Volume 1: Fundamental Aspects and Contaminated Sites
    - Volume 2: Microbial Approaches and Recent Trends
    - Volume 3: Inventive Techniques, Research Methods, and Case Studies
    - Volume 4: Degradation of Pesticides and Polychlorinated Biphenyls
- Dairy Engineering: Advanced Technologies and Their Applications
- Developing Technologies in Food Science: Status, Applications, and Challenges
- Emerging Technologies in Agricultural Engineering
- Engineering Interventions in Agricultural Processing
- Engineering Interventions in Foods and Plants
- Engineering Practices for Agricultural Production and Water Conservation: An Interdisciplinary Approach
- Engineering Practices for Management of Soil Salinity: Agricultural, Physiological, and Adaptive Approaches
- Engineering Practices for Milk Products: Dairyceuticals, Novel Technologies, and Quality
- Enzyme Inactivation in Food Processing: Technologies, Materials, and Applications
- Field Practices for Wastewater Use in Agriculture: Future Trends and Use of Biological Systems
- Flood Assessment: Modeling and Parameterization
- Food Engineering: Emerging Issues, Modeling, and Applications
- Food Process Engineering: Emerging Trends in Research and Their Applications
- Food Processing and Preservation Technology: Advances, Methods, and Applications
- Food Technology: Applied Research and Production Techniques
- Functional Dairy Ingredients and Nutraceuticals: Physicochemical, Technological, and Therapeutic Aspects
- Handbook of Research on Food Processing and Preservation Technologies:
    - Volume 1: Nonthermal and Innovative Food Processing Methods
    - Volume 2: Nonthermal Food Preservation and Novel Processing Strategies

  - Volume 3: Computer-Aided Food Processing and Quality Evaluation Techniques
  - Volume 4: Design and Development of Specific Foods, Packaging Systems, and Food Safety
  - Volume 5: Emerging Techniques for Food Processing, Quality, and Safety Assurance
- Modeling Methods and Practices in Soil and Water Engineering
- Nanotechnology Horizons in Food Process Engineering, 3-volume set:
  - Volume 1: Food Preservation, Food Packaging and Sustainable Agriculture
  - Volume 2: Scope, Biomaterials, and Human Health
  - Volume 3: Trends, Nanomaterials, and Food Delivery
- Nanotechnology and Nanomaterial Applications in Food, Health, and Biomedical Sciences
- Nanotechnology Applications in Agricultural and Bioprocess Engineering: Farm to Table
- Nanotechnology Applications in Dairy Science: Packaging, Processing, and Preservation
- Novel and Alternate Methods in Food Processing: Biotechnological, Physicochemical, and Mathematical Approaches
- Novel Dairy Processing Technologies: Techniques, Management, and Energy Conservation
- Novel Processing Methods for Plant-Based Health Foods: Extraction, Encapsulation and Health Benefits of Bioactive Compounds
- Novel Strategies to Improve Shelf-Life and Quality of Foods: Quality, Safety, and Health Aspects
- Processing of Fruits and Vegetables: From Farm to Fork
- Processing Technologies for Milk and Milk Products: Methods, Applications, and Energy Usage
- Quality Control in Fruit and Vegetable Processing: Methods and Strategies
- Scientific and Technical Terms in Bioengineering and Biological Engineering
- Soil and Water Engineering: Principles and Applications of Modeling
- Soil Salinity Management in Agriculture: Technological Advances and Applications

- State-of-the-Art Technologies in Food Science: Human Health, Emerging Issues and Specialty Topics
- Sustainable Biological Systems for Agriculture: Emerging Issues in Nanotechnology, Biofertilizers, Wastewater, and Farm Machines
- Sustainable Nanomaterials for Biomedical Engineering: Impacts, Challenges, and Future Prospects
- Sustainable Nanomaterials for Biosystems Engineering: Trends in Renewable Energy, Environment, and Agriculture
- Technological Interventions in Dairy Science: Innovative Approaches in Processing, Preservation, and Analysis of Milk Products
- Technological Interventions in Management of Irrigated Agriculture
- Technological Interventions in the Processing of Fruits and Vegetables
- Technological Processes for Marine Foods, from Water to Fork: Bioactive Compounds, Industrial Applications, and Genomics

OTHER BOOKS ON AGRICULTURAL AND BIOLOGICAL ENGINEERING FROM APPLE ACADEMIC PRESS, INC.

Management of Drip/Trickle or Micro Irrigation
Megh R. Goyal, PhD, PE, Senior Editor-in-Chief

Evapotranspiration: Principles and Applications for Water Management
Megh R. Goyal, PhD, PE, and Eric W. Harmsen, Editors

Book Series: Research Advances in Sustainable Micro Irrigation
Senior Editor-in-Chief: Megh R. Goyal, PhD, PE

Volume 1: Sustainable Micro Irrigation: Principles and Practices
Volume 2: Sustainable Practices in Surface and Subsurface Micro Irrigation
Volume 3: Sustainable Micro Irrigation Management for Trees and Vines
Volume 4: Management, Performance, and Applications of Micro Irrigation Systems
Volume 5: Applications of Furrow and Micro Irrigation in Arid and Semi-Arid Regions
Volume 6: Best Management Practices for Drip Irrigated Crops
Volume 7: Closed Circuit Micro Irrigation Design: Theory and Applications
Volume 8: Wastewater Management for Irrigation: Principles and Practices
Volume 9: Water and Fertigation Management in Micro Irrigation
Volume 10: Innovation in Micro Irrigation Technology

Book Series: Innovations and Challenges in Micro Irrigation
Senior Editor-in-Chief: Megh R. Goyal, PhD, PE

Volume 1: Management of Drip/Trickle or Micro Irrigation
Volume 2: Sustainable Micro Irrigation Design Systems for Agricultural Crops
Volume 3: Principles and Management of Clogging in Micro Irrigation
Volume 4: Performance Evaluation of Micro Irrigation Management

Volume 5: Potential Use of Solar Energy and Emerging Technologies in Micro Irrigation
Volume 6: Micro Irrigation Management: Technological Advances and Their Applications
Volume 7: Micro Irrigation Engineering for Horticultural Crops
Volume 8: Micro Irrigation Scheduling and Practices
Volume 9: Engineering Interventions in Sustainable Trickle Irrigation
Volume 10: Management Strategies for Water Use Efficiency and Micro Irrigated Crops
Volume 11: Fertigation Technologies In Micro Irrigation

ABOUT THE EDITORS

Junaid Ahmad Malik, PhD
Lecturer, Department of Zoology, Government Degree College, Bijbehara, Kashmir (J&K), India

Junaid Ahmad Malik, PhD, is a Lecturer with the Department of Zoology at Government Degree College, Bijbehara, Kashmir (J&K), India, and is actively involved with teaching and research activities. He has more than eight years of research experience. His areas of interest are ecology, soil macrofauna, wildlife biology, conservation biology, etc. Dr. Malik has published 21 research articles and technical papers in international peer-reviewed journals from Springer, Elsevier, and T&F, etc. He has authored three books, edited two books, 11 book chapters, and more than 10 popular editorial articles. He is also serving as editor and reviewer of several journals. He has participated in several state, national, and international conferences, seminars, workshops, and symposia, with more than 20 conference papers to his credit. He is a life member of SBBS (Society for Bioinformatics and Biological Sciences).

Dr. Malik received a BSc (2008) from the University of Kashmir, Srinagar, J&K; MSc (2010) in Zoology from Barkatullah University, Bhopal, Madhya Pradesh; and PhD (2015) in Zoology from the same university. He completed his BEd program in 2017 from the University of Kashmir, Srinagar, J&K. He started his career as a Lecturer in the School Education Department, Government of J&K, for two years.

Readers may contact him at: *malik.junaidahmad@gmail.com*.

Megh R. Goyal, PhD, PE

Retired Professor in Agricultural and Biomedical Engineering, University of Puerto Rico, Mayaguez Campus; Senior Acquisitions Editor, Biomedical Engineering and Agricultural Science, Apple Academic Press, Inc.

Megh R. Goyal, PhD, PE, is, currently a retired professor of agricultural and biomedical engineering from the General Engineering Department at the College of Engineering at the University of Puerto Rico–Mayaguez Campus (UPRM); and Senior Acquisitions Editor and Senior Technical Editor-in-Chief for Agricultural and Biomedical Engineering for Apple Academic Press Inc.

During his long career, he has worked as a Soil Conservation Inspector; Research Assistant at Haryana Agricultural University and Ohio State University; Research Agricultural Engineer/Professor at the Department of Agricultural Engineering of UPRM; and Professor of Agricultural and Biomedical Engineering in the General Engineering Department of UPRM. He spent a one-year sabbatical leave in 2002–2003 at the Biomedical Engineering Department of Florida International University, Miami, USA.

Dr. Goyal was the first agricultural engineer to receive the professional license in agricultural engineering from the College of Engineers and Surveyors of Puerto Rico. In 2005, he was proclaimed the "Father of Irrigation Engineering in Puerto Rico for the Twentieth Century" by the American Society of Agricultural and Biological Engineers, Puerto Rico Section, for his pioneering work on micro irrigation, evapotranspiration, agroclimatology, and soil and water engineering.

During his professional career of 52 years, he has received many awards, including Scientist of the Year, Membership Grand Prize for the American Society of Agricultural Engineers Campaign, Felix Castro Rodriguez Academic Excellence Award, Man of Drip Irrigation by the Mayor of Municipalities of Mayaguez/Caguas/Ponce and Senate/Secretary of Agriculture of ELA, Puerto Rico, and many others. He has been recognized as one of the experts "who rendered meritorious service for the development of [the] irrigation sector in India" by the Water Technology Centre of Tamil Nadu Agricultural University in Coimbatore, India, and ASABE who bestowed on him the 2018 Netafim Microirrigation Award. VDGOOD Professional Association of India awarded Lifetime Achievement Award at 12th Annual

Meeting on Engineering, Science and Medicine that was held on 20-21 of November of 2020 in Visakhapatnam, India.

Dr. Goyal has authored more than 200 journal articles and edited more than 100 books. AAP has published many of his books, including *Management of Drip/Trickle or Micro Irrigation; Evapotranspiration: Principles and Applications for Water Management;* ten-volume set on *Research Advances in Sustainable Micro Irrigation*. Dr. Goyal has also developed several book series with AAP, including Innovations in Agricultural & Biological Engineering, Innovations and Challenges in Micro Irrigation; and Innovations in Plant Science for Better Health: From Soil to Fork.

Dr. Goyal received his BSc degree in Engineering from Punjab Agricultural University, Ludhiana, India, and his MSc and PhD degrees from the Ohio State University, Columbus, Ohio, USA. He also earned a Master of Divinity degree from the Puerto Rico Evangelical Seminary, Hato Rey, Puerto Rico, USA.

Readers may contact him at goyalmegh@gmail.com.

CONTENTS

CONTRIBUTORS

Fatima Akmal
Assistant Professor, Institute of Soil and Environmental Sciences, University of Agriculture, Faisalabad – 38040, Pakistan, Mobile: +92-3336266280, E-mail: fatimaakmaluaf@gmail.com

Shahnaz Anjum
Junior Research Fellow, Department of Botany, University of Kashmir, Hazratbal, Srinagar – 190001, Jammu and Kashmir, India, Mobile: +91-9906608135, E-mail: shinaz.anjum@gmail.com

Wajiha Anum
Assistant Research Officer, Department of Agronomy, Regional Agricultural Research Institute, Bahawalpur – 63100, Pakistan, Mobile: +92-3366755784, E-mail: wajiha_anum@live.com

Asad-Ur-Rahman
Research Officer, Department of Pathology, Regional Agricultural Research Institute, Bahawalpur – 63100, Pakistan, Mobile: +92-3053987681, E-mail: sherani639@gmail.com

Tabinda Athar
Visiting Lecturer, Institute of Soil and Environmental Sciences, University of Agriculture, Faisalabad – 38040, Pakistan, Mobile: +92-3405459552, E-mail: athartabinda@gmail.com

Pritisha Barik
PhD Research Scholar, Department of Chemistry, Veer Surendra Sai University of Technology, Burla – 768018, Odisha, India, Mobile: +91-943936304, E-mail: pritishabarik@gmail.com

Chanda Parulekar Berde
Assistant Professor, Department of Marine Microbiology, SEOAS, Goa University, Goa – 403206, India, Mobile: +91-9860360635, E-mail: cvberde27@gmail.com

Vikrant B. Berde
Assistant Professor, Department of Zoology, Arts, Commerce, and Science College, Lanja – 416701, Ratnagiri, Maharashtra, India, Mobile: +91-7387766390, E-mail: berdevikrant31@gmail.com

Trinath Biswal
Associate Professor, Department of Chemistry, Veer Surendra Sai University of Technology, Burla – 768018, Odisha, India, Mobile: +91-943936304, E-mail: biswal.trinath@gmail.com

Pallaval Veera Bramhachari
Assistant Professor, Department of Biotechnology, Krishna University, Machlipatnam – 521001, Andhra Pradesh, India, Mobile: +91-8897019580, E-mail: veerabramha@gmail.com

Suraksha Chanotra
Lecturer, P.G. Department of Sericulture, Poonch Campus, University of Jammu, Jammu and Kashmir – 185101, India, Mobile: + 91-7780986497, E-mail: chanotrasuraksha@gmail.com

Natasha Erum
Assistant Professor, Department of Biochemistry, Faculty of Sciences, University of Agriculture, Faisalabad – 38040, Pakistan, Mobile: +92-3440868611, E-mail: natashaerum@yahoo.com

Subhosh Chandra Muni Ramanna Gari
Assistant Professor, Department of Microbiology, Yogi Vemana University, Vemanapuram, Kadapa – 516005, Andhra Pradesh, India, Mobile: +91-9493372743, E-mail: chandra298@gmail.com

Megh R. Goyal
Senior Editor-in-Chief (Agriculture and Biomedical Engineering) for AAP; and Retired Professor in Agricultural and Biomedical Engineering, University of Puerto Rico-Mayaguez, USA,
Mobile: 001-787-536-0039, E-mail: goyalmegh@gmail.com

Azad Gull
Senior Field Assistant, Silkworm Bivoltine Breeding Laboratory, Central Sericultural Research and Training Institute, Central Silk Board Srirampura, Manandavadi Road, Mysuru – 570008, Karnataka, India, Mobile: +91-7889932126, E-mail: azadgull72@gmail.com

Shazia Iqbal
PhD Research Scholar, Institute of Soil and Environmental Sciences, University of Agriculture, Faisalabad – 38040, Pakistan, Mobile: +92-3326006267, E-mail: shaziaiqbal91@yahoo.com

Naga Raju Maddela
Research Professor, Faculty of Health Sciences, Technical University of Manabí, Portoviejo – 130105, Manabí, Ecuador, Mobile: +593-0962921508, E-mails: mnraju8875@gmail.com; rmaddela@utm.edu.ec

Tariq Mehmood
Post-Doctorate Scholar, College of Environment, Hohai University, Nanjing – 210098, China, Mobile: +8613021112577, E-mail: tariqslamat@hhu.edu.cn

Junaid Ahmad Malik
Lecturer, Department of Zoology, Government Degree College, Bijbehara – 192124, Kashmir, Jammu and Kashmir, India, Mobile: +917006317311, E-mail: malik.junaidahmad@gmail.com

Srinivasulu Mandala
Academic Consultant, Department of Biotechnology, Yogi Vemana University, Kadapa – 516005, Andhra Pradesh, India, Mobile: +91-8333011076, E-mail: seenmahi452@gmail.com

Urfeya Mirza
PhD Scholar, FVSc & AH, SKUAST-K, Shuhama, Ganderbal – 191201, Jammu and Kashmir, India, Mobile: +91-9797807655, E-mail: urfeyamirza@gmail.com

Ghulam Murtaza
Professor, Institute of Soil and Environmental Sciences, University of Agriculture, Faisalabad – 38040, Pakistan, Mobile: +92-3219320153, E-mail: gmurtazauaf@gmail.com

Suresh Yadav Panyam
Academic Consultant, Department of Microbiology, Yogi Vemana University, Vemanapuram, Kadapa – 516005, Andhra Pradesh, India, Mobile: +91-8919763308, E-mail: suriabhi4@gmail.com

Shabir A. Rather
Postdoctoral Fellow, College of Life Sciences, Northwest A&F University, Yangling – 712100, PR China, +Mobile: 86-15529254050, E-mail: rathershabir100@gmail.com

Muhammad Umair Raza
Production Unit Manager, Lok Sanjh Foundation, Bahawalnagar, Pakistan, Mobile: +92-3336262879, E-mail: umair.muhammad240@gmail.com

Umair Riaz
Agricultural/Quality Control Officer, Soil and Water Testing Laboratory for Research, Bahawalpur – 63100, Department of Agriculture, Government of Punjab, Pakistan, Mobile: +92-3126208789, E-mails: umairbwp3@gmail.com; umair.riaz@uaf.edu.pk

Sahil
PhD Research Fellow, Department of Botany, University of Delhi, North Delhi – 110007, India, Mobile: +91-9899279575, E-mail: Sahilc95@gmail.com

Laila Shahzad
Assistant Professor, Sustainable Development Study Center, Government College University, Lahore – 54000, Punjab, Pakistan, Mobile: +92-3227878517, E-mail: lailashahzad@gcu.edu.pk

Faiza Sharif
Professor, Sustainable Development Study Center, Government College University, Lahore – 54000, Punjab, Pakistan, Mobile: +92-3214000818, E-mail: faizasharif@gcu.edu.pk

Kirti Sharma
Assistant Professor, M.K.M Group of Colleges, MDU, Near Bus Stand, Hodal, Palwal – 121006, Haryana, India, Mobile: +91-8860026556, E-mail: kirtisharma3294@gmail.com

Muhammad Irfan Sohail
Assistant Professor, Institute of Soil and Environmental Sciences, University of Agriculture, Faisalabad – 38040, Pakistan, Mobile: +92-3336723714, E-mail: irfansohail9@gmail.com

Khursheed Ahmad Wani
Lecturer, Department of Environmental Science, Government Degree College, Bijbehara – 192124, Kashmir, Jammu and Kashmir, India, Mobile: +919682169453, E-mail: kuurshidevs@gmail.com

Shayista Yousuf
Junior Research Fellow, Department of Zoology, University of Kashmir, Hazratbal, Srinagar – 190001, Jammu and Kashmir, India, Mobile: +91-8082028584, E-mail: shayistayousuf806@gmail.com

Rayees Ahmad Zargar
Assistant Professor, Baba Ghulam Shah Badshah University, Rajouri – 185234, Jammu and Kashmir, India, Mobile: +91-9149502050, E-mail: rayeesphy12@gmail.com

ABBREVIATIONS

3,4-DCA	3,4-dichloroaniline
AA	aromatic amines
ABA	abscisic acid
ABTS	2,2'-azino-bis(3-ethylbenzothiazoline)-6-sulfonic acid
AC	activated carbon
ACC	1-aminocyclopropane-1-carboxylate
ACCD	1-aminocyclopropane-1-carboxylate deaminase
Al	aluminum
AMF	arbuscular mycorrhizal fungi
APU	amphiphilic polyurethane
As	arsenic
Au	gold
BaP	benzo[a]pyrene
BHC	benzene hexachloride
BOD	biological oxygen demand
BSC	biological soil crusts
BTEX	benzene, toluene, ethylbenzene, and xylene
C	carbon
CAM	crassulacean acid metabolism
CAX	cation proton exchanger
Cd	cadmium
CDF	cation diffusion facilitator
cDNA	complementary DNA
CE	chloroethane
CEC	cation exchange capacity
CF	chloroform
Cl	chlorine
CM	chloromethane
CMC	carboxy-methyl-cellulose
CNFs	carbon nanofibers
CNTs	carbon nanotubes
Co	cobalt
CO_2	carbon dioxide
COD	chemical oxygen demand

CP	calcium peroxide
Cr	chromium
CT	carbon tetrachloride
Cu	copper
DDT	dichlorodiphenyltrichloroethane
DNT	dinitrotoluene
DO	dissolved oxygen
DTPA	minepentaacetic acid
E^0	standard potential
EC	environmental conditions
ECS	glutamylcysteine synthetase
EDDS	S-ethylenediaminedisuccinic acid
EDTA	ethylene diamine tetra-acetic acid
EGTA	ethylene glycol tetra-acetic acid
EIL	economic injury level
EK	electrokinetic
EPA	Environmental Protection Agency
EPR	electrochemical potentiokinetic reactivation
EPS	extracellular polysaccharides/exopolysaccharides
ETPs	effluent treatment plants
Fe	iron
FeS	iron sulfide
GEM	genetically engineered microorganisms
GHG	greenhouse gas
gL^{-1}	gram per liter
GS	glutathione synthesis
GSH	glutathione
gshI	glutathione synthetase I
GST	glutathione S-transferases
GTN	glycerol trinitrate
H_2O_2	hydrogen peroxide
HBT	1-hydroxybenzotriazole
HEDTA	hydroxyethylene diamine triacetic acid
Hg	mercury
HM	heavy metal
HMA	heavy metal ATPase
IAS	*in situ* air sparging
IRT1	iron-regulated transporter 1
K	potassium

kDa	kilo Daltons
LAB	lactic acid bacteria
LAC	laccases
LiP	lignin peroxidase
LMEs	lignin modifying enzymes
LNAPLs	light non-aqueous phase liquids
LSF	life support functions
M	molar
MAMP	microbe-associated molecular pattern
MAR	multiple antibiotic resistance
MB	methylene blue
MF	microfiltration
Mg	magnesium
MIC	minimum inhibitory concentration
Mn	manganese
MnO_2	manganese oxide
MnP	manganese peroxidase
mRNA	messenger RNA
MSW	municipal solid waste
MTBE	methyl tertiary butyl ether
MWCNTs	multi-walled carbon nanotubes
N	nitrogen
N_2	dinitrogen
NA	natural attenuation
NAT2	N-acetyltransferase 2
NH_3	ammonia
NH_4^+	ammonium
Ni	nickel
NO_2	nitrogen dioxide
NPs	nanoparticles
NRAMP	natural resistance-associated macrophage protein
NRC	National Research Council
NTA	nitrilotriacetic acid
nZVI	nano-scale zero-valent iron
OFMSW	organic component of MSW
ORC	oxygen releasing compound
P	phosphorous
P450	cytochrome P450 monooxygenases
PAH	polycyclic aromatic hydrocarbon

PAL	phenylalanine ammonia-lyase
PAMAM	poly(amidoamine) or dendrimers
Pb	lead
PCB	polychlorinated biphenyl
PCDDs	polychlorinated dibenzo-dioxins
PCDFs	polychlorinated dibenzofurans
PCE	pentachloroethane/tetrachloroethylene
PCP	pentachlorophenol
PCP	phenylcyclohexyl piperidine
PCR	polymerase chain reaction
PCS	phytochelatin synthase
PGPR	plant growth-promoting rhizobacteria
pH	potential of hydrogen
PHB	polyhydroxybutyrate
PHC	petroleum hydrocarbons
PNCs	polymer nanocomposites
POPs	persistent organic pollutants
POX	peroxidase
PPCPs	pharmaceuticals and personal care products
ppm	parts per million
PPO	polyphenol oxidase
PR	pathogenesis-related
PRB	permeable reactive barrier
PTEs	potentially toxic elements
rDNA	ribosomal DNA
RDX	royal demolition explosive/organic compound-$(O_2NNCH_2)_3$
rGO-Ag	graphene oxide silver nanoparticles
ROS	reactive oxygen species
rRNA	ribosomal RNA
S	sulfur
SDS	sodium dodecyl sulfate
Se	selenium
SOD	superoxide dismutase
Sp.	species
St.	strain
SVE	soil vapor extraction
SWCNTs	single-walled carbon nanotubes
Tc	technetium
TCA	1,1,1-trichloroethane

TCE	trichloroethene
TF	translocation factor
TNT	trinitrotoluene
TPHs	total petroleum hydrocarbons
TrOCs	trace organic contaminants
U	uranium
U.S.	United States
UF	ultrafiltration
UPOs	unspecific peroxygenases
USA	United States of America
VACV	vaccinia virus
VC	vinyl chloride
VOC	volatile organic compounds
VP	versatile peroxidase
VSV	vesicular stomatitis virus
YCF1	yeast cadmium factor protein
ZIP	zinc-iron permease
Zn	zinc
ZVI	zero-valent iron

FOREWORD

Münir Öztürk, PhD, DSc, FIAS
Vice President of the Islamic World Academy of Sciences, Prof. (Emer.) of Ecology and Environmental Sciences, Consultant Fellow, Faculty of Forestry, University Putra Malaysia, Malaysia; Distinguished Visiting Scientist, ICCBS, Karachi University, Pakistan
Books: http://www.amazon.com/-/e/B00JFW5DS8

Soil, for a layman, is the homogenous substance often known as mud or dirt or soil. Live soils are not only the roots of our food system but, in fact, of our civilization. Knowing that our food is from the soils, it becomes imperative that we keep the soil healthy.

The increased population size has led to the need for increased food production and chemical compounds that use huge amounts of water. In addition to residential and agricultural pollution, manufacturing waste gasses and the processing of toxic coal comprising heavy metals (HMs) have risen and are being pumped through the bodies of water and thereby into the nearby land. The excessive input of hazardous waste ends up polluting our soils, air, and waters.

The bioremediation and phytoremediation issues discussed in this book highlight the biological agents used to clean contaminated ecosystems. The latest approaches come as positive forms to the edge. Many of these approaches include standardization as a practical method for researching plants in regions polluted by heavy metals. Likewise, the usage of different microbes and other species along with microalgae is recommended for remediation of polluted natural resources. The soils are part of the terrestrial ecosystem and require complete protection of their quality against cycles of depletion.

This edited volume 2 entitled *Bioremediation and Phytoremediation Technologies in Sustainable Soil Management: Microbial Approaches and Recent Trends* covers all these aspects within 12 chapters grouped under three parts, mostly written by researchers from India.

Part I include the titles related to "Rhizoremediation and Archaea." There are two chapters in this section, namely: "Role of Rhizobacteria in Phytoremediation of Metal Contaminated Lands" and "Role of Archaea in the Remediation of Contaminated Soils." Chapter 1, prepared by researchers from Pakistan, discusses the role of rhizobacteria in phytoremediation of metal-impacted lands. The authors review the processes through which rhizobacteria transfer and make the metals available to the plants. They also discuss the need to isolate the selected bacterial isolates from the metal-polluted rhizosphere of higher plants for the further improvement of phytoremediation projects. Chapter 2 presented here includes information on the role of archaea in the remediation of contaminated soils. The aim here has been to provide an overview of the role of Archaea in bioremediation with information on their distribution, bioremediation mechanisms, applications in bioremediation, and future research possibilities. It provides an overall view of Archaea as key players in the remediation process.

Part II has been assembled under the title "Role of Microalgae and Fungi in Bioremediation" and includes four chapters. This part starts with Chapter 3, titled "Microalgae and Cyanobacteria: Potential Biofertilizers," which discusses the role of micro-algae as biofertilizers for sustainable plant and soil health and argues about the importance of microalgae for soil and plant sustenance and their utilization as a biofertilizer in substitution of inorganic fertilizers. This is followed by Chapter 4, titled "Microalgae as Biofertilizer for Sustainable Agricultural Productivity," which focuses on the utility of various microalgal species as substitutes for chemical fertilizers in the biofertilizer industry. A group of authors from India and Ecuador have prepared Chapter 5, which presents an evaluation of the fungal enzymes in soil bioremediation with a microbial biotechnological perspective. They have discussed the main principle in the degradation and transformation of pollutants, carried out in an enzymatic way. Different methods and strategies applied in the area at the global level have been discussed, including stimulation, bioaugmentation, bioventing, biopiles, and bioattenuation with their advantages and disadvantage. The chapter "Soil Bioremediation Strategies Based on Fungal Enzymes" has been written by authors from Pakistan in Chapter 6. They have presented details on the importance of fungal enzymes with respect to soil bioremediation techniques together with advantages and limitations in their use as well as the prospects for future research.

Part III has been brought together under the heading "Recent Trends and Approaches in Bioremediation" and includes six chapters. This part starts with Chapter 7, titled "Assessment of Current Soil Bioremediation

Technologies." The chapter presents an account of some modern bioremediation techniques adopted for effective bioremediation, such as electrokinetic (EK) process, rhizoremediation, smart bioremediation technology, metagenomics, and biofilm in bioremediation. It stresses that nanotechnology used in the process of bioremediation makes it eco-friendly and cost-effective. Chapter 8, prepared by researchers from Pakistan, presents information on "Sustainable Agroecosystems: Recent Trends and Approaches in Phytoremediation and Rhizoremediation." It gives an insight into the concepts of phytoremediation and rhizoremediation, their mode of action, recent trends and practices for enhancing soil health, and their suitability in adaptation. The chapter further discusses the headings like prevailing issues in the sustainability of agroecosystems, how the world is coping with sustainability problem, introduction to phytoremediation and rhizoremediation together with advantages and limitations.

In Chapter 9, recent trends in bioremediation tools for sustainable soil health and crop improvement have been pooled together. The authors have prepared a comprehensive review on various sources of bioremediation agents and their limitations in treating pollutants present in the environment, which include bioremediation, bioattenuation, biostimulation, bioaugmentation, PGPR as bioremediators, and bioremediation with microalgae. In Chapter 10, the topic reviewed is microbial indicators of soil health for the assessment of remediation efficiency. It introduces the microbial indicators and their significance for improving soil health, together with their bioremediation potential. Topic 11 deals with the transgenic plants for enhanced phytoremediation. It discusses the use of plants for detoxification and volatilization of organic and inorganic pollutants from contaminated soils, sediments, air, and groundwater. The efficacy of transgenic plants for cleaning various soil contaminants has been highlighted, and their uses debated for environmental clean-up.

Rayees Ahmad Zargar, Khursheed Ahmad Wani, and Junaid Ahmad Malik have prepared the title "Nanotechnology and Nanobioremediation: An Integrated Approach for Management of Contaminated Sites."

The authors have discussed this interesting and latest technology topic saying that the utilization of nanotechnology and nanocomposites emerge as a promising option with strategies for soil nanoremediation. The role of nanotechnology has been highlighted and discussed at length in this chapter as a suitable technology for preventing soil pollution from the potential wastes and chemicals, especially heavy metals.

This volume 2 is a continuum of the 4-volume set *Bioremediation and Phytoremediation Technologies in Sustainable Soil Management,* published within the series Innovations in Agricultural and Biological Engineering is an unique and ideal source of scientific information for this research scholars, faculty members, and scientists involved in the studies on environmental aspects. It will prove highly beneficial for environmental engineering students.

I deeply appreciate the contributions of the editors Junaid Ahmad Malik and Megh R. Goyal in getting this amazing volume together successfully.

PREFACE

Soil is a dynamic ecosystem as all the soil parameters (like pH, temperature, nutrient availability, etc.) fluctuate very rapidly, and it is contaminated very easily by different types of pollutants. The different types of pollutants (like pesticides, heavy metals (HMs), aromatic hydrocarbons, etc.) have contaminated the soils to a dangerous level as today our foods contain traces of these pollutants and have also been detected in human blood, urine, and saliva. Sometimes, a single pollutant has a capacity to fluctuate the physical, chemical, and biological properties of the soil to a level that is quite harmful to the whole soil ecosystem. As such there is a need to prevent soil contamination, and if already contaminated, urgent measures need to be taken to remediate the soil. The different approaches to remediate the contaminated soil are to degrade, transform, and detoxify the pollutants to the levels that are harmless to plants as well as animals.

The Romans were the first who are considered to implement bioremediation around 600 B.C. However, the official credit was given to George Robinson in the 1960s when he experimented with microbes present inside a glass jar, and in 1972, real experiments of bioremediation started almost in every part of the world. Since then, the science of bioremediation and phytoremediation has progressed by leaps and bounds, and it has recovered contaminated sites to a level that are quite harmless and now used for various purposes, however initially deemed unfit for the same. Scientists are now in a position to use nanoparticles (NPs) to remediate contaminated soils, which is an example of advances in the science of bioremediation and phytoremediation.

The volume II of the present book has been categorized under three subsections; Rhizoremediation and Archae, Role of Microalgae and Fungi in Bioremediation; and Recent Trends and Approaches in Bioremediation. These three parts comprise 12 chapters that cover the role of rhizobacteria; the role of archaea, microalgae, and cyanobacteria; microalgae as biofertilizers; microbial biotechnology; soil bioremediation; assessment of soil bioremediation technologies; sustainable ecosystems; and sustainable soil health and crop improvement through bioremediation and phytoremediation. Each chapter provides in-depth coverage of the mechanisms, advantages, and disadvantages of the technologies used. The book highlights the use of different microbial enzymes that are used in the process of bioremediation and phytoremediation to clean up different pollutants without causing any damage to the natural environment. This book brings diverse topics together, representing the novel resource for researchers to solve problems in bioremediation.

This book is intended for academicians, environmentalists, practitioners, NGOs, and industrialists who are working in the field of bioremediation and phytoremediation. There is vast information to aid researchers in continue their research beyond the technologies that are currently available; as such, the information has been presented in every chapter under a separate heading, "Future Strategies." The research gaps identified in this section will open new avenues for future research. We hope this volume will meet the demands of researchers all over the world, and we welcome the suggestions from all the readers to be incorporated in our next volume.

— *Editors*

PART I
Rhizoremediation and Archaea

CHAPTER 1

ROLE OF RHIZOBACTERIA IN PHYTOREMEDIATION OF METAL CONTAMINATED LANDS

UMAIR RIAZ, TARIQ MEHMOOD, MUHAMMAD IRFAN SOHAIL, FATIMA AKMAL, and NATASHA ERUM

ABSTRACT

Rhizobacteria having great potential for metal adaption are getting more research interest for improving the phytoremediation methods. This chapter focuses on some advanced techniques to growing higher plants having hyper-accumulating ability. After the discovery of hyperaccumulating plants, phytoextraction is also getting more attention. This study also reviews the processes, through which rhizobacteria transfer and makes the metals available to the plants. Moreover, research is needed to isolate the selected bacterial isolates from the metal-polluted rhizosphere of higher plants for the further improvement of phytoremediation projects.

1.1 INTRODUCTION

The modern developments in industrial processing and advanced material synthesis, population explosions as well as alternative fuel consumptions for meeting energy requirements have intensely deteriorated the environment [67]. Among the various other environmental consequences of development in the modern era, the heavy metal (HM) pollution in the soil is becoming more and more severe over time. Specific activities including extensive mining, intensive agricultural practices, sewage water irrigation, application of fertilizers and metal-enriched feed to livestock have either directly or indirectly became a major source of pollution in agricultural soils with a wide range of contaminants [91].

HM or trace elements is a term alternatively used for toxic metals [62]. These elements have both natural as well as anthropogenic sources. The natural sources include: volcano eruption, weathering of parent material, groundwater, forest fire, etc. Mineral ores like the dissolution process of mineral ores arsenopyrite, cerussite, cassiterite, and galena also release several metals in the environment that remained as their constituents in normal circumstances. Acid rains, dust storm is also included in natural routes of various pollutants including HMs in the environment [1]. However, a substantial amount of these toxic metals originates from human activities known as anthropogenic sources [67].

Agricultural practices (such as fertilizer application, sewage water irrigation, pesticides, and fungicides) are major contributors to HM pollution in soil. Similarly, construction activities, paints, and other building material, industries, smelting, and/or metallurgy coating, textile processing, tanneries are also producing excessive HMs [92]. HMs are also generated from the transport industry, such as heavy vehicle transportation activities and emission [67].

These metals are constituents of rocks and remain non-hazardous until they remain adsorbed or in unavailable forms. However, several geochemical processes and human activities make them in available forms and thus become an environmental hazard. Living organisms require several essential metals for performing their metabolic activities. Generally, these metals are recognized as micronutrients and/or essential elements, e.g., manganese (Mn), chromium cobalt, iron, zinc, selenium, and copper, etc. The deficiencies of these essential nutrients result in numerous health problems, even life-threatening situations, and excess of any of these elements could change the normal physiological function of the body. Other metals, such as Sb, As, Ti, Pb, Ba, Ag, Cd, and Hg are non-essential metals and have the potential to enter into the food chain, and can biomagnify in the animal bodies and can lead to severe health effects [1].

Among these metals, As, Zn, Cd, Pb, Hg, and Cr are more toxic and have been widely reported as potential toxic soil contaminants. These metals cannot be degraded naturally but can transfer from one to another form by changing the oxidation state and or redistributed into other organic complexes [11]. In urban lands, plants can directly uptake these metals from the soil via their roots, and the metals get deposited in the shoot and leaves by wet and dry deposition from atmospheric dust and particulate matter [38, 66].

The treatment of metals in soil is a challenging task, where several strategies have been proposed for optimum treatment of contamination of these metals. For instance, landfill, soil excavation, electroreclamation, acid wash, and thermal treatment are effective but not conveniently applicable. High cost, land degradation, damaging the soil structure and properties, dangerous to soil macro and microflora, low efficiency, and non-sustainable are major drawbacks of these strategies [36]. The past 30 years have seen increasingly rapid advances in the field of phytoremediation.

Several plants (known as hyperaccumulators) have an excellent affinity to uptake these metals and store in above-ground biomass. Some other specific characteristics of these plants include: high biomass productions, tolerate high polluted environments and have a short-grown period [65]. Phytoremediation is *in-situ*, low cost, ecologically benevolent, intrusive, sustainable, green, and publicly as well as esthetically successful strategy; for which it is mandatory to make these metals more bioavailable for plant uptakes. The plant-rhizobacteria interactions have been well known to be promising for efficient and effective phytoremediation.

Rhizobacteria play a crucial role in the detoxification of pollutants, such as make them less toxic and more bioavailable via several biological, chemical, and physical processes. The incorporation of plants and microbes can enhance soil fertility, preserve biodiversity, improve nutrient cycling; and has several advantages over conventional remediation practices [36]. However, the behavior of rhizobacteria and plants and their functional interactions under different metals and contamination level are needed to be studied in detail.

This chapter attempts to highlight the global concern of metal pollution and provide insights for rhizobacteria-assisted phytoremediation approach. Phytoremediation processes along with the potential and scope of rhizobacteria in HM treatment are discussed. Moreover, several interactions between plants, microbes, and metals in soil are also discussed.

1.2 METAL CONTAMINATION: A GLOBAL THREAT

Metal contamination refers to the excessiveness of special metallic properties bearing elements, whose atomic mass is >20. Predominantly As, Zn, Cd, Pb, Hg, and Cr are known major environmental hazards [11, 65]. Several modern practices (such as urbanization, the industrial revolution, population explosion, and intensified agriculture, food processing, etc.), have led to an increase in HM pollution around the globe [67]. The elevated level of HMs in soil and water has continually degraded the quality of these mediums and make them more toxic, unfit, and dangerous for living bodies. Besides, HM continuously enters the food chain from polluted water and soil; and makes food consumption of such food potentially unsafe for animal and human health. Thus, HM contamination is increasingly recognized as a serious, worldwide public health concern [43].

Recent reports have indicated that >10 million sites are potentially contaminated and more than 50% of these have elevated metal pollution levels worldwide. The distribution of these HM pollution sites showed that the USA (United States of America) has the highest contaminated site where more than 100,000 sites reported have been reported as contaminated and from these sites >70% have elevated metals pollution. Europe comes next to the USA (80,000 sites) followed by Australia (50,000 sites) polluted sites and the corresponding metal pollution sites are 37 and 60%, respectively [42].

The metal impacted lands are more in developed countries due to more industries, mining, and agriculture practices. Several other topographical characteristics are also a major reason for excessive land pollution in these areas. The metal-enriched rocks (e.g., black shale and serpentine) are also inherent sources of toxic metals in these metal-impacted lands. HM contamination is a global concern and needs joint efforts of researchers, policymakers, and government at local, regional, national, and global levels.

1.3 PHYTOREMEDIATION TECHNOLOGIES FOR COMBATING THE CONTAMINATION SITES

Phytoremediation approach is easy, economical, and ecologically pleasant with the least environmental disturbance [23]. It is also advantageous from an agricultural point of view, since it is related to other agronomical practices. Furthermore, this method has esthetic appeal and normally gains public acceptance at all sites and locations. Compared to other existing methods, phytoremediation is a solar energy-driven process that presents the benefits of low energy cost and maintenance expenses [34].

Phytoremediation is a self-driven process that normally performs several geochemical processes without any additional inputs; and being a plant-based approach, the process is economically acceptable than other remediation methods [65]. Interestingly, phytoremediation has a high recovery rate of valuable metals, once plants uptake and accumulate metals in harvestable parts, these metals can be recovered by a process called phytomining. Specific plant species, particularly deep-rooted species with high metal translocation rates from roots to above-ground plant parts are recommended to be exploited for this process [35].

1.4 BASIC CONCEPTS OF RHIZOSPHERIC PHYTOREMEDIATION

The rhizosphere phytoremediation is mainly the process, in which plant uptake, accumulate, adsorb, and stabilize the metal in the rhizosphere area that is surrounded by plant roots and is a highly active area from which plants directly uptake water and nutrients. This area is also biologically very important due to the presence of several microorganisms in this zone. The rhizosphere offers a dynamic and multifunctional microenvironment, wherein association with microbes, plants develop exceptional communities with capacity for metal detoxification [18, 89]. Plants develop several interactions with these microbial communities. Predominantly several kinds of bacterial genera are proliferated in this zone, generally called rhizospheric bacteria. Many of these bacterial species promote plant growth under pollutant stress by offering several geochemical functions for plants known as plant growth-promoting rhizobacteria (PGPR) [44].

The bacteria are ubiquitous in soils even many of them has a noticeable tolerance against severe pollution. Such a diverse group of bacteria also help to reduce the toxicity of metal pollution and transfer them into less toxic form as well as promote their phytoextraction by making them more bioavailable. They release siderophores and iron chelators for ensuring iron availability; modulate soil pH, the solubility of metal phosphates and make them more available for plants, thus can improve the phytoremediation process [21, 88]. Rhizobacteria also play a crucial role in the formation and development of root architectural characteristics. An increase in root exudation might lead to a decrease in the rate of phyto-siderophore release, which can establish a symbiotic interaction with roots and can accelerate metal uptake [64].

Several studies have reported an improvement in the uptake of Zn, Ni, Pb, Cr or Cu, in plants in the presence of rhizobacteria [24, 77, 95]. *Alyssum muralea*, Ni-hyperaccumulator, also showed improved Ni-phytoextraction with the addition of rhizobacteria [19]. Another study conducted by Abou-Shanab et al. [3] demonstrated that some bacteria like *M. arabinogalactanolyticum, S. macrogoltabidus* and *M. liquefaciens* can significantly enhance the *A. murale* efficiency for Ni accumulation by 32.4%, 17% and 24%, respectively, compared to the uninoculated plants. Similarly, rhizobacteria-assisted phytoremediation can improve performance of *A. murale*, *B. juncea* and *T. caerulescens* in Se, Zn, and Ni contaminated soils. Siderophore development can also be induced by the involvement of certain HMs for Fe absorption, which may influence the bioavailability of other metals. For instance, *Azotobacter vinelandii*, siderophore production, is increased in the presence of Zn (II).

The production of iron-chelating hydroxamic acid in *Bacillus megaterium* is improved by sensitivity to Cr, Cu, Cd, Zn, and Al, in which *Pseudomonas aeruginosa* has been shown to increase siderophore development [4]. The rhizobacteria population is also reported to be impacted under elevated levels of toxic metals. Studies showed the HM pollution can change microbial community structure, their activities, and their population size [41, 53].

For groundwater metal impacted land, rhizofiltration is the most suitable type of phytoremediation. It can work under *in situ* and ex *situ* experimental conditions. Plants with deep and extensive root structures are further recommended for this process. The approach has been successful in the case of Zn, Cd, Ni, Pb, Cr, Cu, and U [33, 75].

In rhizofiltration, plants grown in clean water (once they develop a long root structure) are acclimated with metals in the control environment. After the development of tolerance against specific pollutants, they are transported to polluted sites, where these plants grow up for a certain time. Once the plants are saturated with metals, they are harvested. This process is performed several times to achieve the desired safe level of pollutants.

1.5 SCOPE OF RHIZOBACTERIA IN PHYTOREMEDIATION

Rhizobacteria have the potential to improve phytoremediation through several mechanisms. These bacteria have inherent traits that are beneficial for plant growth and development. For instance, the bacteria being root colonizers can survive in diverse environmental stress and provide a conducive environment for plant growth as well as establish several other multifunctional interactions with other microbial flora till they exhibit their plant protection, growth, and development activities [4].

Rhizobacteria or PGPR are the special bacterial classes, which have their functional performances (such as bacterial enhanced fertility (known as biofertilizers), phytostimulating capacity (provide phytohormones to enhance plant growth) and metal solubilization capacity (make metals more phytoavailabe)), are called rhizoremediators. Moreover, the bacterial strains, which support the plants against several diseases and pathogens, attack by secreting the lytic enzymes and metabolic compounds are called biopesticides [61].

These bacteria facilitate plants by specific and non-specific means of action. For instance, in a specific mechanism, they promote plant growth by cell division and improve metabolic stabilization, enhance nutrient availability, and stimulate roots networking. Whereas in a non-specific mechanism, they favor plants by suppressing metal toxicity, kill the pathogens and induce the resistance potential against diseases in plants. Several factors significantly contribute to rhizobacteria-assisted phytoremediation including bacterial activities, the structure of the microbial community, metal concentration and bioavailability, plant growth and inherent traits of tolerance against specific metal toxicity [53, 61].

Generally, depending on the chemical speciation of HMs bound with various inorganic and organic constituents are less bioavailable [65]. Rhizobacteria secrete diverse metabolic molecules, such as biosurfactants, siderophores, and EPS (extracellular polysaccharides/exopolysaccharides), etc., in the rhizosphere; and these metabolites tend to change the nature of metals by oxidation, reduction, and acidification as well as modulate their bioavailability by chelation/immobilization, mineralization, and precipitation, which ultimately impact the phytoremediation, phytostabilization, phytoaccumulation, and phytovolatilization efficiency of plants [2, 4].

Mousavi et al. [71] investigated the effect of siderophore producing bacterial strains on the plant growth and metal accumulation, in which microbial inoculations solubilize and increase the accumulation of Pb and Zn. Similarly, Ali et al. [8]; and Mishra et al. [69] reported that the inoculation of siderophore producing PGPR strains was significantly able to increase the accumulation of Zn, Pb, and Fe in their respective host plants.

1.5.1 MICROBE-ASSISTED PHYTOEXTRACTION

The process of metal uptake from roots and translocation to the areal parts is known as phytoextraction. Several factors impact on the phytoextraction ability of plants and the bacteria play an important role to improving phytoextraction of metals. The bacteria have the potential to alter the bioavailability of metals in soil by changing their oxidation state, making them less toxic and chelating them with siderophores [70]. This procedure when repeated at frequent times, produces better results due to less existing contamination than the previous cultivation. Therefore, the remediation and extraction efficiency of plants is improved, and finally, a desired level of a specific pollutant in the soil can be achieved. The phytoextraction can remove metals from 1–3 \m of soil surface depending on the root structure [34, 54].

Till now, about 400 plant species have been known for the phytoextraction process, many of which have excellent metal accumulation capacity [31]. More prominently plants belonging to Scrophulariaceae, Euphorbiaceae, Asteraceae, Brassicaceae, Lamiaceae, and Fabaceae families are more efficient and are widely suited for the phytoremediation purpose [4].

Ahemad [6] studied the impacts of microbes on growth and metal uptake in *H. annuus* that inoculation of siderophores significantly enhanced Pb and Zn accumulation in *H. annuus*. Mishra et al. [69]; and Ali et al. [8] also advocated high accumulation of Zn, Fe, and Pb in plants assisted with siderophores inoculation. Several other low molecule weight organic acids (such as gluconic, 2-ketoglutaric, acetic, oxalic, citric, malic, and succinic) play crucial roles in metal solubilization via complexation reduction process. Whereas, the bacterial biosurfactants intensively affect the bioavailability and mobility of metals in the soil. Several strains of PGPR (e.g., Alasan, bidisperse, emulsan, and nanoemulsion, belonging to genus *Acinetobacter*, glycolipid biosurfactant in *Pseudomonas*, surfactin in *Bacillus* as well as a cyclic lipopeptide biosurfactant serrawettin in *Serratia* genus) are major biosurfactant producing strains [61].

The interaction of the biosurfactants develop complexes with insoluble metals on soil binding sites, metabolize, and desorb them from soil particles and modulate the liability and transport these metals in the soil solution. Besides biosurfactants, bioleaching is also the main characteristic of these bacteria that mobilizes the metals from mineral deposits. Collectively these studies indicated that rhizobacterial mediation in phytoremediation improves the overall process by making metals more bioavailable and supporting plants in a stressed environment.

1.5.2 MICROBE-ASSISTED PHYTOSTABILIZATION

Opposite to phytoextraction process, some rhizobacteria play an important role in reducing the bioavailability of metals by specific mechanisms, such as, complexation, precipitation, alkalization, biotransformation, biosorption, bioaccumulation, and adsorption of extracellular capsules, EPS, and some anionic functional groups on cell surfaces including sulfonate, amine, and amide sulfhydryl, hydroxyl, and carboxyl; the bacterial groups also play an important role in retaining the metals for bioavailability [61].

Similarly, functional groups (such as amine, sulfhydryl carboxyl hydroxyl, and phosphate) also negatively change the microbial EPS. These functional groups are mainly effective for cationic changed metals, make a complex with these cations and make them unavailable for plant uptake. Sinha and Mukherjee [85] addressed that negatively charged bacterial cell walls offer defensive support to *Hordeum vulgare* plant against Cd toxicity. Pratush and Hu [76] also advocated that rhizobacteria reduced the mobility of metals via bioaccumulation and biosorption process.

Metals can be chelated/fixed and/or accumulated on plant roots in soil by phytostabilization. In this process, metals from sludge, soil, and sediments can be retained by complexation and make them unavailable for plant and microbial community. In the rhizosphere, plants adsorb, precipitate, and accumulate the metals. However, this is a limited-time process and these metals again leach down in groundwater through several other biological and physicochemical processes. Thus, plant reduces or limits the mobility of metals in soil and retains their entrance into the food chain. This practice can also be employed to the restoration of plant community on land, where these were previously denuded due to metal toxicity. Once a species achieve tolerance against metal toxicity, then they grow in masses, and land cover becomes less venerable for wind erosion; hence spread of metals reduces [79].

Microorganisms, including bacteria, can reduce the bioavailability of metals by sequestration, transformation, precipitation, and accumulation [57]. Also, microorganisms can change the ionic state of Hg, Se, Fe, and Mn by oxidizing or reducing process and convert these metals into less toxic forms. Krithika and Balachandar [48] explains the biotransformation of Cr(VI) in bacteria and reported four major stages of conversion: adsorption of Cr(VI) on the surface functional group, transportation in cytosolic solution via phosphate/sulfate transporters, reduction in cytosolic solution and finally Cr(III) immobilization.

Srivastava and Thakur [87] found *Serratia* sp. mediated Cr(VI) conversion into reduced and immobilized form Cr(III) in complex mechanisms, such as nonexchange, complexation, co-precipitation, and immobilization. Majumder et al. [60] reported that oxidizing bacteria (e.g., *Geobacillus* sp. and *Bacillus* sp.) can successfully convert As(III) into less toxic As(V) form. Similarly, Gregorio et al. [27] also addressed that *Stenotrophomonas maltophilia* can reduce and immobilize Se in soil. Thus, it can be concluded that the rhizobacteria can successfully reduce the bioavailability of metals and reduce their environmental hazard severity.

1.5.3 MICROBE-ASSISTED PHYTOVOLATILIZATION

The phytovolatilization is a mechanism by which plants can uptake pollutants from soil, transport them to shoot and leaves, and finally release into the atmosphere via transpiration and volatilization. Plants metabolize these metals in leaves and other tissues and make them soluble in water, which further excretes out from the plant body with water from open parts of the plant [103]. One prominent demonstration of this process is the conversion of toxic mercuric ions into less toxic elemental mercury. Genetically modified microorganisms for selenium have been successfully used for the volatilization of these metals [4]. However, this method is often discussed as only a transformation and is less admired in the scientific community.

Rhizobacteria can also facilitate the phytovolatilization process. For example, organomercurial lyase enzymes encoding with MerB gene convert organomercurials into mercuric ion (HgII) [63]. Silver and Hobman [84] reported that encoding of MerA gene by an intracellular flavoenzyme (mercuric reductase) excellently transformed toxic ionic Hg(II) into less toxic volatile Hg(0). These microbial functions dynamically reduce the toxicity of metals and improve their phytovolatilization. The incorporation of these microbial genes into plants also has been deliberated in the literature.

1.6 POTENTIAL OF RHIZOBACTERIA IN PHYTOREMEDIATION

1.6.1 PLANT GROWTH

It is widely reported that elevated metal concentration in soil is toxic even for hyperaccumulators, which can only perform their physiological process to a certain level. Plants grown under metal stress normally encounter iron deficiencies and disturbed growth of chloroplast and inhibit chlorophyll biosynthesis, thus becoming chlorotic. However, rhizobacteria produce siderophores, which are iron-enriched and readily available for plants under stressed environment [4]. Thus, inoculation of plants with siderophores producing bacteria prevents the plants from iron deficiencies grown under metal impacted soils. These bacterial strains have been reported for significant growth-promoting factors in plants and support in root elongation and early establishment under elevated levels of Zn, Ni, and Pb contaminated soils [70, 76, 85].

On successful establishment of plants at contaminated lands, these bacteria facilitate plants for nutrient uptake and improve iron availability. Another study showed the pseudomonads isolated from paint industry effluents reduce uptake of Cr and enable *Triticum aestivus* seed germination under potassium bichromate soil contamination conditions [31, 40]. Several bacteria help the plant to accumulate toxic metals in different tissues, for instance, under axenic conditions, rhizosphere bacteria accumulated up to 80 and 60% Se in roots and shoots of saltmarsh bulrush plants [40]. Rhizobacteria can also regulate metal bioavailability to plants by changing their speciation in soil and influence metal solubility in the rhizosphere [31].

The physicochemical properties of rhizosphere differ from those of the bulk soil, as a result of several mechanisms that are performed by the rhizobacteria and/or by plant roots [9, 13, 24, 98]. Plants and the bacterial combination produce several metabolites that can potentially alter the soil conditions in the rhizospheric zone. For instance, the metal uptake efficiency of *Thalaspi caerulescens* can be significantly enhanced due to acidification in the rhizosphere. It was assumed that plants and bacteria together decreased the soil pH, which makes Hg more available for plant uptake [4].

Several bacteria that need metals as terminal electron acceptors in anaerobic respiration are known as assimilatory metal-reducing bacteria, which are physiologically and phylogenetically diverse and mainly use Fe^{3+} and S(0) as terminal electron acceptors [55]. The bacterial-assisted reduction of Cr^{6+} to Cr^{3+} has been intensively studied in different forms of metal bioremediation [41].

1.6.2 INHIBITION OF PLANT PATHOGENS

Rhizobacteria can also induce resistance in plants against viral, fungal, and bacterial diseases as well as against insects and nematode pests. Several rhizobacteria offer a variety of mechanisms to support plants against pathogens. For instance, they compete with pathogens for space, food, and habitats, and also produce several antibiotics (such as pyocyanine, pyrrolnitrin, and 2,4-diacetyl phloroglucinol). Secretion of siderophores, e.g., pseudo action, can limit the availability of iron necessary for the growth of pathogens [18, 50].

Other main means include releasing of lytic enzymes (such as chitinases and β-1,3-glucanases), which breakdown chitin and glucan present in the cell wall of fungi [94], HCN production [26], and digestion of toxin produced by a pathogen [4].

1.6.3 HIGH SURFACE-AREA-TO-VOLUME RATIO

The high surface-area-to-volume ratio of rhizobacteria enhances its contact with metals and makes it an efficient bio-chelator. Inoculation of Indian mustard plants with bacterial enhanced root hairs helps to accumulate more Se compared to plants germinated from axenic seeds [25]. However, bacterial mediated improved root hairs are not the only process of metal accumulation in plants as many plants successfully accumulate higher metal concentration in the absence of bacteria as well. Evidence from antibacterial application in As and Cr contaminated soil revealed no change in metal accumulation in both treated and untreated soil; suggesting that bacteria are not only contributors in metal accumulation or plant uptake [4, 23].

1.6.4 TRANSFORMATION OF TOXIC HEAVY METALS (HMS)

The phytoremediation performance directly relates to the bioavailability of metals to plants in soil. Rhizobacteria transform toxic metals into less toxic and more bioavailable forms. For instance, bacterial mediated reduction of Selenate in organic Se enhanced its bioavailability to plants. A relative change in percentage of speciation distribution in inoculated and bulk soil is also obvious in provisional studies, which indicated that the inoculated soil showed more changes (+5%, +3%, +23%) in organic bound Cu, Pb, and Zn compared to bulk soil (0.8%, –2%, –3%). Rhizobacteria can also alter the chemical properties of metals by changing their redox state, pH, and organic contents in soil and promote leaching of these metals in soil. Studies suggest that organic matter contents and pH are a major factor contributing to the solubility of metals, which further define their bioavailability. For instance, *Pseudomonas maltophilia* impacted the bioavailability of Cd^{2+}, Hg^{2+} and Pb^{2+} in the soil and reduced toxic Cr^{6+} to nontoxic and immobile Cr^{3+} forms [16].

1.6.5 STIMULATION OF TRANSPORT PROTEINS

Rhizobacteria proliferation and survival in a variety of environments is mainly dependent on sequestration and consuming transitional elements, such as Mn, Fe, and Zn. For example, cells may rigorously adjust intracellular zinc concentration; since elevated levels of zinc are toxic and inhibit cellular functions, and have generated a variety of proteins associated with accumulation and transportation of Zn [22]. Plant root plasma membrane contains several transport proteins, such as sulfate/phosphate transport protein. Rhizobacteria stimulate these proteins to transport metals, e.g., stimulation of sulfate protein for transports selenate [4].

1.7 INTERACTIONS IN RHIZOSPHERE

The phytoremediation ability relies on the associations between water, toxic metals, microbes, and plants. Several factors (such as soil conditions, plant health, pollution level, etc.), dictate the working conditions of these interactions in metal impacted lands.

1.7.1 PLANT-MICROBE INTERACTIONS ENHANCE THE HEAVY METAL (HM) PHYTOREMEDIATION

Plant-microbe interactions promote bacterial proliferation in the rhizosphere and also facilitate plant growth under stress conditions, including remediation of metal impacted land. Plants under these interactions significantly manage HM toxicity by metal accumulation and remediate metal impacted land. Several bacteria live in the root zone to promote plant growth and provide a conducive environment to plants for phytoremediation. The major advantage of these interactions includes: biomass production improvement, metal translocation in the plant, enhancing plant tolerance for metals, etc. Bacterial species in the rhizosphere in particular are considered to influence the mobility and supply of HMs to plants [3].

A substantial amount of metals remains unavailable to plants under natural conditions, rhizobacteria enhance dissolution and solubilization of metals, which previously were bound to soil constituents or present as insoluble precipitates. Several organic chemicals released by living and intact root systems in certain stages of plant growth are called root exudates that offer both allelopathic and detoxification functions for plants: (i)

Allelopathic functions of root exudates enhance the microbial growth in the root zone which results in enhanced plant metal uptake efficiency [57]; and (ii) detoxification functions of root exudates include chelation, inactivation of metals, adsorption, and transformation.

Evidence suggests that organic root exudates act as a natural chelator and enhance the phytoextraction efficiency of plants [65]. The organic acids are released from *Echinochloa crus galli* (i.e., oxalic acid and citric acid), which have a high ability of bioaccumulating and translocation of Cu, Pb, and Cd in soil. Depending on rhizospheric conditions and composition of root exudates, every plant has specific microflora in the root zone [57].

For better phytoremediation, it is necessary to select special plants according to soil and pollutant characteristics. Furthermore, inoculation of bacteria also requires special consideration while selecting the mechanism of phytoremediation (e.g., phytoextraction, phytoaccumulation, and phytovolatilization). Advanced research on plant-bacteria interactions in metal impacted lands to elucidate the phytoremediation practices has shown improved metal tolerance as well as proper geochemical functioning are still needed to be studied.

1.7.2 PLANT-BACTERIA INTERACTION

There are several functional interactions prevailing between plants and rhizobacteria, which play a main determinant role in the successful phytoremediation process. Both micro partners and their associated bacteria significantly impacted the performance of associative plant-bacterial symbioses. Rhizobacteria significantly contribute to marinating the soil structure, detoxification of noxious chemicals, nutrient recycling, and killing the pests, and induce resistance in plants against diseases [4, 57].

Therefore, bacteria are highly necessary for supporting the phytoremediation under stressed environment attributed to metal toxicity, nutrient deficiencies, and pest attack. In fact, different interactions among bacteria and plants provide the bacteria with a common supply of carbon that causes the bacteria to reduce the toxicities of the affected soil. On the other hand, nonspecific associations between bacteria and plants can enable plants to stimulate the microbial community, which normally performs degradation activities in soil. Root exudates released from plants enhance the metal solubility. Collectively these findings indicate that bacterial assisted biochemical mechanisms enhance the metal remediation process associated with plant roots.

1.7.3 METALS AND BACTERIAL INTERACTION

The rhizobacteria bacteria have a specific function to modulate metal bioavailability in soil. As discussed earlier, changing of oxidation state, acidification, change in redox potential and pH, chelation, complex formation, uptake, etc., are the main routes, by which bacteria regulate metals in soils [1, 92]. For instance, Abou-Shanab et al. [4] stated that the addition of *M. liquefaciens*, *M. arabinogalactanolyticum* and *S. macrogoltabidus* reduced the soil pH of serpentine soil, which resulted in higher Ni accumulation in *Alyssum murale*. However, HMs in excessive amounts in soil possess severe toxic consequences to plants and microorganisms. Liu et al. [53] addressed those HMs showed detrimental effect to microbial community structure and their activities.

1.7.4 SOIL-PLANT-BACTERIA INTERACTION

The soil conditions are important for plant-metals-bacterial interactions. The change in soil physicochemical properties changes the microbial community structure, metal bioavailability and plant growth. Moreover, plant-bacteria interactions (i.e., specific/nonspecific for metal remediation) may also be dictated by several metabolic requirements. Besides metal toxicity, many other factors (such as moisture content, nutrients level, soil structure, and air conditions) are also known to significantly impact these interactions.

1.8 CHARACTERISTICS AND FACTORS AFFECTING PHYTOREMEDIATION

Despite the fact, the plants used in phytoremediation showed inherent tolerance towards sequestering the metals, yet various biotic and abiotic factors limit their efficiency. In general, these factors are the ones, which need to be supportive of the steady functioning of plants in stressed (metal impacted media) environment. Broadly, these environmental factors are divided into two groups:

- The biotic factors cover all the living aspects, including microbiota of rhizosphere, plants characteristics (morphology, physiology, genomics, etc.), and animals; and

- The abiotic factors cover all nonliving aspects mainly including soil (structure, texture, etc.), temperature, pH, precipitation, radiations, and xenobiotic concentration.

All these factors are mutually linked, because change in one factor can change the whole system [14, 58].

1.8.1 CHARACTERISTICS OF PHYTOREMEDIATION PLANTS

Selection of a plant is highly essential in achieving successful phytoremediation strategy. There are certain characteristics for consideration in phytoremediation, such as high photosynthetic rate, fast-growing, tolerance towards metal toxicity, profuse root system, more vegetative volume, resistant towards pathogens, high hyperaccumulation capacity (bio-concentration and translocation factor (TF) >1) [5, 7, 78, 81, 90].

Among different phytoremediation techniques, phytoextraction has attained the most attention with the documentation of more hyper-accumulators and their potential to adapt to environmental conditions (EC). Different plant species showed different absorption/accumulation capability towards each metal. Even the varieties/cultivars of the same species showed varied metal accumulation capacity under the same growing conditions. Nowadays, in cultivar/varietal selection not only accumulation capacity but also their susceptibility capacity should be considered [86].

Native plants showed better results in coping with stress conditions, because of ease in adaptation, thereby minimizing the influence of environmental factors and generated good results of phytoremediation [30]. Genetically engineered plants are also being introduced for the phytoremediation of metals. These plants have the required genetic trait (phytochelatins, metallothioneins, and metal transporter genes) of hyperaccumulation [72]. Accordingly, success in phytoremediation is determined by the right selection of plant species for specific trace metal while keeping in consideration the cropping site and cropping pattern [89]. Thus, the selection of plants is crucial for both hyper and non-hyperaccumulator plants.

1.8.2 SOIL PHYSICOCHEMICAL PROPERTIES

Soil characteristics play a prime role in phytoremediation by affecting numerous factors, such as aeration, microorganisms, organic matter, mineral composition, cation exchange capacity (CEC), soil pH, soil water, metals specifications/availability. This environmental factor is so diverse that properties varied within close proximities. All these factors have influenced plants directly but also have an immense effect on phytoremediation technique by determining the availability/absorption of metals to plants. As soil is a receptacle of various pollutants, thus the phytoremediation process varies according to the type of pollution. In the case of hydrocarbons, the microbes degrade the organic pollutant into the absorbable metabolites and use them as a source of carbon [97]. In HM cases, hyperaccumulator plants are used to extract metals from soil and translocate them in above-ground tissues [59].

The bioavailability and solubility of metals determine the success of phytoremediation; and their availability is dependent on the soil solution characteristics (metal content, pH, Eh). For example, soil pH directly influences the metal availability for plant uptake. In soil with low pH, higher will be the metal concentration in the soil solution; while in higher pH case, lower will be the metal concentration due to increase in adsorption capacity [83].

Cd, Zn, Hg, Mn, Pb, Fe, and Cu are some metal cations, which showed increased bioavailability with lower pH [98]. Likewise, metal mobilization had increased with an increase in soil moisture [10], organic matter addition. However, metal mobilization decreases with increase in CEC of soil [28]. The increase in CEC increased the binding sites, and thereby immobilization of metals takes place.

Soil texture is another soil factor, which determines the availability of metals. Fine-textured soil (clay, clay loam) showed higher metal concentration than coarse-textured soils (sand and silt). It is reported that more clay content in soil shows the lesser Cd uptake by *Noccaea caerulescens*, thus there exists a negative correlation between metal uptake and clay content of the soil [100].

1.8.3 ROOT ZONE DIVERSITY AND ROOT EXUDATIONS

Root zone plays an essential role in phytoremediation technique by providing channel to metabolize, absorb, and transfer the contaminant into the plant. It provides the site, where plant roots release materials to convert these metals into nontoxic forms via secretions of phytochelatins, protons, amino acids, organic acids, phytosiderophores, enzymes, and metal chelates. These proteins further increase metal solubility/bioavailability by altering their speciation [49]. The increase in metal solubility is probably done by root activities, such as redox potential modification, chelates, and ligands exudation or acidification/alkalinization [57].

The activities of roots are immensely affected by EC, such as soil moisture content, temperature, precipitation, and drought. For example, under drought conditions, root diameter is increased while elongation is decreased due to poor soil permeability. Similarly, under field conditions, the root structure is different than in the greenhouse environment [20, 68]. Consequently, intensive study of root zone activities provides an important indicator for hyperaccumulator identification.

1.8.4 ROLE OF MICROBES IN RHIZOREMEDIATION

Soil is the habitat of a wide range of diverse microbes, which played a quintessential role in metal availability, nutrients supply and physicochemical properties of soil. Among microbes, plant growth-promoting rhizobacteria (PGPR) showed significant role in assisting phytoremediation by releasing organic acids (citric, succinic malic, formic, 5-ketogluconic and oxalic acids), bio-surfactants, enzymes (1-aminocyclopropane-1-carboxylate (ACC) deaminase), antibiotics, siderophore (Fe ligand) production and phytohormones secretion especially indole-3-acetic acid and cytokinin [45, 46].

Microorganisms (mostly bacteria and fungi) exhibit a significant role in phytoremediation process by increasing metal bioavailability and improving plant growth under stress. Many studies reported that microbial biomass in soil readily elevates the labile metal pool that is taken up by plants; and this microbe-plant synergism is very profitable in ameliorating the polluted soil conditions [7, 32, 104]. These microbes increase metal mobility via acidification, redox changing, phosphate solubilization, and chelation. Microbes have increased the phytoremediation potential either by enhancing metal uptake (assisting phytoextraction) or reducing metal mobility by immobilizing metals (phytostabilization) [45].

Among microorganisms, bacteria have the standout position in phytoremediation as it improves plant growth, reduces phytotoxicity of soil and activate siderophore production. These siderophores make stable complexes with metals which further facilitate the metal uptake to plants [17]. Furthermore, bacteria acidify the environment and secrete organic acids (especially citric acid, oxalates), which make complexes with metals and also played a role in lowering the soil pH [74].

1.8.5 ENVIRONMENTAL FACTORS AND AGRONOMIC PRACTICES

The environmental factors play a very crucial role in achieving phytoremediation success. For better plant growth and functioning, water content plays a primary role. Adequate soil moisture provides superior biomass of plants, which further extract more metals into the plant. Many studies showed the influence of water on the biochemical and physiological aspects of plants [12, 99]. Maintaining water balance is critical not only for maintaining cell structure but also for the other aspects as it directly affects the natural flow of water in soil, plant, and atmosphere [80]. This phenomenon determines the uptake and transport of various substances. Therefore, for phytoremediation plants, transport/uptake of metal is very important.

According to Angle et al. [10], the high-water content resulted in better growth and biomass of hyperaccumulators, while the low moisture content showed inhibited plant growth. Temperature has a direct role in contaminant uptake by controlling plant metabolism, the chemistry of water, and most importantly, transpiration [15]. It is established that there exists a linear relationship between temperature and metal uptake. As temperature rises, there would be an increase in metal uptake. Likewise, with a rise in temperature caused higher Cd uptake in safflower.

1.9 CHEMICALLY-ASSISTED PHYTOREMEDIATION BY RHIZOBACTERIA

The key factor involved in achieving phytoremediation success in the metal-contaminated site is the availability of metals and their high accumulation in plants. Most of the metals present in the soil are either bounded with crystal lattice of minerals or bounded by microbes. Some microbe-plant interactions lead to reduction in metal uptake, consequently making metals unavailable to the plant via [57]:

- In biosorption process, microbes take up the metals into their body or just bind them to their surfaces (like in bacteria, peptidoglycan, phospholipids, and lipopolysaccharides provide binding sites of metals);

- In biopolymer secretion, sequestration of metal takes place, which immobilize metals and lowers their entry to the plants.

To compensate this limiting factor, chelating agents are added in the soil to improve the solubility of metals, their translocation in plants and eventually improve phytoextraction efficiency. The commonly used chelates in metal solubilization are ethylenediaminetetraacetic acid (EDTA), diethylene tri, hydroxyethylene diamine triacetic acid (HEDTA), minepentaacetic acid (DTPA), nitrilotriacetic acid (NTA), S-ethylenediaminedisuccinic acid (EDDS), and ethylene glycol tetra acetic acid (EGTA). These chelates improve metal bioavailability/absorption via making metal-chelate complexes and improve the efficiency of phytoremediation [73].

Among all chelates, EDTA is the most efficient chelator of metals as it influences the chemical speciation of metals and subsequently changes its solubility/availability to the plant [82]. As studied by Hassan et al. [39], EDTA is an organic ligand to facilitate the bioavailability of Pb to roots and their translocation to above-ground parts of bamboo plants by making stable and soluble Pb-EDTA complexes. In addition, EDTA application in soil proved to increase Cu uptake in chickpea [47] and Cd availability by 233% in *Atriplex lentiformis* [29].

Due to persistence and poor degradable nature of EDTA, its application may pose a potential threat to the surrounding environment. Therefore, easily bio-degradable, and fewer toxic chelators are selected for phytoextraction either alone or in combination with EDTA. Luo et al. [56] showed effective role of EDDS in enhancing mental concentration as well as it showed no residual effect in the next season in comparison to EDTA. Likewise, NTA have been used to accumulate and translocate Cd in plants, and it showed positive effects in enhancing the phytoextraction efficiency [101]. EGTA is another important synthetic chelate that has been successfully used to improve phytoextraction techniques. Various studies have showed its effective role as it increased Pb concentration in *Cicer arietinum* by 80%, increased Cd content in *A. rosea* by 72% [51] and in *Calendula officinalis* by 217% [52].

The effectiveness of this strategy largely depends on the concentration of metal, type of metal, type of chelate, the concentration of chelate, plant species and physicochemical properties of soil [39]. Although, these synthetics chelates improve metal bioavailability but also cause metal phytotoxicity by inducing oxidative stress and exert a negative effect on soil biochemical activities. This hazardous effect is alleviated by rhizobacteria application, which improved enzymatic activities in soil and enhanced plant tolerance towards phytotoxicity. Thus, combining the application of chelates and bacteria has a substantial role in phytoremediation. There exist some potential chelates, which have attained immense interest as a promising technique for assisting phytoremediation coupled with rhizobacteria.

The combined application of chelate and bacteria showed enhanced dissolution of metals and effective translocation from soil solution into plants. The co-application of chelate (EDDS) and microbes (rhizobium strain and PGPR strain) has improved plant growth, soil enzymes activities, nutrient content of soil and Cu accumulation in roots and shoots of Alfalfa in Cu contaminated soil. Co-application of PGPR inoculated pots with EDTA showed increased metal (Cd, Pb, and Zn) in roots (2 folds higher) and shoots (1.3 folds higher) of maize in comparison to untreated soil [91].

1.10 MECHANISMS OF RHIZOBACTERIA IN PHYTOREMEDIATION

Soil contamination, particularly of HMs is a worldwide environmental threat [96], and phytoremediation is an economical approach taking advantage of plants' ability to clean and detoxify the environment in which they live. Rhizobacteria are the most significant and abundant symbiotic partners of plants. Synergistic use of these symbiotic partners is well known to enhance the effectiveness of phytoremediation. Rhizobacteria move from the surrounding soil to the rhizosphere and roots of the plant and colonize themselves [57].

Plant growth-promoting rhizobacteria (PGPR) consist of an assorted group of free-living bacteria. The soil rhizosphere offers a complex and dynamic microenvironment where PGPRs in connotation with plant's roots yields a significant potential for detoxification of xenobiotics [3]. The rhizobacteria can alleviate the metal stress in plants and remediate the polluted soil through various mechanisms, which mainly are [57]:

- By improving the synthesis of phytohormones (gibberellins, auxins, cytokinins), enzymes, nitric oxide;
- Phosphate solubilization;
- Nitrogen fixation; and
- Disease control by producing antifungal and antibacterial metabolites.

Since the phytoremediation efficiency considerably depends on the growth of plants, PGPRs can thus improve the process of phytoremediation [37]. Rhizobacteria and plant growth relationship is fundamentally controlled by several biotic and abiotic factors (such as pH, salinity, temperature, and organic matter). Growth stimulation-effect also differs with different bacterial species [57].

In contaminated soils, HMs are usually bound with certain inorganic or organic constituents and difficult to absorb by plants [57]. The major constraints for successful phytoremediation of metal-polluted lands are the meager or non-bioavailability of metal to plants roots. A range of biochemical dynamics in the rhizosphere is influenced by PGPRs to enhance the bioavailability. Numerous species of rhizobacteria are proficient to withstand the raised concentrations of metals.

The PGPRs secrete certain organic acids, biosurfactants, osmolytes, and siderophores, which help in the bioavailability of HMs for plants. These metabolites affect the mobility (by chelation and precipitation) and nature (by acidification and redox reactions) of HMs in rhizosphere; hence by improving their phytoextraction, phytostabilization, and phytovolatilization efficiencies. The *Achromobacter, Arthrobacter, Azobacter, Azospirillum, Bacillus, Enterobacter, Pseudomonas, Serratia, and Streptomyces* spp. are widely acknowledged rhizobacteria for their metal detoxification from the environment [37, 96].

The biosurfactant-producing PGPRs significantly promote the mobilization and subsequent uptake of metals by plants [11, 102]. Release of numerous biochemicals from PGPRs results in enhanced desorption of metals from immobilized states and thus enhance the solubility. Similarly, the inoculant *Microbacterium liquefaciens* and *Microbacterium arabinogalactanolyticum* decreased pH and increased Ni intake in sweet alyssum (*Alyssum murale* L.), which is itself a Ni hyperaccumulator [3, 4]. The hyperaccumulator plant *Noccaea caerulescens* increased the bioaccumulation of Zn, Cd, and Ni by four times compared to control when it was augmented with a consortium of *Enterobacter cancerogenes, Pseudomonas monteilii,* and *Microbacterium saperdae* [93].

The Phytohormones are among major groups of biochemicals released by PGPRs. These phytohormones even exert a pronounced effect on plant physiology even at meager availability to plants. Most important hormones include auxins, cytokines, gibberellic acid, 1-aminocyclopropane-1-carboxylate deaminase (ACCD), abscisic acid (ABA), polyamines, brassinosteroids, jasmonate, salicylic acid, strigolactones, etc. Such microbial substances affect plant development significantly by attenuating the plant's different biochemical and physiological functions under metal toxicity [61]. *Pseudomonas fluorescen* is a cytokinin producing rhizobacteria and improves tolerance level of the host plant to biotic and abiotic stress. Similarly, *Bacillus megaterium* is another cytokinin producing strain known to induce root morphogenesis in *Arabidopsis thaliana* and to regulate the transcriptional level of root and shoot cytokinin receptor (AHK3 and AHK4).

1.11 SUMMARY

The phytoremediation/rhizoremediation brings plant and microbes into action for the persistence of human living conditions. The phytoremediation assayed with rhizosphere microbes is characterized by its low-cost budget, eco-friendly, effective for a wide range of metals, effectively eliminate the exposure of pollutant organisms, supports the biodiversity and equivocally effective for various media (soil, surface water groundwaters and sediments). The mechanisms of phytoremediation are phytoextraction, phytostabilization, phytotransformation, phytovolatilization, and rhizodegradation. The plants used in phytoremediation show inherent tolerance towards the accumulated metals, yet the efficiency is limited by various biotic and abiotic factors.

Screening from native fauna seems to be the most successful strategy. Genetically engineered plants have also shown promising potential for remediation. As the remediation efficiency can be limited by the poor availability of target metal(s) in media and this could be resolved with the addition of various bioremediation techniques or some chelating agents. The numerous benefits associated with phytoremediation and the potential of microbes and plants have led this technology to be utilized on commercial scales in its two largest markets: Europe and America. However, there are meager commercial reports on cost: benefit ratio available for bioremediation projects.

KEYWORDS

- **contamination**
- **ethylene glycol tetra acetic acid**
- **metal-sites**
- **phytoremediation**
- **plant growth-promoting rhizobacteria**
- **rhizobacteria**

REFERENCES

1. Abdul, N., Abdullahi, A. A., & Abdulkadir, A., (2017). Heavy metals and soil microbes. *Environmental Chemistry Letters, 15*, 65–84.
2. Abdullah, S. R. S., Al-Baldawi, I. A., Almansoory, A. F., Purwanti, I. F., Al-Sbani, N. H., & Sharuddin, S. S. N., (2020). Plant-assisted remediation of hydrocarbons in water and soil: Application, mechanisms, challenges, and opportunities. *Chemosphere, 247*, 125932.
3. Abou-Shanab, R., Delorme, T., Angle, J., Chaney, R., Ghanem, K., Moawad, H., & Ghozlan, H., (2003). Phenotypic characterization of microbes in the rhizosphere of *Alyssum murale*. *International Journal of Phytoremediation, 5*, 367–379.
4. Abou-Shanab, R. A., El-Sheekh, M. M., & Sadowsky, M. J., (2010). Role of rhizobacteria in phytoremediation of metal-impacted sites. In: Bharagava, R. M., & Chowdhary, P., (eds.), *Emerging and Eco-Friendly Approaches for Waste Management* (pp. 299–328). Springer Nature, Singapore.
5. Adesodun, J. K., Atayese, M. O., Agbaje, T., Osadiaye, B. A., Mafe, O., & Soretire, A. A., (2010). Phytoremediation potentials of sunflowers (*Tithonia diversifolia* and *Helianthus annuus*) for metals in soils contaminated with zinc and lead nitrates. *Water, Air and Soil Pollution, 207*(1–4), 195–201.
6. Ahemad, M., (2015). Phosphate-Solubilizing bacteria-assisted phytoremediation of metalliferous soils: A review. *3 Biotech, 5*, 111–121.
7. Ali, H., Khan, E., & Sajad, M. A., (2013). Phytoremediation of heavy metals-concepts and applications. *Chemosphere, 91*(7), 869–881.
8. Ali, D. G., Mahar, A., Ma, F., Li, R., Shen, F., Wang, P., & Zhang, Z., (2017). *Streptomyces pactum* assisted phytoremediation in Zn/Pb smelter contaminated soil of Feng county and its impact on enzymatic activities. *Scientific Reports, 7*, 9. Article ID: 46087.
9. Al-Sid-Cheikh, M., Pedrot, M., Dia, A., Guenet, H., Vantelon, D., Davranche, M., & Delhaye, T., (2015). Interactions between natural organic matter, sulfur, arsenic and iron oxides in re-oxidation compounds within riparian wetlands: Nano SIMS and x-ray adsorption spectroscopy evidences. *Science of the Total Environment, 515*, 118–128.
10. Angle, J. S., Baker, A. J. M., Whiting, S. N., & Chaney, R. L., (2003). Soil moisture effects on uptake of metals by Thlaspi, Alyssum, and Berkheya. *Plant and Soil, 256*(2), 325–332.
11. Ashraf, M. A., Hussain, I., Rasheed, R., Iqbal, M., Riaz, M., & Arif, M. S., (2017). Advances in microbe-assisted reclamation of heavy metal contaminated soils over the last decade: A review. *Journal of Environmental Management, 198*, 132–143.
12. Asrar, A. A., & Elhindi, K. M., (2011). Alleviation of drought stress of marigold (*Tagetes erecta*) plants by using arbuscular mycorrhizal fungi. *Saudi Journal of Biological Sciences, 18*(1), 93–98.
13. Balapure, K., Jain, K., Bhatt, N., & Madamwar, D., (2016). Exploring bioremediation strategies to enhance the mineralization of textile industrial wastewater through sequential anaerobic-microaerophilic process. *International Biodeterioration & Biodegradation, 106*, 97–105.
14. Beard, K. H., Wang, D., Waite, C. E., Decker, K. L. M., Hawley, G. J., DeHayes, D. D., Hughes, J. W., & Cumming, J. R., (2005). Quantifying ecosystem controls and their contextual interactions on nutrient export from developing forest mesocosms. *Ecosystems, 8*(2), 210–224.

15. Bhargava, A., Carmona, F. F., Bhargava, M., & Srivastava, S., (2012). Approaches for enhanced phytoextraction of heavy metals. *Journal of Environmental Management, 105,* 103–120.
16. Blake, R. C., Choate, D. M., Bardhan, S., Revis, N., Barton, L. L., & Zocco, T. G., (1993). Chemical transformation of toxic metals by a pseudomonas strain from a toxic waste site. *Environmental Toxicology and Chemistry: An International Journal, 12,* 1365–1376.
17. Braud, A., Geoffroy, V., Hoegy, F., Mislin, G. L. A., & Schalk, I. J., (2010). The siderophores pyoverdine and pyochelin are involved in *Pseudomonas aeruginosa* resistance against metals: Another biological function of these two siderophores. *Environmental Microbiology Reports, 2,* 419–425.
18. Brereton, N., Gonzalez, E., Desjardins, D., Labrecque, M., & Pitre, F., (2020). Co-cropping with three phytoremediation crops influences rhizosphere microbiome community in contaminated soil. *Science of The Total Environment, 711*, 135067.
19. Broadhurst, C. L., Chaney, R. L., Angle, J. S., Erbe, E. F., & Maugel, T. K., (2004). Nickel localization and response to increasing Ni soil levels in leaves of the Ni hyperaccumulator *Alyssum murale. Plant and Soil, 265*, 225–242.
20. Brunner, I., Herzog, C., Dawes, M. A., Arend, M., & Sperisen, C., (2015). How tree roots respond to drought. *Frontiers in Plant Science, 6,* 547–558.
21. Chen, D. W., Liu, C., Lu, J., Mehmood, T., & Ren, Y. Y., (2020). Enhanced phycocyanin and don removal by the synergism of H_2O_2 and micro-sized ZVI: Optimization, Performance, and mechanisms. *Science of The Total Environment,* 140134.
22. Claverys, J. P., (2001). New family of high-affinity ABC manganese and zinc permeases. *Research in Microbiology, 152,* 231–243.
23. Cristaldi, A., Copat, C., Conti, G. O., Zuccarello, P., Grasso, A., & Ferrante, M., (2020). Phytoremediation. Chapter 10. In: Chaudhery, M. H., (ed.), *The Handbook of Environmental Remediation* (pp. 268–298). Royal Society of Chemistry, London -UK.
24. Dąbrowska, G., Hrynkiewicz, K., Trejgell, A., & Baum, C., (2017). The effect of plant growth-promoting rhizobacteria on the phytoextraction of Cd and Zn by *Brassica napus* L. *International Journal of Phytoremediation, 19,* 597–604.
25. De Souza, M., Huang, C., Chee, N., & Terry, N., (1999). Rhizosphere bacteria enhance the accumulation of selenium and mercury in wetland plants. *Planta, 209,* 259–263.
26. Defago, G., & Haas, D., (1990). *Pseudomonads as Antagonists of Soilborne Plant Pathogens: Mode of Actions and Genetic Analysis, 4*, 249–291.
27. Di Gregorio, S., Lampis, S., & Vallini, G., (2005). Selenite precipitation by a rhizospheric strain of *Stenotrophomonas* sp. isolated from the root system of *Astragalus bisulcatus*: A biotechnological perspective. *Environment International, 31,* 233–241.
28. Egene, C. E., Poucke, V. R., Ok, Y. S., Meers, E., & Tack, F. M. G., (2018). Impact of organic amendments (biochar, compost and peat) on Cd and Zn mobility and solubility in contaminated soil of the Campine region after three years. *Science of the Total Environment, 626,* 195–202.
29. Eissa, M. A., (2017). Phytoextraction mechanism of Cd by *Atriplex lentiformis* using some mobilizing agents. *Ecological Engineering, 108,* 220–226.
30. Elekes, C. C., (2014). Eco-technological solutions for the remediation of polluted soil and heavy metal recovery. Chapter 10. In: Hernandez-Soriano, M. C., (ed.), *Environmental Risk Assessment of Soil Contamination* (pp. 309–335). InTech Open, Rijeka.

31. Eslamian, S., (2016). *Urban Water Reuse Handbook* (1st edn., p. 1141). Boca Raton - FL: CRC Press.
32. Franchi, E., Rolli, E., Marasco, R., Agazzi, G., Borin, S., Cosmina, P., Pedron, F., et al., (2017). Phytoremediation of a multi contaminated soil: Mercury and arsenic phytoextraction assisted by mobilizing agent and plant growth-promoting bacteria. *Journal of Soils and Sediments, 17*(5), 1224–1236.
33. Gardea-Torresdey, J., De La Rosa, G., & Peralta-Videa, J., (2004). Use of phytofiltration technologies in the removal of heavy metals: A review. *Pure and Applied Chemistry, 76*, 801–813.
34. Gaurav, G. K., Mehmood, T., Cheng, L., Klemeš, J. J., & Shrivastava, D. K., (2020). Water hyacinth as a biomass: A review. *Journal of Cleaner Production,* 122214.
35. Guerrero-Zúñiga, L., López-López, E., Rodríguez-Tovar, A. V., & Rodríguez-Dorantes, A., (2020). Functional diversity of plant endophytes and their role in assisted phytoremediation. In: Bharagava, R. N., & Saxena, G., (eds.), *Bioremediation of Industrial Waste for Environmental Safety* (Vol. 2, pp. 237–255). Springer Nature, Singapore.
36. Haldar, S., & Ghosh, A., (2020). Microbial and plant-assisted heavy metal remediation in aquatic ecosystems: A comprehensive review. *3-Biotech, 10,* 1–13.
37. Hamidpour, M., Nemati, H., Dahaji, P. A., & Roosta, H. R., (2019). Effects of plant growth-promoting bacteria on EDTA-assisted phytostabilization of heavy metals in a contaminated calcareous soil. *Environmental Geochemistry and Health,* 1–11.
38. Hanfi, M. Y., Mostafa, M. Y. A., & Zhukovsky, M. V., (2019). Heavy metal contamination in urban surface sediments: Sources, distribution, contamination control, and remediation. *Environmental Monitoring and Assessment, 192*, 32–40.
39. Hasan, M., Uddin, M., Ara-Sharmeen, I. F., Alharby, H., Alzahrani, Y., Hakeem, K. R., & Zhang, L., (2019). Assisting phytoremediation of heavy metals using chemical amendments. *Plants, 8*(9), 295–303.
40. Hasnain, S., & Sabri, A. N., (1997). Growth stimulation of *Triticum aestivum* seedlings under Cr-stresses by non-rhizospheric pseudomonad strains. *Environmental Pollution, 97,* 265–273.
41. He, Z. I., & Yang, X. E., (2007). Role of soil rhizobacteria in phytoremediation of heavy metal contaminated soils. *Journal of Zhejiang University Science B, 8,* 192–207.
42. He, Z. I., Shentu, J., Yang, X. E., Baligar, V. C., Zhang, T., & Stofella, P. J., (2015). Heavy metal contamination of soils: Sources, indicators and assessment. *Journal of Environmental Indicators, 9*, 17–18.
43. Hembrom, S., Singh, B., Gupta, S. K., & Nema, A. K., (2020). A comprehensive evaluation of heavy metal contamination in foodstuff and associated human health risk: A global perspective. In: Singh, P., Singh, R. P., & Srivastava, V., (eds.), *Contemporary Environmental Issues and Challenges in Era of Climate Change* (pp. 33–63). Springer Nature, Singapore.
44. Horstmann, J. L., Dias, M. P., Ortolan, F., Medina-Silva, R., Astarita, L. V., & Santarém, E. R., (2020). *Streptomyces* sp. CLV45 from *Fabaceae* rhizosphere benefits growth of soybean plants. *Brazilian Journal of Microbiology, 6*, 1–11.
45. Hrynkiewicz, K., Złoch, M., Kowalkowski, T., Baum, C., & Buszewski, B., (2018). Efficiency of microbially assisted phytoremediation of heavy-metal contaminated soils. *Environmental Reviews, 26*(3), 316–332.

46. Jebara, S. H., Ayed, S. A., Chiboub, M., Fatnassi, I. C., Saadani, O., Abid, G., & Jebara, M., (2019). Phytoremediation of cadmium-contaminated soils by using legumes inoculated by efficient and cadmium-resistant plant growth-promoting bacteria. Chapter 19. In: Hasanuzzaman, M., Prasad, M. N. V., & Fujita, M., (eds.), *Cadmium Toxicity and Tolerance in Plants* (pp. 479–493). London: Academic Press.
47. Kambhampati, M. S., & Vu, V. T., (2013). EDTA enhanced phytoremediation of copper contaminated soils using chickpea (*Cicer aeritinum* L.). *Bulletin of Environmental Contamination and Toxicology, 91*(3), 310–313.
48. Krithika, S., & Balachandar, D., (2016). Expression of zinc transporter genes in rice as influenced by zinc-solubilizing *Enterobacter cloacae* strain ZSB14. *Frontiers in Plant Science, 7,* 446–452.
49. Laghlimi, M., Baghdad, B., Hadi, H. E., & Bouabdli, A., (2015). Phytoremediation mechanisms of heavy metal contaminated soils: A review. *The Open Ecology Journal, 5,* 375–388.
50. Lemanceau, P., Bakker, P., De Kogel, W. J., Alabouvette, C., & Schippers, B., (1992). Effect of Pseudobactin 358 production by *Pseudomonas putida* WCS358 on suppression of fusarium wilt of carnations by nonpathogenic *Fusarium oxysporum* Fo47. *Applied and Environmental Microbiology, 58,* 2978–2982.
51. Liu, J. N., Zhou, Q. X., Sun, T., Ma, L. Q., & Wang, S., (2008). Identification and chemical enhancement of two ornamental plants for phytoremediation. *Bulletin of Environmental Contamination and Toxicology, 80*(3), 260–265.
52. Liu, J., Zhou, Q., & Wang, S., (2010). Evaluation of chemical enhancement on phytoremediation effect of Cd-contaminated soils with *Calendula officinalis* L. *International Journal of Phytoremediation, 12*(5), 503–515.
53. Liu, C., Lu, J., Liu, J., Mehmood, T., & Chen, W., (2020). Effects of lead (Pb) in stormwater runoff on the microbial characteristics and organics removal in bioretention systems. *Chemosphere*, 8. Article ID: 126721.
54. Lone, M. I., He, Z. I., Stoffella, P. J., & Yang, X. E., (2008). Phytoremediation of heavy metal polluted soils and water: Progresses and perspectives. *Journal of Zhejiang University Science B, 9,* 210–220.
55. Lovley, D. R., (1997). Microbial Fe (III) reduction in subsurface environments. *FEMS Microbiology Reviews, 20,* 305–313.
56. Luo, C., Shen, Z., Lou, L., & Li, X., (2006). EDDS and EDTA-enhanced phytoextraction of metals from artificially contaminated soil and residual effects of chelant compounds. *Environmental Pollution, 144*(3), 862–871.
57. Ma, Y., Oliveira, R. S., Freitas, H., & Zhang, C., (2016). Biochemical and molecular mechanisms of plant-microbe-metal interactions: Relevance for phytoremediation. *Frontiers in Plant Science, 7,* 918–925.
58. Magdziak, Z., Gąsecka, M., Goliński, P., & Mleczek, M., (2015). Phytoremediation and environmental factors. Chapter 4. In: Ansari, A. A., Gill, S. S., Gill, R., Lanza, G. R., & Newman, L., (eds.), *Phytoremediation* (pp. 45–55). Springer Nature, Cham.
59. Mahar, A., Wang, P., Ali, A., Awasthi, M. K., Lahori, A. H., Wang, Q., Li, R., & Zhang, Z., (2016). Challenges and opportunities in the phytoremediation of heavy metals contaminated soils: A review. *Ecotoxicology and Environmental Safety, 126,* 111–121.
60. Majumder, K. B., Bhattacharyya, S., & Kole, S., (2013). Arsenic-tolerant, arsenite-oxidizing bacterial strains in the contaminated soils of West Bengal, India. *Science of the Total Environment, 463,* 1006–1014.

61. Manoj, S. R., Karthik, C., Kadirvelu, K., Arulselvi, P. I., Shanmugasundaram, T., Bruno, B., & Rajkumar, M., (2020). Understanding the molecular mechanisms for the enhanced phytoremediation of heavy metals through plant growth-promoting rhizobacteria: A review. *Journal of Environmental Management, 254*, 110–118.
62. Markert, B., & Friese, K., (2000). *Trace Elements: Their Distribution and Effects in the Environment* (1st edn., p. 596). New York: Elsevier.
63. Matsui, K., Yoshinami, S., Narita, M., Chien, M. F., Phung, L. T., Silver, S., & Endo, G., (2016). Mercury resistance transposons in *bacilli* strains from different geographical regions. *FEMS Microbiology Letters, 363*, 5–12.
64. Mazhar, R., Ilyas, N., Arshad, M., Khalid, A., & Hussain, M., (2020). Isolation of heavy metal-tolerant PGPR strains and amelioration of chromium effect in wheat in combination with biochar. *Iranian Journal of Science and Technology, Transactions A: Science, 44*, 1–12.
65. Mehmood, T., Bibi, I., Shahid, M., Niazi, N. K., Murtaza, B., Wang, H. L., Ok, Y. S., et al., (2017). Effect of compost addition on arsenic uptake, morphological and physiological attributes of maize plants grown in contrasting soils. *Journal of Geochemical Exploration, 178*, 83–91.
66. Mehmood, T., Tianle, Z., Ahmad, I., & Li, X., (2019). Integration of AirQ+ and particulate matter mass concentration to calculate health and ecological constraints in Islamabad, Pakistan. *16th International Bhurban Conference on Applied Sciences and Technology (IBCAST)-2019*, pp. 248–254.
67. Mehmood, T., Zhu, T., Ahmad, I., & Li, X., (2020). Ambient PM2.5 and PM10 bound PAHs in Islamabad, Pakistan: Concentration, source and health risk assessment. *Chemosphere, 257,* 8. Article ID: 127187.
68. Merkl, N., Schultze-Kraft, R., & Infante, C., (2005). Phytoremediation in the tropics-influence of heavy crude oil on root morphological characteristics of graminoids. *Environmental Pollution, 138*(1), 86–91.
69. Mishra, V., Gupta, A., Kaur, P., Singh, S., Singh, N., Gehlot, P., & Singh, J., (2016). Synergistic effects of arbuscular mycorrhizal fungi and plant growth-promoting rhizobacteria in bioremediation of iron contaminated soils. *International Journal of Phytoremediation, 18,* 697–703.
70. Mohwinkel, D., Kleint, C., & Koschinsky, A., (2014). Phase associations and potential selective extraction methods for selected high-tech metals from ferromanganese nodules and crusts with siderophores. *Applied Geochemistry, 43,* 13–21.
71. Mousavi, S. M., Motesharezadeh, B., Hosseini, H. M., Alikhani, H., & Zolfaghari, A. A., (2018). Root-induced changes of Zn and Pb dynamics in the rhizosphere of sunflower with different plant growth-promoting treatments in a heavily contaminated soil. *Ecotoxicology and Environmental Safety, 147,* 206–216.
72. Nahar, N., Rahman, A., Nawani, N. N., Ghosh, S., & Mandal, A., (2017). Phytoremediation of arsenic from the contaminated soil using transgenic tobacco plants expressing ACR2 gene of *Arabidopsis thaliana. Journal of Plant Physiology, 218,* 121–126.
73. Najeeb, U., Ahmad, W., Zia, M. H., Malik, Z., & Zhou, W., (2017). Enhancing the lead phytostabilization in wetland plant *Juncus effusus* L. through somaclonal manipulation and EDTA enrichment. *Arabian Journal of Chemistry, 10,* S3310–S3317.
74. Panhwar, Q. A., Jusop, S., Naher, U. A., Othman, R., & Razi, M. I., (2013). Application of potential phosphate-solubilizing bacteria and organic acids on phosphate solubilization from phosphate rock in aerobic rice. *The Scientific World Journal, 2013*, 1–10.

75. Pratas, J., Paulo, C., Favas, P. J., & Venkatachalam, P., (2014). Potential of aquatic plants for phytofiltration of uranium-contaminated waters in laboratory conditions. *Ecological Engineering, 69,* 170–176.
76. Pratush, A. K., & Hu, Z., (2018). Adverse effect of heavy metals (As, Pb, Hg, and Cr) on health and their bioremediation strategies: A review. *International Microbiology, 21,* 97–106.
77. Rau, N., Mishra, V., Sharma, M., Das, M. K., Ahaluwalia, K., & Sharma, R. S., (2009). Evaluation of functional diversity in rhizobacterial taxa of a wild grass (*Saccharum ravennae*) colonizing abandoned fly ash dumps in Delhi urban ecosystem. *Soil Biology and Biochemistry, 41,* 813–821.
78. Sakakibara, M., Ohmori, Y., Ha, N. T., Sano, S., & Sera, K., (2011). Phytoremediation of heavy metal contaminated water and sediment by *Eleocharis acicularis*. *Clean-Soil, Air, Water, 39*(8), 735–741.
79. Santibáñez, C., Verdugo, C., & Ginocchio, R., (2008). Phytostabilization of copper mine tailings with biosolids: Implications for metal uptake and productivity of *Lolium perenne*. *Science of the Total Environment, 395,* 1–10.
80. Seagraves, M. P., Riedell, W. E., & Lundgren, J. G., (2011). Oviposition preference for water-stressed plants in *Orius insidiosus*. *Journal of Insect Behavior, 24*(2), 132–143.
81. Shabani, N., & Sayadi, M. H., (2012). Evaluation of heavy metals accumulation by two emergent macrophytes from the polluted soil: An experimental study. *The Environmentalist, 32*(1), 91–98.
82. Shahid, M., Austruy, A., Echevarria, G., Arshad, M., Sanaullah, M., Aslam, M., Nadeem, M., Nasim, W., & Dumat, C., (2014). EDTA-enhanced phytoremediation of heavy metals: A review. *Soil and Sediment Contamination: An International Journal, 23*(4), 389–416.
83. Sheoran, V., Sheoran, A. S., & Poonia, P., (2016). Factors affecting phytoextraction: A review. *Pedosphere, 26*(2), 148–166.
84. Silver, S., & Hobman, J. L., (2007). Mercury microbiology: Resistance systems, environmental aspects, methylation, and human health. In: Nies, D. H., & Silver, S., (eds.), *Molecular Microbiology of Heavy Metals* (pp. 357–370). Springer, Berlin - Heidelberg, AG.
85. Sinha, S., & Mukherjee, S. K., (2008). Cadmium-induced siderophore production by a high Cd-resistant bacterial strain relieved Cd toxicity in plants through root colonization. *Current Microbiology, 56,* 55–60.
86. Sricoth, T., Meeinkuirt, W., Pichtel, J., Taeprayoon, P., & Saengwilai, P., (2018). Synergistic phytoremediation of wastewater by two aquatic plants (*Typha angustifolia* and *Eichhornia crassipes*) and potential as biomass fuel. *Environmental Science and Pollution Research, 25*(6), 5344–5358.
87. Srivastava, S., & Thakur, I. S., (2012). Biosorption and biotransformation of chromium by *Serratia* sp. isolated from tannery effluent. *Environmental Technology, 33,* 113–122.
88. Subiramani, S., Ramalingam, S., Muthu, T., Nile, S. H., & Venkidasamy, B., (2020). Development of abiotic stress tolerance in crops by plant growth-promoting rhizobacteria (PGPR). In: Kumar, M., Kumar, V., & Prasad, V., (eds.), *Phyto-Microbiome in Stress Regulation* (pp. 125–145). Springer Nature, Singapore.
89. Tangahu, V., Sheikh, A. S. R., Basri, H., Idris, M., Anuar, N., & Mukhlisin, M., (2011). A review on heavy metals (As, Pb, and Hg) uptake by plants through phytoremediation. *International Journal of Chemical Engineering, 2011*, 110–118.

90. Tong, Y. P., Kneer, R., & Zhu, Y. G., (2004). Vacuolar compartmentalization: Second-generation approach to engineering plants for phytoremediation. *Trends in Plant Science, 9*(1), 7–9.
91. Ulm, F., Avelar, D., Hobson, P., Penha-Lopes, G., Dias, T., Máguas, C., & Cruz, C., (2020). Sustainable urban agriculture using compost and an open-pollinated maize variety. *Journal of Cleaner Production, 212*, 622–629.
92. Verma, M., (2020). Ecotoxicology of heavy metals: Sources, effects and toxicity. Chapter 2. In: Bhat, R. A., Hakeem, K. R., & Dervash, M. A., (eds.), *Bioremediation and Biotechnology* (Vol. 2, pp. 13–23). Springer Nature, Switzerland AG.
93. Whiting, S. N., Desouza, M. P., & Terry, N., (2001). Rhizosphere bacteria mobilize Zn for hyperaccumulation by *Thlaspi caerulescens*. *Environmental Science Technology, 35,* 3144–3150.
94. Wu, C. T., Leubner-Metzger, G., Meins, F., & Bradford, K. J., (2001). Class I β-1,3-glucanase and chitinase are expressed in the micropylar endosperm of tomato seeds prior to radicle emergence. *Plant Physiology, 126,* 1299–1313.
95. Wu, S., Peng, X., Cheung, K., Liu, S., & Wong, M. H., (2009). Adsorption kinetics of Pb and Cd by two plant growth-promoting rhizobacteria. *Bioresource Technology, 100,* 4559–4563.
96. Wyke, S., Pena-Fernandez, A., Brooke, N., & Duarte-Davidson, R., (2014). The importance of evaluating the physicochemical and toxicological properties of a contaminant for remediating environments affected by chemical incidents. *Environment International, 72,* 109–118.
97. Xia, W., Du, Z., Cui, Q., Dong, H., Wang, F., He, P., & Tang, Y. C., (2014). Biosurfactant produced by novel *Pseudomonas* sp. WJ6 with biodegradation of n-alkanes and polycyclic aromatic hydrocarbons. *Journal of Hazardous Materials, 276,* 489–498.
98. Yadav, B. K., Siebel, M. A., & Van, B. J. J. A., (2011). Rhizofiltration of a heavy metal (Lead) containing wastewater using the wetland plant *Carex pendula. Clean-Soil Air Water, 39*, 467–474.
99. Yadav, K. K., Gupta, N., Kumar, A., Reece, L. M., Singh, N., Rezania, S., & Khan, S. A., (2018). Mechanistic understanding and holistic approach of phytoremediation: A review on application and future prospects. *Ecological Engineering, 120,* 274–298.
100. Yanai, J., Zhao, F. J., McGrath, S. P., & Kosaki, T., (2006). Effect of soil characteristics on Cd uptake by the hyperaccumulator *Thlaspi caerulescens*. *Environmental Pollution, 139*(1), 167–175.
101. Yin, Y., Wang, Y., Liu, Y., Zeng, G., Hu, X., Hu, X., Zhou, L., Guo, Y., & Li, J., (2015). Cadmium accumulation and apoplastic and symplastic transport in *Boehmeria nivea* (L.) gaudich on cadmium-contaminated soil with the addition of EDTA or NTA. *RSC Advances, 5*(59), 47584–47591.
102. Zhang, X., Su, C., Liu, X., Liu, Z., Liang, X., Zhang, Y., & Feng, Y., (2020). Effect of plant-growth-promoting rhizobacteria on phytoremediation efficiency of *Scirpus triqueter* in pyrene-Ni co-contaminated soils. *Chemosphere, 241,* 125–127.
103. Zhu, Y. G., & Rosen, B. P., (2009). Perspectives for genetic engineering for the phytoremediation of arsenic-contaminated environments: From imagination to reality. *Current Opinion in Biotechnology, 20*(2), 220–224.
104. Złoch, M., Kowalkowski, T., Tyburski, J., & Hrynkiewicz, K., (2017). Modeling of phytoextraction efficiency of microbially stimulated *Salix dasyclados* L. *International Journal of Phytoremediation, 19*(12), 1150–1164.

CHAPTER 2

ROLE OF ARCHAEA IN THE REMEDIATION OF CONTAMINATED SOILS

CHANDA PARULEKAR BERDE and VIKRANT B. BERDE

ABSTRACT

Bioremediation has emerged as an effective and alternative method of treating contaminated soils, and it mainly depends on microorganisms. Besides bacteria, archaea can also be used for the remediation purpose to remove hydrocarbons, because archaea can work in very harsh conditions, such as halophilic and acidophilic environments. The industries, which release their effluents with heavy metals (HMs) in soil, rivers, or ocean, can use Archaea for pre-treatment of waste effluents. In the present chapter, the use of various Archaeal domain organisms in remediation of contaminated natural resources has been discussed.

2.1 INTRODUCTION

Industrial pollution and its impact on soil, water, and sediments are becoming a main threat to the environment and health of living organisms. Biostimulation (involves nutrient addition or adjusting the conditions) and bioaugmentation (involves the adding of microbes) both can help in complete removal of contaminants. Bacterial domain plays a crucial role in bioremediation, and it has some limitations in extreme thermophilic and halophilic environments.

Many of *Archae*bacteria are inhabitable to environments, where most other organisms cannot survive. If such extreme environments get contaminated, it is a difficult task to remove pollutants from these environments. The chemical effluents or wastewaters from many industries have high salt contents, high temperatures, extremely alkaline or acidic waters, or even heavy metals (HMs) released from the production process. Under such conditions, extremophilic archaebacteria possess the ability to function in the elimination of pollutants.

Bioremediation involves mechanisms of degradation, eradication, immobilization, or detoxification of chemicals and hazardous materials that pollute the environment. Bioremediation technology involves the use of microorganisms to breakdown toxic compounds to harmless ones, such as CO_2, H_2O, and other inorganic complexes [86]. It is thus the application of microorganisms to degrade and transform the pollutants including hydrocarbons, oil, HMs, pesticides, dyes, etc. The rate of biodegradation depends on biotic and abiotic factors, such as pollutant concentration, nature of the pollutant, environmental conditions (EC), the physicochemical parameters, etc. [33]. The microorganisms that harbor enzymes required for degradation of the pollutants can utilize or degrade these compounds.

Bioremediation of organic pollutants is a very effective biological method that is less costly and does not harm the environment. The microorganisms found in normal EC cannot function efficiently under adverse environments. Their metabolic and degradation abilities are not optimal in extreme habitats. Whereas, the extremophilic microorganisms are adapted to survive and metabolize hydrocarbons under stressed conditions.

Such microorganisms are called extremophiles belonging to the archaea group, and they can grow in conditions unfavorable to eubacteria. Therefore, for the bioremediation and removal of contaminated soils, the focus is on an archaebacterial domain.

This chapter provides a holistic overall view of the role of Archaea in bioremediation. It explicitly covers archaeal distribution, bioremediation mechanisms, and applications in bioremediation and the future of research to reinforce our understanding of the applications of Archaea in the field.

2.2 BACKGROUND INFORMATION ON ARCHAE DOMAIN

Extremities of temperature, pressure, salinity, dryness, radiation, pH, or concentrations of HMs make the survival of organisms difficult. Microorganisms that are adapted and can tolerate the extreme conditions can survive in these ecosystems. These microorganisms are extremophiles classed under Archaebacteria. Depending on the extreme conditions, the archaea may be termed as: halophiles (salt), barophiles (pressure), thermophiles/

psychrophiles (temperature), acidophiles/alkalophiles (pH), etc. Extremophiles harbor unique cell walls and extremozymes that help them in the survival process [6, 15]. Thus, being acquainted with the harsh conditions, the extremophiles with pollutant degradation abilities can be utilized for bioremediation of contaminated soils in such conditions. These archaea have great capacity as new platforms, which can exploit the extreme environments and thrive.

Hypersaline environments constitute high salt concentrations, which create unfavorable stressful conditions for the survival and existence of the majority of the living organisms [52]. Natural hypersaline environments comprise of salty lakes, pans, marshes, flats, and effluents of production units of oil and gas. They are contaminated with crude oil and effluents of various industries; and are highly saline [65]. The biological communities that exist in such environments are the ones that can tolerate high salt concentrations. These organisms are called halophiles, and the presence of the high salts is a must for their survival [81], while the halotolerant strains can survive in the absence of high salt stress [69].

Members of Archaea are found to dwell in these extreme environments, and these are grouped as haloarchaea that are clustered in one class (Halobacteria in the phylum Euryarchaeota), requiring high salt concentrations of 1.8–5.0 M of NaCl [5, 30]. Along with high salinity, some hypersaline environments (such as salterns also experience high temperature). The haloarchaea found in such niches are thus called "thermophilic halophiles."

The thermophilic microorganisms grow optimally at 40°C or above; while the extreme thermophiles and hyperthermophiles grow best at >70°C and >80°C, respectively. Thermophiles belong to eubacteria and archaea [102].

2.3 ROLE OF ARCHAE IN BIOREMEDIATION

Coastal marine sediments are subjected to anthropogenic activities leading to the accumulation of contaminants in these environments. Hydrocarbon degrading archaea are ideal strains for bioremediation of habitats that are contaminated with oils and hydrocarbons. The conditions in these environments are hypersaline, extremely high temperature, or strictly anaerobic. There are very few studies on the application of bacteria from archaeal populations in pollution management.

2.3.1 *REMEDIATION OF HYPERSALINE ENVIRONMENTS*

Hypersaline environments have saturating salt concentrations of about 350 gL^{-1} as compared to seawater with 35 gL^{-1} [63]. These environments support the organisms that can survive the high salinity. These are the halophilic and halotolerant microorganisms belonging to haloarchaea [82]. Oil pollution in hypersaline environments thus is dependent on the archaeal components. Numerous reports have indicated the members of class Gamma-proteobacteria [12, 37, 40, 63, 73, 76, 99, 111] with degradative activities under saline stress. However, studies also show the negative influence of increase in salinity of the habitat to affect the breakdown of hydrocarbons by the archaea, thus affecting the bioremediation process [1, 28, 96].

The exact reason for the decreased rate of degradation cannot be pinpointed yet. Unavailability of hydrocarbons in higher salt concentrations due to the salting-out effect may result in reduced biodegradation under hypersaline environments [68, 72, 105]. The hydrocarbon metabolizing ability of halophiles is seen in isolates from hydrocarbon-contaminated and uncontaminated environments (Tables 2.1 and 2.2).

The Archaebacterial strains not exposed to oils or aromatic compounds (being isolated from uncontaminated environments) show good degradation properties. The presence of growth supplements along with the oil or aromatic compound was found to have a beneficial effect on the degradation by haloarchael communities [2, 19]. Surfactants (such as Tween 20, Tween 80) play a significant role in the degradation of oils; and are used in remediation of oil spills and hydraulic fracturing mixtures [77, 95].

Large diversity of haloarchael strains in hypersaline environments through requiring high temperature as well as requirements of growth supplements have tremendous metabolic degradation capabilities for oil, crude oil, hydrocarbons [3, 4, 10, 30]. Due to high aromatic degradation rates of these communities, they are excellent and promising candidates for bioremediation of hydrocarbons in hypersaline environments.

TABLE 2.1 Oil Degradation by Archaea Isolated from Contaminated Environments

Bacteria	Carbon Source	Salinity	References
Dietzia maris	Hydrocarbons, paraffin	58 to 175 $g.L^{-1}$ NaCl	[5]
Fusarium lateritium Drechslera sp.	Crude oil	10% NaCl	[78]

TABLE 2.1 *(Continued)*

Bacteria	Carbon Source	Salinity	References
Gammaproteobacteria	Monoaromatic	140 to 230 $g.L^{-1}$	[76]
	–	290 $g.L^{-1}$	[99]
Haloarchaea	Mixture of polyaromatic hydrocarbons	200 $g.L^{-1}$	[10]
Haloarcula Haloferax	Heptadecane	225 $g.L^{-1}$	[104]
Haloarcula vallismortis	Hydrocarbons	310 $g.L^{-1}$ (31%)	[104]
Haloferax *Halobacterium Halococcus*	Aliphatic and aromatic hydrocarbons	58 to 175 $g.L^{-1}$ NaCl	[5]
Halomonas strain	Crude oil, diesel, lubricant oil, or hexadecane	Saline water	[73]
Halophilic bacteria	Phenanthrene	5% NaCl	[107]
Marinobacter hydrocarbon oclasticus	Intermediate chain length aliphatics	35 $g.L^{-1}$ NaCl	[40]
Planococcus	BTEX	0.5% to 25% NaCl	[66]

TABLE 2.2 Aromatic Compound Degraders from Uncontaminated Environments

Bacteria	Carbon Source	Source of Isolation	References
33 strains of *Halobacteriaceae*	Crude oil, tween 80	Hypersaline lakes in Turkey	[83]
Haloarcula st. D1	4-Hydroxybenzoic acid	Industrial wastewater	[36]
Haloarcula st. MSNC 2, *Haloferax* st. MSNC 2, *Haloferax* st. MSNC 14, *Haloferax* st. MSNC 16	Heptadecane	Salt crystallization pond in Camargue, France	[104]
Haloarcula vallismortis	Aliphatic and aromatic hydrocarbons	Saltmarsh in France	[104]
Halobacteriaceae L1	Benzoic acid	Dead sea	[20]
Halobacterium	Alkanes	Hypersaline wastewater in Russia	[61]

TABLE 2.2 *(Continued)*

Bacteria	Carbon Source	Source of Isolation	References
Haloferax st. MSNC 14,	n-Heptadecane, pristane, and phenanthrene	Salt crystallization pond in Camargue, France	[31]
Haloferax mediterranei st. M-11	Hydrocarbons	Kalamkass oil field [Mangyshlak, Kazakhstan]	[115]
Haloferax sp., *Halobacterium piscisalsi*, *Halobacterium salinarum*, *Halorubrum ezzemoulense*, *Halorubrum* sp.	Naphthalene, phenanthrene, and pyrene	Çamaltı Saltern, Turkey	[35]
Haloferax st. HA-1, *Haloferax* st. HA-2, *Halobacterium* st. HA-3, *and Halococcus* st. HA-4	Alkane and aromatic degradation abilities	Coast of Arabian Gulf	[05]
Haloferax strains	Polycyclic aromatic hydrocarbons and crude oil	Salt marshes, salterns, salt flats, crystallizer ponds and the dead sea	[30]
Haloferax volcanii strain D1227	Monoaromatic carboxylic acids, 3-phenylpropionate	Saline oil-brine from Michigan	[39]
Halopenitus	4-Hydroxybenzoate	Great Salt Lake [Utah, USA]	[21]
Haloterrigena mahii sp. H13	1,2-Dichloroethane, naphthalene/anthracene, γ-hexachlorocyclohexane, 1-/2-methylnapthalene, and benzoate	Saltern pond in San Diego, CA, USA	[29, 30]
Natrialba sp. st. C21	Phenol, pyrene, and naphthalene,	Contaminated saline water in Ain Salah, Algeria	[56]

2.3.2 ROLE OF ARCHAEA UNDER THERMOPHILIC CONDITIONS

The majority of oil degraders being mesophilic are unable to efficiently bring about oil degradation at elevated temperatures (i.e., under thermophilic conditions). Therefore, thermophilic archaea are of abundant significance in the bioremediation of oil contamination in these regions. Some workers have reported the application of thermophilic microorganisms in the bioremediation process.

The research study by Jiang et al. [50] on bacterial isolates from soil contaminated with oil at Daqing Oilfield, China showed the ability of the bacterial isolates to degrade crude oil at 52 to 80°C. Audrius et al. [7] studied thermophilic bacteria isolated from an oilfield in Lithuania, having high temperature. These isolates could degrade naphthalene, anthracene, benzene, phenol, benzene-1,3-diol, protocatechuic acid at high temperatures. Thermophilic strains *Sulfolobus solfataricus* P2 and *Sulfolobus solfataricus* P1 degrading phenol were studied for the degradative pathways involved in the degradation at 80°C and highly acidic pH [16–18].

2.4 ROLE OF ARCHAEA IN OIL DEGRADATION: OCEANIC WATERS AND MARINE SEDIMENTS

The oil-degrading bacteria found in the oceanic waters and sediments are subjected to temperature and salinity variations. Mostly the oil-degrading organisms isolated are from the domain bacteria [14]. Few groups of Archaea were found including Marine Group II Archaea, Euryarchaeota, and Thaumarchaeota [90].

Archaeal members were indicated in marine sediments near Rio de Janeiro (Brazil) when heptadecane, naphthalene, or crude oil was added. The archaea detected belonged to uncultivable strains of Euryarchaeota [51]. Similar findings were also reported from oil-contaminated mangrove sediments [27, 49]. These findings suggest that uncultivable Archaea are involved in the removal of oil contamination in most of the marine environments, and this includes the haloarchaea and methanogens also [114]. These studies indicated that the increase in methanogens of the groups Methanosarcinales and Methanomicrobiales occurred on the addition of stimulants (such as methanol, acetate, etc.) [113]. There are similar reports of an increase in the haloarchaeal strains such as *Haloferax*, on addition of oil droplets and nitrate [13, 108, 109, 112]. The role of the haloarchaea in oil degradation in sediments has been documented from these results, which is important because the sediments being more contaminated as compared to the ocean waters.

2.5 ROLE OF ARCHAEA IN HEAVY METAL (HM) TREATMENT

HM contamination even at low concentration as well as depending on its oxidation/reduction state poses a serious problem to the environment, due to its toxicity to biotic factors in the environment [94]. Hence, it is crucial to remove the metal contamination. Several physical methods (such as ion exchange, precipitation, filtration, electrochemical treatment, or reverse osmosis) are being employed for the purpose [110]. However, being costly, there is a need for cost-effective and safe methods. Bioremediation using metal tolerant haloarchaea is an alternative.

Ranawat and Rawat [89] have reviewed the metal-tolerant thermophiles in bioremediation. Kashefi et al. [54] have described the role of *P. islandicum* in the Fe[III] oxide reduction and conversion U[VI] to magnetite and uraninite. This conversion takes place in hyperthermophilic habitats along with the reduction of several other metals, such as Tc[VII], Cr[VI], Co[III], Mn[IV], and Au[III] [53, 55]. The Kashefi and his colleagues also reported a reduction of Gold and Fe[III] by archaea *Pyrococcus furiosus* [55]. *Methanobacterium bryantii* was reported to exhibit copper chelating properties [57].

Some thermophilic archaea could cause oxidation of Arsenite [AsIII] to arsenate [AsV], which is a less toxic form. Some of the strains reported belong to Crenarchaeota and Euryarchaeota [48], *Sulfolobus acidocaldarius* st. BC [98], *Aeropyrum pernix* st. K1, *Pyrobaculum calidifontis* strain JCM 11548, and *Sulfolobus tokodaii* strain 7 [43, 64], species belonging to genus *Halorubrum* [47, 79]. Mercuric mercury [HgII] is highly toxic that is reduced to zero-valent mercury [Hg0] (which is volatile) by Crenarchaeota and Euryarchaeota [11], *Sulfolobus solfataricus* [97], *Halococcus*, *Halobacterium*, *Haloferax* [2].

There are several reports of archaeal strains involved in the reduction of U[VI] to U[IV], which resulted in precipitation of Uranium [75]. Some of the examples are hyperthermophile *Pyrobaculum* sp. [54], *Sulfolobus acidocaldarius* [91]. Haloarchaea has been reported to bioconvert cadmium. Examples of these haloarchaea are *Halobacterium noricense* [100] and *Haloferax* st. BBK2 [23].

2.6 ROLE OF ARCHAEA IN SOIL AND DRAINAGE AFFECTED BY THE MINING INDUSTRY

The mines are responsible for the release of large amounts of metals in the environment. These sites are habitat for the metal tolerant microbial communities that react with the metals leading to the generation of acid mine drainage. It is seen that microorganisms involved in mining also have utility in bioremediation of metal-contaminated sites, due to their higher tolerance levels to the metals. The bioremediation process involves the conversion of toxic metals into insoluble non-toxic or less toxic metal precipitates, resulting in the deposition of minerals of the corresponding metal ores. This is indeed an indication of presence of microbial activity.

Studies carried out at mining sites have shown the presence of microbial communities that are involved in the transformation of metals in these environments. Baker and Banfield [8] described thermoplasmatales that were found associated with environments, such as the iron mines. Ferroplasma, Thermoplasma. The five groups A-plasma, B-plasma, C-plasma, D-plasma, and E-plasma were found to colonize these sites. Archael communities were also reported by Druschel et al. [32] from Iron Mountain in California, an iron mining site. Archaea belonging to Thermoplasmatales was found to constitute the microbial community.

The presence of Thermoplasmatales was also reported by Golyshina et al. [41] from bioleaching pilot plants as well as shallow submarine and continental solfataras. Studies on acid mine drainage and the incidence of archaea have also been reported by other workers [103, 106]. Isolation of Archaeal iron reducers from very diverse environments has been reported. One such site is the marine geothermal regions. Both culturable and non-culturable archaebacterial have been encountered, such as, strains belonging to Crenarchaeota and Euryarchaeota [54] and *Aciduliprofundum boonei* strain T469 [93].

2.7 ROLE OF ARCHAEA IN THE TREATMENT OF EFFLUENTS FROM INDUSTRIAL PLANTS

There are numerous reports on the use of extremely thermophilic archaeal in industrial effluent treatment. Thermophilic enzymes find application in the removal of various contaminants from the industrial effluents before their discharge in the water bodies or wastelands. These wastewaters are characterized by high temperatures and high salinities. Removal of phosphate and arsenate, nitrogen, organophosphorus compounds by thermophilic enzymes have been described by several investigators [24, 44, 92].

2.8 ADVANTAGES AND DISADVANTAGES OF ARCHAE IN REMEDIATION

There are several disadvantages to the application of archaea for bioremediation purposes. Firstly, for using thermophilic archaeal strains, the temperature at the site should be optimal for growth. At this temperature, the hydrocarbons are volatile and oxygen availability is less.

Secondly, there are still lacunae in the information about anaerobic degradation by archaeal strains. In the hypersaline habitats, anaerobic conditions exist. Hence, contamination of these sites with hydrocarbons or oils will require the knowledge of tolerant organisms. According to Ramos et al. [88], having halo tolerance or temperature tolerance may still restrict the bioremediation of contaminants by this microflora. Extremophilic microbes can be employed in the cleanup of metal and radionuclide contaminants in hypersaline ecosystems. The metal and radionuclide recycling and recovery ability of archaea can be exploited for remediation purposes [71].

The use of archaea for bioremediation of hypersaline environments is cost-effective as compared to the other treatment methods of oil-contaminated sites [74]. It is a completely eco-friendly and sustainable method that makes use of natural inhabitants of the hypersaline ecosystem. Either indigenous or organisms that can survive in the harsh conditions are used for the purpose [26]. Another advantage of the use of archaea for bioremediation is that it is non-intrusive and has implementation ease [62].

2.9 COMMERCIAL APPLICATIONS

Halophilic archaea find applications in industrial processes, hypersaline wastewater treatment, evolutionary studies, fossil studies, possible involvement in petroleum product synthesis, etc. [63]. Microbial bioremediation technology has been successfully used for the elimination of environmental pollutants converting toxic to non-toxic compounds at low cost [86]. According to Holden et al. [45], the natural attenuation (NA) process was utilized for bioremediation of 25% of petroleum-contaminated land.

Bioremediation of hypersaline environments makes it mandatory to use halophilic or halotolerant microorganisms [85]. Application of the non-halophiles in bioremediation of contaminated hypersaline regions would require dilution of salt-laden soil and water, which increases the cost of the process.

Xenobiotic bioremediation is the most important application of halophilic archaea. Oil and gas recovery processes result in the production of saline wastewater as effluents [80]. Treatment of these wastewaters biologically requires the agents to be tolerant to salinity as well as the aromatic components in the waste. Hence, haloarchaea find application in bioremediation of contaminated saline soils, toxic effluents, and hydrocarbon-polluted environments [29]. *Hfx. mediterranei* is able to grow using oil as a carbon and energy source, while *Natrialba* sp. strain C21 can utilize hydrocarbons (such as phenol, naphthalene, and pyrene) as the sole source of carbon [56]. Similarly, several halophiles isolated from saline habitats with oil and hydrocarbon-degrading properties are enlisted in Tables 2.1 and 2.2. These haloarchaeal members are apt microbial agents for application in the biodegradation of hydrocarbon-polluted saline environments.

2.10 FUTURE PERSPECTIVES FOR USE OF ARCHAEA IN BIOREMEDIATION

The presence of archaebacteria is widespread in environments, where the conditions are unfavorable (such as hypersaline regions, polar regions, hydrothermal vents, etc.), for the growth of eubacteria [34, 59]. They are involved in the transformation processes in these habitats that influence the biogeochemical cycles, including degradation of organic compounds, and metal transformations, etc.

Most of the in-depth studies of these processes involving bacteria are numerous, while there are lacunae in the case of haloarchaea and archaea in general. There is a need to apply gene analysis and expression methods to understand the metabolism and utilization of transformation processes.

Studies on the unculturable archaea will help in generating the complete picture of the entire archaeal communities in the extreme habitats. The information will be applicable in the understanding of the enzymatic pathways involved in biodegradation and biotransformation of contaminants. Some workers have reported microflora with hydrocarbon-degrading abilities using culture-independent molecular techniques [22, 42, 46, 70]. These studies show that aerobic and anaerobic organisms together bring about the hydrocarbon biodegradation in the extreme habitats [58].

Several investigators have researched on the utilization of various aromatic contaminants including oils. These research reports indicate abundance of hydrocarbon-degrading archaea in these habitats focusing on their degradation potential [38, 42]. However, most of the research is laboratory-based. It is high time to have field-based (*in-situ*) research. It is thus a challenge to carry out *in situ* bioremediation of oil and petroleum-contaminated waste containing aromatics, aliphatics, and HMs [84].

There are few reports on the *in-situ* bioremediation of contaminated sites using the indigenous microorganisms, describing the difficulties encountered [67, 101]. Bioremediation carried out with biostimulation results in the growth of the required microflora and successful removal of contaminants has been reported [9, 24, 101, 114]. Thus, the indigenous microbial flora with bioremediation properties needs to be characterized to develop the bioremediation technology.

In most of the studies, there are reports on the less abundance of detection of Archaea. Since the methods used for non-culturable and culturable archaea are PCR (polymerase chain reaction) based and the primers used are universal primers, which are more apt for eubacteria. There is a very high possibility of missing archaeal strains. Hence the amplification post PCR should be done more cautiously not to miss the clades of Archaea [87]. Thus, more research to monitor and to study the degradation process and detect the application of haloarchaea in *in-situ* degradation is required. This will increase the understanding of biodegradation processes of archaeal strains in the contaminated hypersaline environments.

2.11 SUMMARY

Archaea are extremophiles that can live in environments that do not allow other organisms to survive. These are uninhabitable extreme environments with high salt concentrations, high and low temperatures, highly acidic and alkaline pH, etc. The remediation of these environments requires microorganisms that can survive the hostile conditions. Furthermore, effluents of many industries have high salinity, temperature, metal content; and the pH is either too acidic or too alkaline. Thus, when the effluents enter water bodies or are discharged on land, the environment gets polluted. It is the extremophilic Archaea that can survive the conditions as well as have the potential to remove the contaminants.

KEYWORDS

- **archaea**
- **bioremediation**
- **contamination**
- **extremophiles**
- ***Methanobacterium bryantii***
- **microorganism**

REFERENCES

1. Abed, R. M. M., Al-Thukair, A., & De Beer, D., (2006). Bacterial diversity of a *Cyanobacteria*l mat degrading petroleum compounds at elevated salinities and temperatures. *FEMS Microbiology and Ecology, 57*, 290–301.
2. Al-Mailem, D. M., Al-Deieg, M., Eliyas, M., & Radwan, S. S., (2017). Biostimulation Of indigenous microorganisms for bioremediation of oily hypersaline microcosms from the Arabian gulf Kuwaiti coasts. *Journal of Environmental Management, 193*, 576–583.
3. Al-Mailem, D. M., Al-Awadhi, H., Sorkhoh, N. A., Eliyas, M., & Radwan, S. S., (2011). Mercury resistance and volatilization by oil utilizing haloarchaea under hypersaline conditions. *Extremophiles, 15*, 39–44.
4. Al-Mailem, D. M., Eliyas, M., & Radwan, S., (2014). Enhanced bioremediation of oil-polluted, hypersaline, coastal areas in Kuwait via vitamin-fertilization. *Environmental Science and Pollution Research, 21*, 3386–3394.
5. Al-Mailem, D. M., Sorkhoh, N. A., Al-Awadhi, H., Eliyas, M., & Radwan, S. S., (2010). Biodegradation of crude oil and pure hydrocarbons by extreme halophilic archaea from hypersaline coasts of the Arabian Gulf. *Extremophiles, 14*(3), 321–328.
6. Amoozegar, M. A., Siroosi, M., Atashgahi, S., Smidt, H., & Ventosa, A., (2017). Systematics of haloarchaea and biotechnological potential of their hydrolytic enzymes. *Microbiology, 163*(5), 623–645.
7. Audrius, B., Gražina, G., Kalėdienė, L., & Nivinskiene, O., (2008). Degradation of naphthalene by thermophilic bacteria via a pathway, through protocatechuic acid. *Central European Journal of Biology, 3*(1), 61–68.
8. Baker, B. J., & Banfield, J. F., (2003). Microbial communities in acid mine drainage. *FEMS Microbiology and Ecology, 44*, 139–152.
9. Bell, T. H., Stefani, F. O. P., Abram, K., Champagne, J., Yergeau, E., Hijri, M., & St-Arnaud, M., (2016). A diverses soil microbiome degrades more crude oil than specialized bacterial assemblages obtained in culture. *Applied Environmental Microbiology, 82*, 5530–5541.
10. Bonfá, M. R. L., Grossman, M. J., Mellado, E., & Durrant, L. R., (2011). Biodegradation of aromatic hydrocarbons by haloarchaea and their use for the reduction of the chemical oxygen demand of hypersaline petroleum produced water. *Chemosphere, 84*(11), 1671–1676.

11. Boyd, E. S., & Barkay, T., (2012). The mercury resistance operon: From an origin in geothermal environment to an efficient detoxification machine. *Frontiers in Microbiology, 3*, 349–354.
12. Bruns, A., & Berthe-Corti, L., (1999). *Fundibacter jadensis*: New slightly halophilic bacterium, isolated from intertidal sediment. *Int. J. Syst. Bacteriol., 49*, 441–448.
13. Campeão, M. E., Reis, L., Leomil, L., De Oliveira, L., Otsuki, K., Gardinali, P., Pelz, O., et al., (2017). The deep-sea microbial community from the Amazonian basin associated with oil degradation. *Frontiers in Microbiology, 8,* 8. Article ID: 1019.
14. Catania, V., Cappello, S., Di Giorgi, V., Satisi, S., Di Maria, R., Mazzola, A., Vizzini, S., & Quatrini, P., (2018). Microbial communities of polluted sub-surface marine sediments. *Marine Pollution Bulletin, 131*(A), 396–406.
15. Cavicchioli, R., (2011). Archaea-timeline of the third domain. *Nature Reviews Microbiology, 9,* 51–61.
16. Christen, P., Vega, A., Casalot, L., Simon, G., & Auria, R., (2012). Kinetics of aerobic phenol biodegradation by the acidophilic and hyperthermophilic archaeon *Sulfolobus solfataricus* 98/2. *Biochemical Engineering Journal, 62,* 56–61.
17. Christen, P., Davidson, S., Combet-Blanc, Y., & Auria, R., (2011). Phenol biodegradation by the thermoacidophilic archaeon *Sulfolobus solfataricus* 98/2 in a fed-batch bioreactor. *Biodegradation, 22*(3), 475–484.
18. Comte, A., Christen, P., Davidson, S., Pophillat, M., Lorquin, J., Auria, R., Simon, G., & Casalot, L., (2013). Biochemical, transcriptional and translational evidences of the phenolmeta- degradation pathway by the hyperthermophilic *Sulfolobus solfataricus* 98/2. *PLoS One, 8*(12), 9. Article ID: e82397.
19. Corsellis, Y. Y., Krasovec, M. M., Sylvi, L. L., Cuny, P. P., & Militon, C. C., (2016). Oil removal and effects of spilled oil on active microbial communities in close to salt saturation brines. *Extremophiles, 20*(3), 235–250.
20. Cuadros-Orellana, S., Pohlschröder, M., Grossman, M. J., & Durrant, L. R., (2012). Biodegradation of aromatic compounds by a halophilic archaeon isolated from the dead sea. *Chemical Engineering Transactions, 27,* 13–18.
21. Dalvi, S., Youssef, N. H., & Fathepure, B. Z., (2016). Microbial community structure analysis of a benzoate-degrading halophilic archaeal enrichment. *Extremophiles, 20*(3), 11–321.
22. Das, D., Salgaonkar, B. B., Mani, K., & Braganca, J. M., (2014). Cadmium resistance in extremely halophilic archaeon *Haloferax* strain BBK2. *Chemosphere, 112*, 385–392.
23. Das, R., & Kazy, S. K., (2014). Microbial diversity, community composition and metabolic potential in hydrocarbon-contaminated oily sludge: Prospects for *in situ* bioremediation. *Environmental Science and Pollution Research, 21*(12), 7369–7389.
24. Dashti, N., Ali, N., Eliyas, M., Khanafer, M., Sorkhoh, N. A., & Radwan, S. S., (2015). Most hydrocarbon oclastic bacteria in the total environment are diazotrophic, which highlights their value in the bioremediation of hydrocarbon contaminants. *Microbes and Environments, 30,* 70–75.
25. Del, G. I., Coppolecchia, R., Merone, L., Carusone, T. M., Mandrich, L., Worek, F., & Manco, G., (2016). Efficient thermostable organophosphate hydrolase and its application in pesticide decontamination. *Biotechnology and Bioengineering, 113*(4), 724–734.
26. Dell'Anno, A., Beolchini, F., Rocchetti, L., Luna, G. M., & Danovaro, R., (2012). High bacterial biodiversity increases the degradation performance of hydrocarbons during

bioremediation of contaminated harbor marine sediments. *Environmental Pollution, 167,* 85–92.
27. Dias, A. C. F., Dini-Andreote, F., Taketani, R. G., Tsai, S. M., Azevedo, J. L., & De Melo, I. S., (2011). Archaeal communities in the sediments of three contrasting mangroves. *Journal of Soils Sediments, 11*(8), 1466–1476.
28. Díaz, M. P., Boyd, K. G., Grigson, S. G. W., & Burgess, J. G., (2002). Biodegradation of crude oil across a wide range of salinities by an extremely halotolerant bacterial consortium MPD-M, immobilized onto polypropylene fibers. *Biotechnology and Bioengineering, 79*(2), 145–153.
29. Ding, J. Y., & Lai, M. C., (2010). The biotechnological potential of the extreme halophilic archaea *Haloterrigena* sp. H13 in xenobiotic metabolism using a comparative genomics approach. *Environmental Technology, 31,* 905–914.
30. Ding, J. Y., Chen, S. C., Lai, M. C., & Liao, T. L., (2017). *Haloterrigena Mahii* Sp. Nov.: An extremely halophilic archaeon from a solar saltern. *International Journal of Systematic Evolutionary Microbiology, 67*(5), 1333–1338.
31. Djeridi, I., Militon, C., Grossi, V., & Cuny, P., (2013). Evidence for surfactant production by the haloarchaeon *Haloferax* sp. MSNC14 in hydrocarbon-containing media. *Extremophiles, 17*(4), 669–675.
32. Druschel, G. K., Baker, B. J., Gihring, T. M., & Banfield, J. F., (2004). Acid mine drainage biogeochemistry at iron mountain, California. *Geochemistry, 5*(T), 13–32.
33. El Fantroussi, S., & Agathos, S. N., (2005). Is bioaugmentation a feasible strategy for pollutant removal and site remediation? *Current Opinion in Microbiology, 8,* 268–275.
34. Elshahed, M. S., Najar, F. Z., Roe, B. A., Oren, A., Dewers, T. A., & Krumholz, L. R., (2004). Survey of archaeal diversity reveals an abundance of halophilic archaea in a low-salt, sulfide- and sulfur-rich spring. *Applied Environmental Microbiology, 70,* 2230–2239.
35. Erdoğmuş, S. F., Mutlu, B., Korcan, S. E., Güven, K., & Konuk, M., (2013). Aromatic hydrocarbon degradation by halophilic archaea isolated from Çamaltı Saltern, Turkey. *Water, Air and Soil Pollution, 224*(3), 7. Article ID: 1449.
36. Fairley, D. J., Boyd, D. R., Sharma, N. D., Allen, C. C. R., Morgan, P., & Larkin, M. J., (2002). Aerobic metabolism of 4-hydroxybenzoic acid in archaea via an unusual pathway involving an intramolecular migration. *Applied Environmental Microbiology, 68*(12), 6246–6255.
37. Fathepure, B. Z., (2014). Recent studies in microbial degradation of petroleum hydrocarbons in hypersaline environments. *Frontiers in Microbiology, 5,* 173–180.
38. Fowler, S. J., Toth, C. R., & Gieg, L. M., (2016). Community structure in methanogenic enrichments provides insight into syntrophic interactions in hydrocarbon-impacted environments. *Frontiers in Microbiology, 7,* 562–569.
39. Fu, W., & Oriel, P., (1999). Degradation of 3-phenyl propionic acid by *Haloferax* sp. D1227 *Extremophiles, 3*(1), 45–53.
40. Gauthier, M. J., Lafay, B., Christen, R., Fernandez, L., Acquaviva, M., Bonin, P., & Bertrand, J. C., (1992). *Marinobacter hydrocarbon oclasticus*: New, extremely halotolerant, hydrocarbon-degrading marine bacterium. *International Journal of Systematic Bacteriology, 42*(4), 568–576.
41. Golyshina, O. V., Yakimov, M. M., Lunsdorf, H., Ferrer, M., Nimtz, M., Timmis, K. N., Wray, V., et al., (2009). *Acidiplasma Aeolicum*: Novel euryarchaeon of the family ferroplasmaceae isolated from a hydrothermal pool, and transfer of *Ferroplasma*

cupricumulans to *Acidiplasma cupricumulans* comb. *International Journal of Systematic Evolutionary Microbiology, 59,* 2815–2823.

42. Head, I. M., Gray, N. D., & Larter, S. R., (2014). Life in the slow lane; biogeochemistry of biodegraded petroleum containing reservoirs and implications for energy recovery and carbon management. *Frontiers in Microbiology, 5*, 566–573.
43. Heinrich-Salmeron, A., Cordi, A., Brochier-Armanet, C., Halter, D., Pagnout, C., Abbaszadeh-fard, E., Montaut, D., et al., (2011). Unsuspected diversity of arsenite-oxidizing bacteria as revealed by widespread distribution of the Aoxb gene in prokaryotes. *Applied and Environmental Microbiology, 77*(13), 4685–4692.
44. Hennessy, J. E., Latter, M. J., Fazelinejad, S., Philbrook, A., Bartkus, D. M., Kim, H. K., Onagi, H., et al., (2018). Hyperthermophilic carbamate kinase stability and anabolic *in vitro* activity at alkaline pH. *Applied and Environmental Microbiology, 84*(3), 9. Article ID: e02250–17.
45. Holden, P. A., LaMontagne, M. G., Bruce, A. K., Miller, W. G., & Lindow, S. E., (2002). Assessing the role of *Pseudomonas aeruginosa* surface-active gene expression in hexadecane biodegradation in sand. *Applied and Environmental Microbiology, 68,* 2509–2518.
46. Hu, G., Li, J., & Zeng, G., (2013). Recent development in the treatment of oily sludge from petroleum industry: A review. *Journal of Hazardous Materials, 261,* 470–490.
47. Huber, R., Sacher, M., Vollmann, A., Huber, H., & Rose, D., (2000). Respiration of arsenate and selenate by hyperthermophilic archaea. *Systematic and Applied Microbiology, 23*(3), 305–314.
48. Jackson, C. R., Langner, H. W., Donahoe-Christiansen, J., Inskeep, W. P., & McDermott, T. R., (2001). Molecular analysis of microbial community structure in an arsenite-oxidizing acidic thermal spring. *Environmental Microbiology, 3*(8), 532–542.
49. Jeanbille, M., Gury, J., Duran, R., Tronczynski, J., Ghiglione, J. F., Agogué, H., Saïd, O. B., et al., (2016). Chronic polyaromatic hydrocarbon contamination is a marginal driver for community diversity and prokaryotic predicted functioning in coastal sediments. *Frontiers in Microbiology*, *7*, 7. Article ID: 1303.
50. Jiang, Y., Ma, H. X., Zong, L., Rong, Y., & Wang, Y., (2016). The screening and capability of heat resistant oil-degrading bacteria. *Chinese Journal of Microecology, 28*(6), 674–677.
51. Jurelevicius, D., De Almeida, C. C. R., Alvarez, V. M., Vollú, R. E., De Almeida, D. F., & Seldin, L., (2014). Response of the archaeal community to simulated petroleum hydrocarbon contamination in marine and hypersaline ecosystems. *Water, Air & Soil Pollution, 225*(2), 10. Article ID: 1871.
52. Kargi, F., & Dincer, A. R., (1996). Effect of salt concentration on biological treatment of saline wastewater by fed-batch operation. *Enzyme and Microbial Technology, 19,* 529–537.
53. Kashefi, K., & Lovley, D. R., (2000). Reduction of Fe[III], Mn[IV], and toxic metals at 100 °C by *Pyrobaculum islandicum. Applied and Environmental Microbiology, 66*(3), 1050–1056.
54. Kashefi, K., Moskowitz, B. M., & Lovley, D. R., (2008). Characterization of extracellular minerals produced during dissimilatory Fe[III] and U[VI] reduction at 100 °C by *Pyrobaculum islandicum. Geobiology, 6,* 147–154.

55. Kashefi, K., Tor, J. M., Nevin, K. P., & Lovley, D. R., (2001). Reductive precipitation of gold by dissimilatory Fe[III]-reducing bacteria and archaea. *Applied and Environmental Microbiology, 67*, 3275–3279.
56. Khemili-Talbi, S., Kebbouche-Gana, S., & Akmoussi, S., (2015). Isolation of an extremely halophilic archaeon *Natrialba* Sp. C21 is able to degrade aromatic compounds and to produce stable biosurfactant at high salinity. *Extremophiles, 19*, 1109–1120.
57. Kim, B. K., Pihl, T. D., Reeve, J. N., & Daniels, L., (1995). Purification of the copper response extracellular proteins secreted by the copper-resistant methanogen *Methanobacterium bryantii* BKYH and cloning, sequencing, and transcription of the gene encoding these proteins. *Journal of Bacteriology, 177*(24), 7178–7185.
58. Kolukirik, M., Ince, O., & Ince, B. K., (2011). Increment in anaerobic hydrocarbon degradation activity of halic bay sediments via nutrient amendment. *Microbial Ecology, 61*(4), 871–884.
59. Konneke, M., Bernhard, A. E., De La Torre, J. R., Walker, C. B., Waterbury, J. B., & Stahl, D. A., (2005). Isolation of an autotrophic ammonia-oxidizing marine archaeon. *Nature, 437*, 543–546.
60. Krzmarzick, M. J., Taylor, D. K., Xiang, F., & McCutchan, A. L., (2018). Diversity and niche of archaea in bioremediation. *Archaea, 2018*, 17. Article ID: 3194108.
61. Kulichevskaya, I. S., Milekhina, E. I., Borzenkov, I. A., Zvyagintseva, I. S., & Belyaev, S. S., (1991). Oxidation of petroleum hydrocarbons by extremely halophilic archeobacteria. *Microbiologia, 60*(5), 596–601.
62. Kumar, A., Bisht, B. S., Joshi, V. D., & Dhewa, T., (2011). Review on bioremediation of polluted environment: Management tool. *International Journal of Environmental Sciences, 1*(60), 1079–1093.
63. Le Borgne, S., Paniagua, D., & Vazquez-Duhalt, R., (2008). Biodegradation of organic pollutants by halophilic bacteria and archaea. *Journal of Molecular Microbiology and Biotechnology, 15*, 74–92.
64. Lebrun, E., Brugna, M., Baymann, F., et al., (2003). Arsenite oxidase, an ancient bioenergetic enzyme. *Molecular Biology and Evolution, 20*(50), 686–693.
65. Lefebvre, O., & Moletta, R., (2006). Treatment of organic pollution in industrial saline wastewater: A literature review. *Water Research, 40*(20), 3671–3682.
66. Li, H., Liu, Y. H., Luo, N., Zhang, X. Y., Luan, T. G., Hu, J. M., Wang, Z. Y., et al., (2006). Biodegradation of benzene and its derivatives by a psychrotolerant and moderately haloalkaliphilic *Planococcus* sp. strain ZD22. *Research in Microbiology, 157*(7), 629–636.
67. Lu, L., Huggins, T., Jin, S., Zuo, Y., & Ren, Z. J., (2014). Microbial metabolism and community structure in response to bioelectro-chemically enhanced remediation of petroleum hydrocarbon-contaminated soil. *Environmental Science and Technology, 48*, 4021–4029.
68. Mackay, D., Shiu, W. Y., Ma, K. C., & Lee, S. C., (2006). *Handbook of Physical-Chemical and Environmental Fate for Organic Chemicals* (2nd edn., p. 4216). Boca Raton - FL: CRC Press, Taylor & Francis.
69. Madigan, M. T., Martinko, J. M., Dunlap, P. V., & Clark, D. P., (2009). Brock - *Biology of Microorganisms* (12th edn., p. 1061). New York: Pearson Benjamin Cummings.
70. Maheshwari, D. K., & Saraf, M., (2015). *Halophiles: Biodiversity and Sustainable Exploitation* (p. 456). Combs. – Switzerland: Springer International Publishing.

71. Marques, C. R., (2018). Extremophilic microfactories: Applications in metal and radionuclide bioremediation. *Frontiers in Microbiology, 9*, 1191–1201.
72. McGenity, T. J., (2010). Halophilic hydrocarbon degraders. In: Timmis, K. N., (ed.), *Handbook of Hydrocarbon and Lipid Microbiology* (pp. 1939–1951). Berlin: Springer-Verlag.
73. Mnif, S., Chamka, M., & Sayadi, S., (2009). Isolation and characterization of *Halomonas* sp. strain C2SS100, a hydrocarbon-degrading bacterium under hypersaline conditions. *Journal of Applied Microbiology, 107*, 785–794.
74. Montagnolli, R. N., Matos, L. P. R., & Bidoia, E. D., (2015). Assessing *Bacillus subtilis* biosurfactant effects on the biodegradation of petroleum products. *Environmental Monitoring and Assessment, 187*(4116), 1–17.
75. Newsome, L., Morris, K., & Lloyd, J. R., (2014). The biogeochemistry and bioremediation of uranium and other priority radionuclides. *Chemical Geology, 363*, 164–184.
76. Nicholson, C. A., & Fathepure, B. Z., (2004). Biodegradation of benzene by halophilic and halotolerant bacteria under aerobic conditions. *Applied and Environmental Microbiology, 70*(12), 1222–1225.
77. Nyankson, E., Rodene, D., & Gupta, R. B., (2016). Advancements in crude oil spill remediation research after the Deepwater Horizon oil spill. *Water, Air, and Soil Pollution, 227*(1), 29.
78. Obuekwe, C. O., Badrudeen, A. M., Al-Saleh, E., & Mulder, J. L., (2005). Growth and hydrocarbon degradation by three desert fungi under conditions of simultaneous temperature and salt stress. *International Biodeterioration and Biodegradation Journal, 56,* 197–205.
79. Ordoñez, O. F., Rasuk, M. C., Soria, M. N., Contreras, M., & Farías, M. E., (2018). Haloarchaea from the Andean Puna: Biological role in the energy metabolism of arsenic. *Microbial Ecology, 76*, 695–705.
80. Oren, A., (2010). Industrial applications of halophilic microorganisms. *Environmental Technology, 31*, 825–834.
81. Oren, A., (2008). Microbial life at high salt concentrations: phylogenetic and metabolic diversity. *Saline Systems, 4*(1), 2.
82. Oren, A., (2009). Saltern evaporation ponds as model systems for the study of primary production processes under hypersaline conditions. *Aquatic Microbiology and Ecology, 56*, 193–204.
83. Ozcan, B., Ozcengiz, G., Coleri, A., & Cokmus, C., (2007). Diversity of halophilic archaea from six hypersaline environments in Turkey. *Journal of Microbiology and Biotechnology, 17*(6), 985–992.
84. Pal, S., Banat, F., Almansoori, A., & Haijra, M. A., (2016). Review of technologies for biotreatment of refinery wastewaters: Progress, challenges and future opportunities. *Environmental Technology Reviews, 5*, 12–38.
85. Pernetti, M., & Di Palma, L., (2005). Experimental evaluation of inhibition effects of saline wastewater on activated sludge. *Environmental Technology, 26*, 695–703.
86. Philip, J. C., Bamforth, S. M., Singleton, I., & Atlas, R. M., (2005). Environmental pollution and restoration: A role for bioremediation. In: Atlas, R. M., & Philip, J., (eds.), *Bioremediation: Applied Microbial Solutions for Real-World Environmental Cleanup* (pp. 1–48). Washington - DC: ASM Press.
87. Pinto, A. J., & Raskin, L., (2012). PCR biases distort bacterial and archaeal community structure in pyrosequencing datasets. *PLoS One, 7*(8), 7. Article ID: e43093.

88. Ramos, J. L., Marqués, S., & Van, D. P., (2011). Laboratory research aimed at closing the gaps in microbial bioremediation. *Trends in Biotechnology, 29*, 641–647.
89. Ranawat, P., & Rawat, S., (2018). Metal-Tolerant thermophiles: Metals as electron donors and acceptors, toxicity, tolerance, and industrial applications. *Environmental Science and Pollution Research, 25*(5), 4105–4133.
90. Redmond, M. C., & Valentine, D. L., (2012). Natural gas and temperature structured a microbial community response to the Deepwater Horizon oil spill. *Proceedings of National Academy of Science - USA, 109*(50), 20292–20297.
91. Reitz, T., Merroun, M. L., & Selenska-Pobell, S., (2008). Interactions of *Paenibacillus* sp. and *Sulfolobus acidocaldarius* strains with U[VI]. In: Merkel, B. J., & Hasche-Berger, A., (eds.), *Uranium, Mining, and Hydrogeology* (pp. 703–710). Berlin - Heidelberg: Springer.
92. Remy, B., Plener, L., Poirier, L., Elias, M., Daude, D., & Chabriere, E., (2016). Harnessing hyperthermostable lactonase from *Sulfolobus solfataricus* for biotechnological applications. *Scientific Reports, 6*, 11. Article ID: 37780.
93. Reysenbach, A. L., Liu, Y., Banta, A. B., Beveridge, T. J., Kirshtein, J. D., Schouten, S., Tivey, M. K., et al., (2006). A Ubiquitous thermoacidophilic archaeon from deep-sea hydrothermal vents. *Nature, 442,* 444–447.
94. Rocchetti, L., Beolchini, F., Hallberg, K. B., & Johnson, D. B., (2012). Effects of prokaryotic diversity changes on hydrocarbon degradation rates and metal partitioning during bioremediation of contaminated anoxic marine sediments. *Marine Pollution Bulletin, 64*(8), 1688–1698.
95. Rogers, J. D., Burke, T. L., Osborn, S. G., & Ryan, J. N., (2015). A framework for identifying organic compounds of concern in hydraulic fracturing fluids based on their mobility and persistence in groundwater. *Environmental Sci Technol Lett, 2*(6), 158–164.
96. Santos, H. F., Carmo, F. L., Paes, J. E. S., Rosado, A. S., & Peixoto, R. S., (2011). Bioremediation of mangroves impacted by petroleum. *Water Air Soil Pollution, 216*(1–4), 329–350.
97. Schelert, J., Dixit, V., Hoang, V., Simbahan, J., Drozda, M., & Blum, P., (2003). Occurrence and characterization of mercury resistance in the hyperthermophilic archaeon *Sulfolobus solfataricus* by use of gene disruption. *Journal of Bacteriology, 186*(2), 427–437.
98. Sehline, H. M., & Linström, E. B., (1992). Oxidation and reduction of arsenic by *Sulfolobus acidocaldarius* strain BC. *FEMS Microbiology Letters, 93*(1), 87–92.
99. Sei, A., & Fathepure, B. Z., (2009). Biodegradation of BTEX at high salinity by an enrichment culture from hypersaline sediments of rozel point at great Salt Lake. *Journal of Applied. Microbiology, 107*, 2001–2008.
100. Showalter, A. R., Szymanowski, J. E. S., Fein, J. B., & Bunker, B. A., (2016). An x-ray absorption spectroscopy study of Cd binding onto a halophilic archaeon. *Journal of Physics, Conference Series, 712*. Article 012079.
101. Smith, E., Thavamani, P., Ramadass, K., Naidu, R., Srivastava, P., & Meghraj, M., (2015). Remediation trials for hydrocarbon-contaminated soils in arid environments: Evaluation of bioslurry and biopiling techniques. *International Biodeterioration and Biodegradation, 101*, 56–65.
102. Stetter, K. O., (1998). Hyperthermophiles: Isolation, classification, and properties. In: Horikoshi, K., & Grant, W. D., (eds.), *Extremophiles: Microbial Life in Extreme Environments* (pp. 1–24). New York: Wiley-Liss.

103. Tan, G. L., Shu, W. S., Zhou, W. H., Li, X. L., Lan, C. Y., & Huang, L. N., (2009). Seasonal and spatial variations in microbial community structure and diversity in the acid stream draining across an ongoing surface mining site. *FEMS. Microbiology and Ecology, 70,* 121–129.
104. Tapilatu, Y. H., Grossi, V., Acquaviva, M., Militon, C., Bertrand, J. C., & Cuny, P., (2010). Isolation of hydrocarbon-degrading extremely halophilic archaea from an uncontaminated hypersaline pond (Camargue, France). *Extremophiles, 14*(2), 225–231.
105. Turner, A., & Rawling, M. C., (2001). The influence of salting-out on the sorption of neutral organic compounds in estuaries. *Water Research, 35*(18), 4379–4389.
106. Tyson, G. W., Chapman, J., Hugenholtz, P., Allen, E. E., Ram, R. J., Richardson, P. M., Solovyev, V. V., et al., (2004). Community structure and metabolism through reconstruction of microbial genomes from the environment. *Nature, 428*, 37–43.
107. Wang, C., Huang, Y., Zhang, Z., & Wang, H., (2018). Salinity effect on the metabolic pathway and microbial function in phenanthrene degradation by a halophilic consortium. *AMB Express, 8*, 1–13.
108. Wang, L., Huang, X., & Zheng, T. L., (2016). Responses of bacterial and archaeal communities to nitrate stimulation after oil pollution in mangrove sediment revealed by Illumina sequencing. *Marine Pollution Bulletin, 109*(1), 281–289.
109. Wang, X., Han, Z., Bai, Z., Tang, J., Ma, A., He, J., & Zhaung, G., (2011). Archaeal community structure along a gradient of petroleum contamination in saline-alkali soil. *Journal of Environmental Sciences, 23*(11), 1858–1864.
110. Williams, G. P., Gnanadesigan, M., & Ravikumar, S., (2012). Biosorption and bio-kinetic studies of halobacterial strains against Ni^{2+}, Al^{3+} and Hg^{2+} metal ions. *Bioresource Technology, 107*(1), 526–529.
111. Yakimov, M. M., Golyshin, P. N., Lang, S., Moore, E. R. B., Abraham, W. R., Lünsdorf, H., & Timmis, K. N., (1998). *Alcanivorax borkumensis*: New, hydrocarbon-degrading and surfactant producing bacterium. *International Journal Systematic Bacteriology, 48,* 339–348.
112. Yan, L., Yu, D., Hui, N., Naanuri, E., Viggor, S., Gafarov, A., Sokolov, S. L., Heinaru, A., & Romantschuk, M., (2018). Distribution of archaeal communities along the coast of the Gulf of Finland and their response to oil contamination. *Frontiers in Microbiology, 1*(9), 15–22.
113. Zhang, Z., & Lo, I. M. C., (2015). Biostimulation of petroleum hydrocarbon-contaminated marine sediment with co-substrate: Involved metabolic process and microbial community. *Applied Microbiology and Biotechnology, 99*(13), 5683–5696.
114. Zhang, Z., Lo, I. M., & Yan, D. Y., (2015). An integrated bioremediation process for petroleum hydrocarbons removal and odor mitigation from contaminated marine sediment. *Water Research, 83,* 21–30.
115. Zviagintseva, I. S., Beliaev, S. S., Borzenkov, I. A., Kostrikina, N. A., Milekhina, E. I., & Evanov, M. V., (1995). Halophilic archaebacteria from the kalamkass oil field. *Microbiologia, 64*(1), 83–87.

PART II

Role of Microalgae and Fungi in Bioremediation

CHAPTER 3

MICROALGAE AND CYANOBACTERIA: POTENTIAL BIOFERTILIZERS

JUNAID AHMAD MALIK

ABSTRACT

The utilization of different chemicals to improve the soil and agricultural productivity has led to serious consequences and instabilities within the soil and its biota, which may lead to ecological imbalance. This chapter explores the role of microalgae and cyanobacteria in maintaining soil health and its fertility. Microalgae can benefit the ecological aspects of the soil by producing biological soil crusts (BSC), extracellular polysaccharides (EPS), and antimicrobial compounds to enhance plant growth and fertility by increasing the soil carbon (C). Numerous microalgae are also helpful in heavy metal (HM) removal through the process of biosorption and bioaccumulation. They can also detoxify chemical agents (such as herbicides, insecticides, etc.), by manufacturing favorable hormones and enzymes.

3.1 INTRODUCTION

Expansion of soil toxins upsets soil microbial interaction to a greater extent [85]. Harmful impacts of substantial metals are widely noticeable in plants, humans, and animals. Stress by heavy metals (HMs) present serious dangers to agrarian crops by restraining plant development and harvest as recognized by numerous scientists [21, 30, 36, 47, 53].

Numerous agrarian soils need significant supplements (e.g., nitrogen, phosphorus, potassium, and iron), bringing about the utilization of about 200 million metric tons of chemical manures [29, 72], thus requiring a huge investment to manufacture such quota of fertilizers besides causing air contamination. Different conventional physiochemical strategies have been utilized in financially devolved nations for HM remediation and removal. At the same time, these techniques are costly apart from instigating other destructive and poisonous impacts prompting contamination of natural environs [3].

Along these lines, new practical and ecologically well-disposed strategies are prerequisites for the remediation of HMs. Microorganisms have the capacity to remove harmful contaminants without producing poisonous side effects [49]. Soil serves itself as the main pool of organic carbon regulating the fluctuations of greenhouse gases (GHGs). Therefore, it has importance for the sustainability of an ecological unit and habitat restorations [16].

Biofertilizers are chief alternatives for chemical manures, as they are green, vigorous, and contamination-free. Numerous investigations have indicated that biofertilizers can successfully progress the use of N, P, K, and other components within the soil, upgrade the stress resistance of plants, enhance the crop yield and essentially repress the noxiousness of pathogens. The application of photosynthetic microalgae to the soil increases the carbon content, enhances soil organic matter and upsurges the deployment of macro and micronutrients.

Microalgae have turn out to be appealing for their shares in promoting plant growth and repelling metal build-up in polluted sites [31]. Khan et al. [48] mentioned that expansion of microalgae recovers plant developmental characteristics and yet in addition can remove sullied metals present in soil. Microalgae also assist plants to conquer harmful impacts of metals and eventually improve the parameters of growth, development, and phytoremediation potential [94]. Borowitzka [11] reported that microalgae can produce significant products, such as astaxanthin, β-carotene, phycobilin pigments, and polyunsaturated fatty acids. Similarly, Bidyarani et al. [10] reported that *Cicer arietinum* L. yields were improved by the inoculation of microalgae and cyanobacteria, and the same was later confirmed by Renuka et al. [73] for wheat.

The microalgae are microorganisms with basic nourishing prerequisites for development and present significant manufacture of biomass, which in turn is utilized as a sustainable substitute for agricultural practices as biopesticides and other functions [87, 95]. A feature of microalgae generally exploited in horticulture and agriculture is their utilization for improving soil attributes in terms of bioavailability of macro- and micronutrients. Moreover, microalgae can improve the general soil health and prevent the soil erosion and degradation, by means of their passage and course of action [70].

Moreover, microalgae (as biofertilizers) are well known to recuperate the soils [10], aid in biological soil crust formation [74], wastewater treatment and HM removal [55].

Algae are a vast, varied group of photosynthetic eukaryotic life-forms, which are not certainly related (polyphyletic). This group of organisms range from unicellular microalgae (for example, *Chlorella* and the diatoms) to multicellular structures (for example, the giant kelp: about 50 m long). Within favorable settings, several strains of algae can reproduce their cells a number of times per day.

Soil quality enhancement tools used by microalgae/cyanobacteria are:

- Biological resistance of pathogens/pests;
- Breakdown of organic matter and decomposition;
- Contaminated topsoil exclusion;
- Easing microbial associations;
- Extra fertilizer use efficacy;
- Improving plant resistance and defense;
- Lifting physical, chemical, and biological properties of soil;
- Manufacturing phytohormones;
- Nutrient conversions (nitrogen fixation, P-solubilization/ mobilization);
- Soil acclimation;
- Soil recovery and reclamation.

This chapter assesses potential applications of microalgal biofertilizers and their role in allaying HM and toxicity in agricultural soils and promoting plant growth traits.

3.2 ROLE OF MICROALGAE IN SOIL FERTILITY

Synthetic composts contain numerous unproductive elements, which are aggregated in the soil and can harm soil fertility by combining with the metal ions. Microalgae, by assimilation process, are well known to amend CO_2, N_2 and can form symbiotic relationships through producing Extra Cellular Polysaccharides (EPS) to escalate the soil profile (Figure 3.1). Dineshkumar et al. [22] showed that *Chlorella vulgaris* and *Spirulina platensis* biomass (when mixed with cow-dung and presented to soil) could escalate growth, yield, organic content and minerals in onion, which indicates that microalgal biofertilizers are best alternatives to chemical fertilizers.

3.2.1 SEQUESTRATION OF CARBON

The contribution of carbon dioxide (CO_2) is substantial in GHG emissions. The utilization of synthetic manures has expanded the emission of ozone depleting gases by numerous means, for example, crude oil extractions, industrial, and manufacturing processes, mechanical composting; and these together have expanded CO_2 emanations. On the application of chemical fertilizers to the soil, GHGs in the form of CO_2 and NO_2 (nitrogen dioxide) are likewise emitted.

Cyanobacteria and blue-green algae are vital springs of organic matter in agricultural activities because they undoubtedly assimilate CO_2 into organic biomass by means of photosynthesis. Being the primary producers, algae contribute to 50% of the overall photosynthesis on earth. The organic carbon quota of the soil is comprehended to increase through the assimilation of CO_2 and excretion of carbon polysaccharides by algae; thus, encouraging the development of supplementary vegetation and fauna. Yilmaz and Sönmez [99] presented that the algal biofertilizers are essential to amend the soil profile and its carbon quota.

3.2.2 NITROGEN FIXATION

Nitrogen is the second vital supplement for the manufacture of biomass after carbon. Nitrogen fixation is a system, through which molecular nitrogen of the atmosphere is transformed into ammonia (NH_3) or associated nitrogenous combinations by microorganisms within the soil. Nitrogen present in the air is in the N_2 form, which is non-reactive and unusable molecule. This process plays a practical and vital role in continuing the natural nitrogen cycle as all living entities depend on such organic nitrogen amended by the microorganisms. With an aim to improve and maintain soil fertility, study, and application of nitrogen-fixing entities can dig out a foundation of agrarian stimulant and fertilizer.

The specialized heterocyst cells of cyanobacteria fix the atmospheric nitrogen and hence afford the stresses of soil flora and fauna. Numerous investigations have envisaged that the agrarian crops inoculated with microalgae and cyanobacteria have considerably amplified the nitrogen quota of the soil. Around 25–40% of synthetic nitrogen fertilizer can be set aside through the inoculation of soil with cyanobacteria. Pereira et al. [66] showed that around 50% of the fertilizer could be saved by the inoculation of filamentous nitrogen-fixing cyanobacteria in rice crops, not affecting the overall yield and grain quality. *Anabaena* biofilm inoculation in wheat upsurges soil nitrogen content by 40–57% [86].

Osman et al. [64] assessed the potential of *Nostoc entophytum* and *Oscillatoria augustissima* as biofertilizers for the pea plant. The utilization of cyanobacterial biofertilizer is known to escalate the rice harvests besides plummeting expenditure of synthetic fertilizers. The utilization of nitrogen fixing cyanobacteria improves the hay and yield of rice grains to a considerable level, besides subsiding 25% of synthetic fertilizer use [45]. The implementation of *Anabaena variabilis* in the rice fields has boosted the growth and yield of grain crops [81].

3.2.3 UPGRADING SOIL STRUCTURE AND BIOLOGICAL SOIL CRUSTS (BSC)

Soil erosion is mainly caused by the physical forces (for example, rain, storm, and farming practices), which in turn disturb the lushness and efficiency of agricultural soils. Various green algae and cyanobacteria can generate EPS within the nearby environs, which can considerably increase the carbon-based content of soil and check soil degradation and erosion.

The adhesiveness of EPS secreted by green algae and cyanobacteria assists to accumulate soil particles, which provide the smallest pore size for the growth and infiltration of crop roots. The soil masses thus formed play another role of keeping-up soil edifice by resisting to rainwater and its course to a definite mark. The soil combinations, which consist of fibers and EPS, were prepared after six weeks of controlled presentation with green algae and cyanobacteria [56].

Several soil qualities like soil profile and stability are known to remain improving by microalgae and cyanobacteria. Yilmaz and Sönmez [99] presented that the controlled inoculation of *Chlorella* alone has enhanced the strength of soil micro-aggregates. Cyanobacteria decrease soil erosion and runoff by supporting the extension in flinch period. Sadeghi et al. [76] have investigated the influence of *Nostoc, Oscillatoria, and Lyngbya* on the frequency of overflow at abandoned sites and reported that the initiation phase of soil erosion and runoff has prolonged by 38.2–168.8% after the successful cyanobacterial presentation.

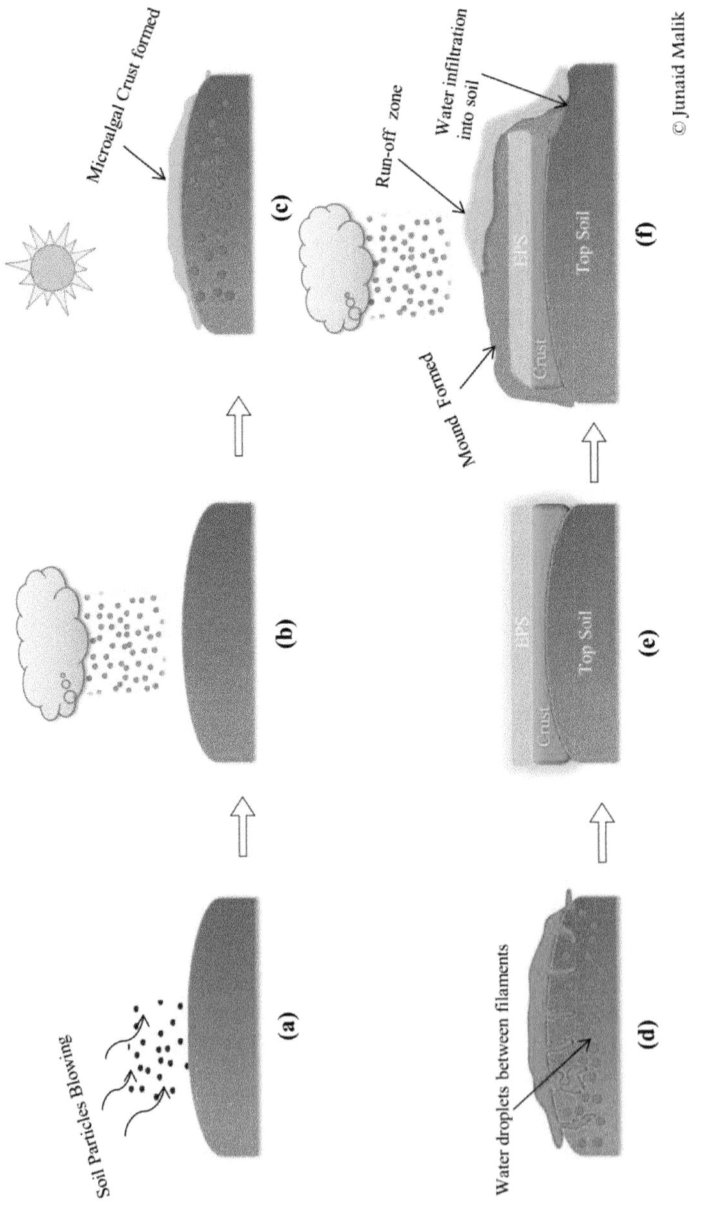

FIGURE 3.1 Formation of biological soil crust (BSC): (a) Upper soil particles driven by wind; (b) rain droplet infiltration into soil; (c) cyanobacteria/microalgae colonization to grow a scabbard; (d) soil/water assortment to setup a crust; (e) extracellular polysaccharides (EPS) created on the crust; and (f) mound formed, an overspill zone to re-route rainwater into soil and augments water infiltration.

Biological soil crust is the result of the integration among the soil microorganisms (such as cyanobacteria, microalgae, fungi, and bacteria), which play its role in soothing and composting the soil and adding carbon-based matter to it [13, 52]. BSCs are also indicators of organic vigor [82]. Numerous scientists have been studying to examine the BSC formation, particularly molded by microalgae and cyanobacteria [8, 9, 27, 35, 43, 44, 59, 79, 93]. Microalgae and cyanobacteria filaments along with their secretion function as bonding and pasting agents for the development of stable soil crust. Figure 3.1 delineates the development of the biological soil crusts (BSC) by microalgae and cyanobacteria and their implication in soil.

Biological soil crust (BSC) is formed by a close relationship among inorganic and organic materials with variable shares of soil microorganisms by secreting extracellular polysaccharides (EPS). The polysaccharides secreted by microalgae and cyanobacteria are structured into a condensed edifice with the soil particles establishing an assorted mass on the soil [59]. The measure and quality of EPS formed by microalgae and cyanobacteria under natural conditions differs through the soil type and its nature [35, 63], and in their growth settings [12, 39, 63]. Moisture availability recovers that activates the metabolic pathways in microalgae and cyanobacteria to improve the physical structure of EPS [50, 77, 78, 88]. Raised-up nutrient amounts by the cyanobacterial soil crusts are known to deliver a viable benefit to the soil flora by providing them promptness and accessibility of nutrients [33].

Biological soil crust formation by microalgae and cyanobacteria can also stabilize the soil surface and supports it to defend from erosion by water permeation [18]. EPS, the high molecular weight biopolymers, are secreted by microalgae and cyanobacteria into the nearby environs through their development and proliferation [19, 96] and remain glued loosely to the cell walls or are oozed out into the surroundings [75, 89]. Numerous microalgae, particularly red algae and cyanobacteria, produce diverse kinds of EPS, thus protecting the cells from adverse pressure [68].

Dunaliella salina is a vivacious microalga, which have abundant tolerance to salt stress, secretes EPS with great polysaccharide proportion, which ultimately adds to the endurance of *D. salina* under changing salt measures [60]. EPS from red microalgae and cyanobacterium *Arthrospira platensis* display antiviral, antibacterial, and antioxidant properties [71].

3.2.4 WASTELAND RECLAMATION

Microalgae and cyanobacteria are pervasive entities, which are well adapted to sustain in the drought, salty lands and metal sullied areas. They have been able to recover the wetlands owing to their special property of tolerating the extreme EC. Reclamation of eroded soils in deserts was studied by Trejo et al. [90] in terms of application and ability of *Chlorella aeruginosa* and *Spirillum Azotobacter*, which exhibited a substantial increase in the organic matter, microbial biomass and carbon content of the soil after three successive inoculations of the bacteria. The soil damaged by heat was known to be recovered through the combined presentation of *Oscillators*, *Candida*, and *Filaria* [2]. Microbial crust degraded in the soils heated with >350°C was recovered with the application of cyanobacteria owing to the intensification in a number of nitrate and nitrite producers.

Abed [1] indicated that microalgae and cyanobacteria have distinctive association in the restoration of lands adulterated with oil, besides supporting their mutual progression. Similar results were obtained by Chaillan et al. [14] while studying the involvement of cyanobacteria in petroleum polluted sites and their associations with other related microbes. Cyanobacteria are reported to trap extra sodium ions with EPS matrix, thereby preventing the sodium ion engagement by plants [83]; and on the degradation of cyanobacterial biofilm, the trapped ions were unconstrained for the environs.

Cyanobacteria and green microalgae are frequently used in the metal-polluted soils because of their greater HM removal ability and renovation of degraded lands.

3.3 HEAVY METAL (HM) REMOVAL THROUGH MICROALGAE

HMs include Arsenic, Cadmium, Chromium, Copper, Iron, Lead, Mercury, Silver, Zinc, etc. Anthropogenic and industrial activities are responsible for the release of HMs into the environment [101], causing serious environmental and health issues owing to their biodegradable nature [97]. Compared to other microorganisms, microalgae are well recognized to grow under the unfavorable conditions of HM stress and high salinity. Table 3.1 portrays some important microalgae species and their status for bioremediation of HMs.

Microalgae are known to involve the significant HMs for their cellular metabolism and enzymatic processes leading to the flocculation and consequently decrease the HM toxicity [7]. Microalgae decrease the effect of

HMs by forming protein-HMs complexes despite maintaining their own metabolism [69]. Microalgae through their metabolic processes can change, detoxify, and volatilize HMs and also pose no threat of inadvertent discharge in the environment owing to their non-pathogenic nature. They use the pollutants in their metabolism and nutrition, thereby degrading them enzymatically. Microalgae are having great metal-binding abilities as their cell-wall surfaces possess polysaccharides, proteins, and lipids, thereby providing more binding sites and functional groups for the attachment of HMs [100]. Moreover, microalgae in association with fungi and other photosynthetic eukaryotes are notorious for producing peptides, which form organometallic complexes after binding with HMs, consequently checking or defusing their possible poisonous influence [17].

Bioremediation of HMs by microalgae is a two-step process:

1. **Biosorption:** This of heavy metals is a reversible, passive, and rapid process whereby the metal ions (biosorbate) are binding to the surface of the microalgae (biosorbent), by forming a covalent bond between the microalgal cell wall and the metal ion.
2. **Bioaccumulation:** On the other hand, bioaccumulation is an active, irreversible, complex but much slower process in which the metal ions amass and accumulate inside the microalgal cells (biosorbent).

The HMs are vigorously carried through the cellular membrane and subsequently bind the proteins and peptides inside the cell [41, 67] as shown in Figure 3.2.

Morelli et al. [61] observed 35% inhibition of arsenate by microalgae *Phaeoductylum tricornutum* at 1 μm of Arsenic. Similarly, Levy et al. [51] envisaged that *Chlorella* sp. is tolerant to arsenic in comparison to *Monoraphidium arcuatum*. Huang et al. [38] comprehended the successful removal of arsenic by binding to extracellular cell-wall and active sites of *Anabaena affinis* and *Oscillatoria tenuis*; while by utilizing the cyanobacterium *Microcystis aeruginosa*, arsenic was amassed in the cytoplasm by intracellular bioaccumulation. Likewise, *Chlorella minutissima* displayed sturdy tolerance to cadmium and adsorption value of 35.65 mg/g [97]. Having a number of acidic functional groups for instance, carboxylic acid, *Parachlorella* sp. has proven an enhanced biosorbent for cadmium [23]. Husein et al. [40] reported that *Chlorella sorokiniana* tolerates 100 ppm of Chromium up to 72 h and reaches elimination capacity of 99.7% after 24 h presentation period.

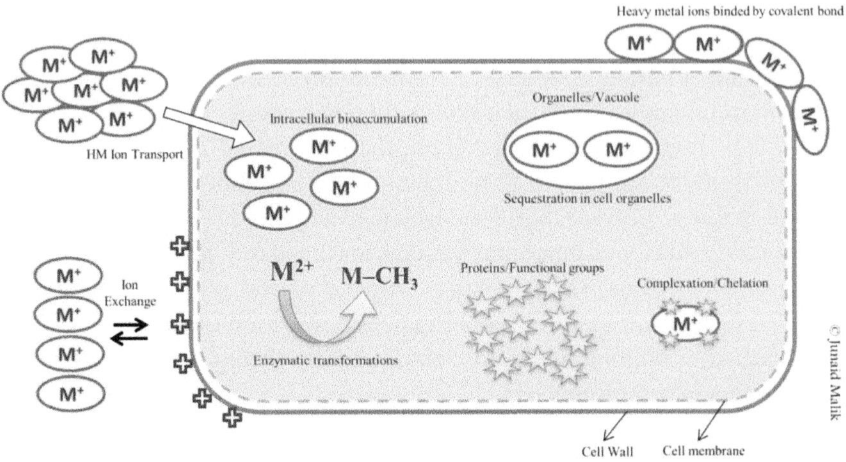

FIGURE 3.2 Procedure of heavy metal removal by microalgae.

TABLE 3.1 Some Microalgae Species and Their Use in Bioremediation of Heavy Metals

Microalgae	Heavy Metal	References
Anabaena sp.	Ni, Cr	[24, 65, 80]
Aulosira sp.	Ni	[24, 65]
Chlorella sp.	Cu, Al, Fe, Mg, Mn, Pb	[4, 5, 91, 98]
Chlorococcum sp.	Hg, Pb	[4, 5, 42]
Closterium sp.	Cu, Pb, Hg	[4, 5]
Fischerella sp.	Hg, Cd, Pb	[4, 5, 42]
Gloeocapsa sp.	Hg, Cd, Pb	[4, 5, 42]
Lyngbya sp.	Hg, Cd, Pb	[4, 5, 42]
Nostoc sp.	Pb, Se, Cr, Cd	[4, 5, 42, 65, 80]
Oscillatoria sp.	Cu, Pb, Ni, Cr, Zn, Co, Cd, Hg	[4, 5, 24, 42, 65, 80, 91]
Phormidium sp.	Hg, Cd, Pb	[4, 5, 42]
Scenedesmus sp.	Cu, Pb, Zn, Co, Cd, Hg	[4, 5, 42, 91]
Spirogyra sp.	Pb, Ni, Se, Co, Cd, Hg, As	[4, 5, 24, 42, 65]
Spirulina sp.	Cu, Pb, Hg	[4, 5, 42, 91, 98]
Stigonema sp.	Hg, Cd, Pb	[4, 5, 42]
Tolypothrix sp.	Hg, Cd, Pb	[4, 5, 42]
Tolypothrix sp.	Hg, Cd, Pb,	[4, 5, 42]
Westiellopsis sp.	Ni, Cr	[24, 65, 80]

3.4 PLANT GROWTH PROMOTERS AND DISEASE CONTROL

Microalgae have also proven imperative for the growth and development of the plants as they secrete important growth hormones. They are also known to improve the immune system of the plants, thus increasing their resistance to the diseases by secreting antibiotics.

3.4.1 PHYTOHORMONES

Phytohormones assume a significant character in the process of growth and development of plants. Externally supplemented synthetic phytohormones, though instigating an environmental risk, are very commonly used for the agronomic farming to improve crop efficiency and harvest. Various types of microalgae and cyanobacteria are well known to produce phytohormones (such as auxins, cytokinins, jasmonic acid, etc.), within their cells and release it into the neighboring environs, where they act as the stimulants for biotic entities. Mazhar et al. [58] envisaged that cyanobacterial auxin levels were increased when symbiotically associated with wheat, which directly indicates the manifestation of signal transduction process between the wheat plant and cyanobacteria. Hence the utilization of potential cyanobacteria excreting phytohormones can be an approach to stimulate plant growth.

Lv et al. [54] found that height, leaf size and number, stem diameter in cucumber was increased to a greater extent by using microalgal biomass as biofertilizer, which is similar to the measures of cucumber cultured through farmyard and inorganic fertilizer [25, 26]. The conclusion was moreover sturdy with past reports demonstrating that *Spirulina platensis* improved height, leaf number and length of roots [32, 95], altogether envisioning the robust evidences of microalgae as premium fertilizers and apt for total plant growth. *Chlorella vulgaris* and *Stichococcus hacillaris* can upsurge the production of extracellular abscisic acid (ABA) by 5–10 folds under stress conditions of salt, acid or drought [57].

3.4.2 IMPROVEMENT OF DEFENSE MECHANISMS

The microalgae and cyanobacteria are known to promote the defense mechanisms as they stimulate and enhance the activity of defense enzymes (such as β-1,3-endoglucanase, chitinase, catalase, peroxidase (POX), polyphenol oxidase (PPO), superoxide dismutase (SOD), phenylalanine ammonia-lyase

(PAL)) involved in different pathways and pathogenic mechanisms. Cyanobacterial inoculants considerably improved the activity of β-1,3-endoglucanase in spice plants [69].

Babu et al. [6] envisaged that treating wheat plants with high sugar have substantially increased the activities of POX, PPO, and PAL. Nonetheless, the practical appliance of green microalgae-based association with plants in boosting resistance mechanisms still needs to be studied, by investigating useful features of various green algae and cyanobacteria as biofertilizers. Numerous species of microalgae produce active metabolites and phytotoxins into the environment, which eventually encourage growth and resistance of the plants [92] (as shown in Table 3.2).

TABLE 3.2 Phytotoxins from Some Microalgal Species

Microalgae	Compound Secreted	Role/Target	References
Anabaena flosaquae	Peptides	Inhibition of *Chlamydomonas reinhardtii*	[46]
Calothrix sp.	Indolo phenanthridine calothrixin A	*Bacillus subtilis*	[62]
Chlorella vulgaris	C18 fatty acids	Growth stimulation Inhibition of *Pseudokirchneriella subcapitata*	[20]
Chlorella vulgaris *Anabaena* PCC 7120	Polysaccharides	Resistance for *Microcystis aeruginosa*	[28]
Isochrysis galbana	Diethyl malonate	Inhibition of: • *Dunaliella salina;* • *Platymonas elliptica;* • *Nitzschia Closterium;* • *Chaetoceros muelleri;* • *Chaetoceros gracilis;* • *Nitzschia closterium;* • *Chlorella minutissima;* • *Phaeodactylum tricornutum.*	[84]

3.4.3 DISEASE CONTROL

The use of biological alternatives to control pests and diseases in agrarian practices signifies an eco-friendly approach. Fungi, bacteria, and micro-algae are supreme common categories of microorganisms for the potential biological regulation of pests, besides increasing the nutrition and crop yields.

Cyanobacteria are reported to manufacture numerous antimicrobial compounds (such as hydrolases, ambigol A benzoic acid, carbamidocyclophane A and majusculonic acid), which inhibit and kill pathogens and pests. Such compounds work by altering the organization of cellular plasma membranes, deactivating enzymes as well as hindering protein synthesis. Bioactive substances from algae substantially inhibit and control soil borne diseases.

Cyanobacteria could inhibit cotton root rot by means of encouraging useful modifications in rhizosphere microbial populations. Cyanobacterial extracts of *Nodularia harveyana* as well as *Nostoc insulare* anti-cyanobacterial have proven to reduce the progression of bacterium *Staphylococcus aureus* and fungus *Candida albicans*. The resistance to diseases in tomato seedlings was increased to a considerable measure using *Anabaena variabilis* and *Anabaena oscillarioides* besides boosting plant growth [15]. Singh and Datta [81] envisaged the beneficial role of *Anabaena* for its resistance to herbicides besides increasing growth, yield, and fecundity of paddy.

Cyanotoxins produced by cyanobacteria are specifically accumulated by the plants, which indirectly pose a risk to humans and animals. Hence, strict control and verification of these toxins and their effects is a necessity in agriculture. The EPS polysaccharides from *P. cruentum* show antiviral, antibacterial, and immunomodulatory properties, hence is known to boost resistance against vesicular stomatitis virus (VSV), vaccinia virus (VACV), and against bacterium *Salmonella enteritidis* [37].

3.5 SUMMARY

This review on application of microalgae along with cyanobacteria and their consortia with various advantageous microbes has proven promising biofertilizers in agrarian practices and maintaining sustainability. Numerous studies reflect the capability of BSC to preserve and recover the biological nature and moistness of the soil. Microalgae are able to bio-transform and biodegrade the soil contaminants and HMs thereby can help in eliminating the toxins and preserves the natural environs. Secretion of enzymes, phytohormones, phytotoxins, and allelochemicals by microalgae and cyanobacteria play a fundamental role for the soil enrichment, plant growth, and establishing associations between other microorganisms.

KEYWORDS

- **biofertilizer**
- **biological soil crusts**
- **cyanobacteria**
- **microalgae**
- **nitrogen fixation**
- **phytohormones**
- **soil fertility**

REFERENCES

1. Abed, R. M., (2010). Interaction between *Cyanobacteria* and aerobic heterotrophic bacteria in the degradation of hydrocarbons. *International Biodeterioration and Biodegradation, 64*(1), 58–64.
2. Acea, M., (2003). *Cyanobacterial* inoculation of heated soils: Effect on microorganisms of C and N Cycles and on chemical composition in soil surface. *Soil Biology and Biochemistry, 35*(4), 513–524.
3. Ahluwalia, S. S., & Goyal, D., (2007). Microbial and plant derived biomass for removal of heavy metals from wastewater. *Bioresour. Technol., 98*, 2243–2257.
4. Ajayan, K. V., Selvaraju, M., & Thirugnanamoorthy, K., (2011). Growth and heavy metals accumulation potential of microalgae grown in sewage wastewater and petrochemical effluents. *Pak. J. Biol. Sci., 14*(16), 805–811.
5. Arunakumara, K. K. I. U., Zhag, X., & Song, X., (2008). Bioaccumulation of Pb2+ and its effects on growth, morphology and pigment contents of *Spirulina* (*Arthrospira*) *platensis*. *J. Ocean Univ. China, 7*(4), 397–403.
6. Babu, S., Bidyarani, N., & Chopra, P., (2015). Evaluating microbe-plant interactions and varietal differences for enhancing biocontrol efficacy in root rot challenged cotton crop. *European Journal of Plant Pathology, 1*(42), 345–362.
7. Balaji, S., Kalaivani, T., Shalini, M., & Gopalakrishnan, M., (2016). Sorption sites of microalgae possess metal-binding ability towards Cr(VI) from tannery effluents: Kinetic and characterization study. *Desalination and Water Treatment, 57*(31), 14518–14529.
8. Belnap, J., (1995). Surface disturbances: Their role in accelerating desertification. *Environmental Monitoring and Assessment, 37*, 39–57.
9. Belnap, J., (2003). World at your feet: Desert biological soil crusts. *Frontiers in Ecology and the Environment, 1*(5), 181–189.
10. Bidyarani, N., Prasanna, R., Babu, S., Hossain, F., & Saxena, A. K., (2016). Enhancement of plant growth and yields in chickpea (*Cicer arietinum* L.) through novel cyanobacterial and biofilm inoculants. *Microbiological Research, 188*, 97–105.
11. Borowitzka, M. A., (2013). High-value products from microalgae: Their development and commercialization. *Journal of Applied Phycology, 25*, 743–756.

12. Brüll, L. P., Huang, Z., & Thomas-Oates, J. E., (2000). Studies of polysaccharides from three edible species of *Nostoc* (*Cyanobacteria*) with different colony morphologies: Structural characterization and effect on the complement system of polysaccharides from *Nostoc* commune. *Journal of Phycology, 36*, 871–881.
13. Cameron, R. E., (1966). Desert algae: Soil crusts and diaphanous substrata as algae habitats. *Jet Propulsion Lab Technical Report, 32*, 1–41.
14. Chaillan, F., Gugger, M., Saliot, A., Coute, A., & Oudot, J., (2006). Role of *Cyanobacteria* in the biodegradation of crude oil by a tropical *Cyanobacterial* mat. *Chemosphere, 62*(10), 1574–1582.
15. Chaudhary, V., Prasanna, R., & Nain, L., (2012). Bioefficacy of novel *Cyanobacteria*-amended formulations in suppressing damping-off disease in tomato seedlings. *World Journal of Microbiology and Biotechnology, 28*(12), 3301–3310.
16. Chen, M. M., Zhu, Y. G., & Su, Y. H., (2007). Effects of soil moisture and plant interactions on the soil microbial community structure. *Eur. J. Soil Biol., 43*, 31–38.
17. Cobbett, C., & Goldsbrough, P., (2002). Phytochelatin and metallothioneins: Roles in heavy metal detoxification and homeostasis. *Annu. Rev. Plant Biol., 53*, 159–182.
18. Colesie, C., Gommeaux, M., Green, A. T. G., & Budel, B., (2014). Biological soil crusts in continental Antarctica: Garwood Valley, southern Victoria land and diamond hill, Darwin mountains region. *Antarct. Sci., 26*(2), 115–123.
19. Delbarre-Ladrat, C., Sinquin, C., & Lebellenger, L., (2014). Exopolysaccharides produced by marine bacteria and their applications as glycosaminoglycan-like molecules. *Front. Chem., 2*, 85–91.
20. Della, G. M., Zarrelli, A., Fergola, P., Cerasuolo, M., Pollio, A., & Pinto, G., (2010). Fatty acids released by *Chlorella vulgaris* and their role in interference with *Pseudokirchneriella subcapitata*: Experiments and modeling. *J. Chem. Ecol., 36*, 339–349.
21. Dheeba, B., Sampathkumar, P., & Kannan, K., (2015). Fertilizers and mixed crop cultivation of chromium tolerant and sensitive plants under chromium toxicity. *J Toxicol., 2015*, 9. Article ID: 367217; online, doi: 10.1155/2015/367217.
22. Dineshkumar, R., Subramanian, J., & Arumugam, A., (2020). Exploring the microalgae biofertilizer effect on onion cultivation by field experiment. *Waste and Biomass Valorization, 11*(1), 77–87.
23. Dirbaz, M., & Roosta, A., (2018). Adsorption, kinetic and thermodynamic studies for the biosorption of cadmium onto microalgae *Parachlorella* Sp. *Journal of Environmental Chemical Engineering, 6*(2), 2302–2309.
24. Diwan, S., (2007). Removal of Ni(II) from aqueous solution by biosorption using two green algal species *Oscillatoria* Sp. and *Spirogyra* sp. In: 5th *WSEAS International Conference on Environment, Ecosystems and Development* (p. 11). Tenerife, Spain, online.
25. Eifediyi, E. K., & Remison, S. U., (2010). Growth and yield of cucumber (*Cucumis sativus* L.) as influenced by farmyard manure (FYM) and inorganic fertilizer. *Journal of Plant Breeding and Crop Science, 2*, 216–220.
26. Eifediyi, E. K., & Remison, S. U., (2009). The effects of inorganic fertilizer on the yield of two varieties of cucumber (*Cucumis sativus* L.). *Report and Opinion, 1*, 74–80.
27. Eldridge, D. J., & Greene, R. S. B., (1994). Biological soil crusts: A view of their roles in soil and ecological processes in the rangelands of Australia. *Aust. J. Soil Res., 32*, 389–415.

28. El-Sheekh, M. M., Khairy, H. M., & El-Shenody, R., (2012). Algal production of extra and intracellular polysaccharides as an adaptive response to the toxin crude extract of *Microcystis aeruginosa*. *Iran. J. Environm. Health Sci. Eng., 9*, 10–18.
29. FAO, (2015). *World Fertilizer Trends and Outlook to 2018: Report* (p. 53). Rome: Food and Agriculture Organization of the United Nations.
30. Gill, R. A., Zang, L., Ali, B., Farooq, M. A., Cui, P., Yang, S., & Zhou, W., (2015). Chromium-induced physicochemical and ultrastructural changes in four cultivars of *Brassica napus* L. *Chemosphere, 120*, 154–164.
31. Glick, B. R., (2010). Using soil bacteria to facilitate phytoremediation. *Biotechnol. Adv., 28*, 367–374.
32. Grzesik, M., Romanowska-Duda, Z., & Kalaji, H. M., (2017). Effectiveness of *Cyanobacteria* and green algae in enhancing the photosynthetic performance and growth of willow (*Salix viminalis* L.) plants under limited synthetic fertilizers application. *Photosynthetica, 55*, 510–521.
33. Harper, K. T., & Belnap, J., (2001). The influence of biological soil crusts on mineral uptake by associated vascular plants. *Journal of Arid Environments, 47*, 347–357.
34. Hu, C., Liu, Y. D., & Zhang, D., (2002). Cementing mechanism of algal crusts from desert area. *Chinese Science Bulletin, 47*(16), 1361–1368.
35. Hu, C., Liu, Y., Song, L., & Zhang, D., (2002). Effect of desert soil algae on the stabilization of fine sands. *Journal of Applied Phycology, 14*, 281–292.
36. Hu, Y. A., Liu, X. P., Bai, J. M., Shih, K. M., Zeng, E. Y., & Cheng, H. F., (2013). Assessing heavy metal pollution in the surface soils of a region that had undergone three decades of intense industrialization and urbanization. *Environ. Sci. Pollut. Res. Int, 20*, 6150–6159.
37. Huang, J., Chen, B., & You, W., (2005). Studies on separation of extracellular polysaccharide from *Porphyridium cruentum* and its anti-HBV activity *in vitro*. *Chin. J. Mar. Drugs, 24*, 18–21.
38. Huang, W. J., Wu, C. C., & Chang, W. C., (2014). Bioaccumulation and toxicity of arsenic in *Cyanobacteria* cultures separated from a eutrophic reservoir. *Environmental Monitoring and Assessment, 186*(2), 805–814.
39. Huang, Z., Liu, Y., Paulsen, B. S., & Klaveness, D., (1998). Studies on polysaccharides from three edible species of *Nostoc* (*Cyanobacteria*) with different colony morphologies: Comparison of monosaccharide compositions and viscosities of polysaccharides from field colonies and suspension cultures. *Journal of Phycology, 34*, 962–968.
40. Husien, S., Labena, A., El-Belely, E. F., Mahmoud, H. M., & Hamouda, A. S., (2019). Absorption of hexavalent chromium by green microalgae *Chlorella sorokiniana*: Live planktonic cells. *Water Practice and Technology, 14*(3), 515–529.
41. Ibuot, A., Dean, A. P., McIntosh, O. A., & Pittman, J. K., (2017). Metal bioremediation by crmtp4 over-expressing *Chlamydomonas reinhardtii* in comparison to natural wastewater-tolerant microalgae strains. *Algal Research, 24*, 89–96.
42. Inthorn, D., Sidtitoon, N., Silapanuntakul, S., & Incharoensakdi, A., (2001). Sorption of mercury, cadmium and lead by microalgae. *Science Asia, 28*, 253–261.
43. Issa, O. M., Bissonnais, Y. L., Defarge, C., & Trichet, J., (2001). Role of a cyanobacterial cover on structural stability of sandy soils in the Sahelian part of Western Niger. *Geoderma, 101*, 15–30.

44. Issa, O. M., Trichet, J., Defarge, C., Couté, A., & Valentin, C., (1999). Morphology and microstructure of microbiotic soil crusts on a tiger bush sequence (Niger, Sahel). *Catena, 37*, 175–196.
45. Jha, M. N., & Prasad, A. N., (2006). Efficacy of new inexpensive cyanobacterial biofertilizer including its shelf-life. *World Journal of Microbiology and Biotechnology, 22*(1), 73–79.
46. Kearns, K. D., & Hunter, M. D., (2001). Toxin-producing Anabaena flos-aquae induces settling of *Chlamydomonas reinhardtii*, a competing motile alga. *Microb. Ecol., 42*, 80–86.
47. Khan, M. U., Shahbaz, N., Waheed, S., Mahmood, A., Shinwari, Z. K., & Malik, R. N., (2016). Comparative health risk surveillance of heavy metals via dietary foodstuff consumption in different land-use types of Pakistan. *Hum. Ecol. Risk Assess an Int. J., 22*, 168–186.
48. Khan, S., Afzal, M., Iqbal, S., & Khan, Q. M., (2013). Plant-bacteria partnerships for the remediation of hydrocarbon contaminated soils. *Chemosphere, 90*, 1317–1332.
49. Kothe, E., Bergmann, H., & Buchel, G., (2005). Molecular mechanisms in biogeo-interactions: From a case study to general mechanisms. *Chemie der Erde Geochem, 65*, 7–27.
50. Lange, O. L., (2003). Photosynthesis of soil crust biota as dependent on environmental factors. In: Belnap, J., & Lange, O. L., (eds.), *Biological Soil Crusts: Structure, Function and Management* (pp. 217–240). Berlin: Springer-Verlag.
51. Levy, J. L., Stauber, J. L., Adams, M. S., Maher, W. A., Kirby, J. K., & Jolley, D. F., (2005). Toxicity, biotransformation, and mode of action of arsenic in two freshwater microalgae (*Chlorella* Sp. and *Monoraphidium arcuatum*). *Environ. Toxicol. Chem., 24*(10), 2630–2639.
52. Li, X. R., Zhou, H. Y., Wang, X. P., Zhu, Y. G., & O'conner, P. J., (2003). The effect of sand stabilization and revegetation on cryptogam species diversity and soil fertility in the tengger Desert, Northern China. *Plant and Soil, 251*, 237–245.
53. Liu, L., Zhang, X., & Zhong, T., (2016). Pollution and health risk assessment of heavy metals in urban soil in China. *Human and Ecological Risk Assessment: An International Journal, 22*(2), 424–434. doi: 10.1080/10807039.2015.1078226.
54. Lv, J., Liu, S., Feng, J., Liu, Q., Guo, J., Wang, L., Jiao, X., & Xie, S., (2020). Effects of microalgal biomass as biofertilizer on the growth of cucumber and microbial communities in the cucumber rhizosphere. *Turkish Journal of Botany, 44*(2), 167–177.
55. MacGregor, J. T., (2006). *Genetic Toxicity Assessment of Microbial Pesticides: Needs and Recommended Approaches* (p. 17). Organization for Economic Cooperation and Development: Paris, France.
56. Malam, O., & Défarge, C., & Le Bissonnais, Y., (2007). Effects of the inoculation of *Cyanobacteria* on the microstructure and the structural stability of a tropical soil. *Plant and Soil, 290*(1), 209–219.
57. Maršálek, B., Zahradnícková, H., & Hronková, M., (1992). Extracellular production of abscisic acid by soil algae under salt, acid or drought stress. *Z. Naturforschung C, 47*, 701–704.
58. Mazhar, S., Cohen, J. D., & Hasnain, S., (2013). Auxin producing non-heterocystous *Cyanobacteria* and their impact on the growth and endogenous auxin homeostasis of wheat. *Journal of Basic Microbiology, 53*, 996–1003.

59. Mazor, G., Kidron, G. J., & Vonshak, A., (1996). The role of cyanobacterial exopolysaccharides in structure desert microbial crusts. *FEMS Microbiology- Ecology, 21*(2), 121–130.
60. Mishra, A., & Jha, B., (2009). Isolation and characterization of extracellular polymeric substances from micro-algae *Dunaliella salina* under salt stress. *Bioresour. Technol., 100*, 3382–3386.
61. Morelli, E., Mascherpa, M. C., & Scarano, G., (2005). Biosynthesis of phytochelatins and arsenic accumulation in the marine microalga *Phaeodactylum tricornutum* in response to arsenate exposure. *Biometals, 18*(6), 587–593.
62. Nga, T. D., Rickards, R. W., Rothschild, J. M., & Smith, G. D., (2000). Allelopathic actions of the alkaloid 12-epihapalindole E isonitrile and calothrixin A from *Cyanobacteria* of the genera *Fischerella* and *Calothrix*. *J. Appl. Phycol., 12*, 409–416.
63. Nicolaus, B., Panico, A., Lama, L., & Romano, I., (1999). Chemical composition and production of exopolysaccharides from representative members of heterocystous and non-heterocystous *Cyanobacteria*. *Phytochemistry, 52*, 639–647.
64. Osman, M. E. H., El-Sheekh, M. M., El-Naggar, A. H., & Gheda, S. F., (2010). Effect of two species of *Cyanobacteria* as biofertilizers on some metabolic activities, growth, and yield of pea plant. *Biology and Fertility of Soils, 46*(8), 861–875.
65. Parameswari, E., Lakshmanan, A., & Thilagavathi, T., (2010). Phytoremediation of heavy metals in polluted water bodies. *Electron. J. Environ. Agric. Food Chem., 9*(4), 808–814.
66. Pereira, I., Ortega, R., Barrientos, L., Moya, M., Reyes, G., & Kramm, V., (2009). Development of a biofertilizer based on filamentous nitrogen-fixing *Cyanobacteria* for rice crops in chile. *Journal of Applied Phycology, 21*(1), 135–144.
67. Pradhan, D., Sukla, L. B., Mishra, B. B., & Devi, N., (2019). Biosorption for removal of hexavalent chromium using microalgae *Scenedesmus* sp. *Journal of Cleaner Production, 209*, 617–629.
68. Prajapat, J., & Patel, A., (2013). Food and health applications of exopolysaccharides produced by lactic acid bacteria. *Adv. Dairy Res., 1*, 107–118.
69. Priatni, S., Ratnaningrum, D., Warya, S., & Audina, E., (2017). Phycobiliproteins production and heavy metals reduction ability of *Porphyridium* sp. *IOP Conf. Series: Earth Environ. Sci., 60*, 11–32.
70. Quadros, P. D., Zhalnina, K., Davis-Richardson, A. G., Drew, J. C., Menezes, F. B., Camargo, F. A. O., & Triplett, E. W., (2016). Coal mining practices reduce the microbial biomass, richness, and diversity of soil. *Appl. Soil Ecol., 98*, 195–203. doi: 10.1016/j.apsoil.2015.10.016.
71. Rafika, C., Lamia, T., Rym, B. D., & Omeya, E. A., (2011). Evaluation of cytotoxicity and biological activities in extracellular polysaccharides released by cyanobacterium *Arthrospira platensis*. *Braz. Arch. Biol. Technol., 54*, 831–838.
72. Rashid, M. I., Mujawar, L. H., Shahzad, T., Almeelbi, T., Ismail, I. M., & Oves, M., (2016). Bacteria and fungi can contribute to nutrients bioavailability and aggregate formation in degraded soils. *Microbiological Research, 183*, 26–41.
73. Renuka, N., Prasanna, R., Sood, A., & Ahluwalia, A. S., (2016). Exploring the efficacy of wastewater-grown microalgal biomass as a biofertilizer for wheat. *Environmental Science and Pollution Research, 23*, 6608–6620.

74. Rossi, F., Olguın, E. J., Diels, L., & Philippis, R. D., (2015). Microbial fixation of CO2 in water bodies and in drylands to combat climate change, soil loss and desertification. *New Biotechnology, 32*, 109–120. doi: 10.1016/j.nbt.2013.12.002.
75. Ruas-Madiedo, P., Hugenholtz, J., & Zoon, P., (2002). An overview of the functionality of exopolysaccharides produced by lactic acid bacteria. *Int. Dairy J., 12*, 163–171.
76. Sadeghi, S. H., Kheirfam, H., Homaee, M., Darki, B. Z., & Vafakhah, M., (2017). Improving runoff behavior resulting from direct inoculation of soil microorganisms. *Soil and Tillage Research, 171*, 35–41.
77. Satoh, K., Hirai, M., Nishio, J., Yamaji, T., Kashino, Y.; & Koike, H., (2002). Recovery of photosynthetic systems during rewetting is quite rapid in a terrestrial cyanobacterium, *Nostoc commune. Plant and Cell Physiology, 43*, 170–176.
78. Scherer, S., Ernst, A., Chen, T. W., & Böger, P., (1984). Rewetting of drought-resistant blue-green algae: Time course of water uptake and reappearance of respiration, photosynthesis and nitrogen fixation. *Oecologia, 62*, 418–423.
79. Schulten, J. A., (1985). Soil aggregation by cryptogams of a sand prairie. *J. Bot., 72*(11), 1657–1661.
80. Singh, S. K., Bansal, A., Jha, M. K., & Dey, A., (2011). An integrated approach to remove Cr(VI) using immobilized *Chlorella minutissima* grown in nutrient rich sewage wastewater. *Bioresour. Technol., 104*, 257–265.
81. Singh, S., & Datta, P., (2007). Outdoor evaluation of herbicide resistant strains of *Anabaena variabilis* as biofertilizer for rice plants. *Plant and Soil, 296*(1, 2), 95–102.
82. Stoddart, L. A., Smith, A. D., & Box, T. W., (1943). *Range Management* (p. 532). New York: McGraw-Hill.
83. Subhashini, D., & Kaushik, B., (1981). Amelioration of sodic soils with blue-green algae. *Soil Research, 19*(3), 361–366.
84. Sun, Y., Xu, S., Li, W., Zhang, J., & Wang, C., (2012). Antialgal substances from *Isochrysis galbana* and its effects on the growth of *Isochrysis galbana* and six species of feed microalgae. In: Zhu, E., & Sambath, S., (eds.), *Information Technology and Agricultural Engineering* (pp. 211–223). Berlin/Heidelberg, Germany: Springer.
85. Swain, H., & Abhijita, S., (2013). Nitrogen fixation and its improvement through genetic engineering. *J. Global Biosci., 2*(5), 98–112.
86. Swarnalakshmi, K., Prasanna, R., & Kumar, A., (2013). Evaluating the influence of novel *Cyanobacterial* biofilm biofertilizers on soil fertility and plant nutrition in wheat. *European Journal of Soil Biology, 55*, 107–116.
87. Tassara, C., Zaccaro, M. C., Storni, M. M., Palma, M., & Zulpa, G., (2008). Biological control of lettuce white mold with *Cyanobacteria. Int. J. Agric. Biol, 10*, 487–492.
88. Thomas, A. D., Hoon, S. R., & Linton, P. E., (2008). Carbon dioxide fluxes from *Cyanobacteria* crusted soils in the Kalahari. *Applied Soil Ecology, 39*, 254–263.
89. Trabelsi, L., Chaieb, O., Mnari, A., Abid-Essafi, S., & Aleya, L., (2016). Partial characterization and antioxidant and antiproliferative activities of the aqueous extracellular polysaccharides from the thermophilic microalgae *Graesiella* sp. *BMC Complementary and Alternative Medicine, 16*, 212–221.
90. Trejo, A., De-Bashan, L. E., Hartmann, A., Hernandez, J. P., Rothballer, M., Schmid, M., & Bashan, Y., (2012). Recycling waste debris of immobilized microalgae and plant growth-promoting bacteria from wastewater treatment as a resource to improve fertility of eroded desert soil. *Environmental and Experimental Botany, 75*, 65–73.

91. Wang, L., Min, M., Li, Y., Chen, P., Chen, Y., Liu, Y., Wang, Y., & Ruan, R., (2010). Cultivation of green algae *Chlorella* sp. in different wastewaters from municipal wastewater treatment plant. *Applied Biochemistry and Biotechnology, 162*(4), 1174–1186.
92. Wang, R., Xiao, H., Zhang, P., Qu, L., Cai, H., & Tang, X., (2007). Allelopathic effects of *Ulva pertusa*, *Corallina pilulifera* and *Sargassum thunbergii* on the growth of the dinoflagellates *Heterosigma akashiwo* and *Alexandrium tamarense*. *J. Appl. Phycol., 19*, 109–121.
93. West, N. E., (1990). Structure and function of microphytic soil crusts in wildland ecosystems of arid to semi-arid regions. *Advances in Ecological Research, 20*, 179–223.
94. Weyens, N., Truyens, S., Dupae, J., & Newman, L., (2010). Potential of the TCE-degrading endophyte pseudomonas putida W619-TCE to improve plant growth and reduce TCE phytotoxicity and evapotranspiration in poplar cuttings. *Environmental Pollution, 158*(9), 2915–2919.
95. Wuang, S. C., Khin, M. C., Chua, P. Q. D., & Luo, Y. D., (2016). Use of *Spirulina* biomass produced from treatment of aquaculture wastewater as agricultural fertilizers. *Algal Research, 15*, 59–64.
96. Xiao, R., & Zheng, Y., (2016). Overview of microalgal extracellular polymeric substances (EPS) and their applications. *Biotechnology Advances, 34*(7), 1225–1244.
97. Yang, J., Cao, J., Xing, G., & Yuan, H., (2015). Lipid production combined with biosorption and bioaccumulation of cadmium, copper, manganese and zinc by oleaginous microalgae *Chlorella minutissima* UTEX2341. *Bioresource Technology, 175*, 537–544.
98. Yao, J., Li, W., Xia, F., Zheng, Y., Fang, C., & Shen, D., (2011). Heavy metals and PCDD/solid waste incinerator fly ash in Zhejiang province, China: Chemical and bioanalytical characterization. *Environ. Monit. Assess, 184*(6), 3711–3720.
99. Yilmaz, E., & Sönmez, M., (2017). The role of organic/biofertilizer amendment on aggregate stability and organic carbon content in different aggregate scales. *Soil and Tillage Research, 168*, 118–124.
100. Yu, Q., Matheickal, J. T., & Kaewsarn, P., (1999). Heavy metal uptake capacities of common marine macro-algal biomass. *Water Res., 33*, 1534–1537.
101. Zakhama, S., Dhaouadi, H., & Mhenni, F., (2011). Nonlinear modulization of heavy metal removal from aqueous solution using *Ulva lactuca* algae. *Bioresource Technology, 102*(2), 786–796.

CHAPTER 4

MICROALGAE AS BIOFERTILIZER FOR SUSTAINABLE AGRICULTURAL PRODUCTIVITY

SAHIL, KIRTI SHARMA, and SHABIR A. RATHER

ABSTRACT

The unicellular or multicellular microorganisms inhabiting both freshwater and marine habitats are collectively referred to as microalgae that are rich in bioactive compounds and efficient photo-synthesizers. This chapter focuses on the utility of various microalgal species biofertilizers. They can enhance plant growth by facilitating nutrient availability and increasing soil fertility by stimulating soil microbial activity. Moreover, as they are already grown at an industrial scale for commercial purposes, protocols to grow them efficiently have already been established. Hence, microalgae present a very sustainable and viable option for the fertilizer industry.

4.1 INTRODUCTION

Food requirements for increasing population growth can be met by upgrading agricultural feasibility [22, 50] through the continuous efforts of farmers, who put in efforts to grow particular crops under varying climate conditions; and researchers with their new innovative ideas, techniques, and products can help to improve the crop quality and quantity [13, 16, 95]. Although some of these products include insecticides, pesticides, fertilizers, yet bioinsecticides, biopesticides, biostimulants, and biofertilizers are being developed.

Biostimulants can improve a plant's physiological processes and hence can lead to increased crop quality, quantity, nutrient uptake, and abiotic stress tolerance. Biofertilizers are basically organic compounds obtained from various organisms (such as bacteria, algae) that can promote plant growth or soil microorganism growth by providing essential nutrients (such as nitrogen, phosphorus) and other essential nutrients [5, 70, 90]. Biofertilizers can also enhance plant growth by activating microbial processes that convert non-accessible nutrients into more accessible forms that can be used by the plants. Biofertilizers are cheaper, renewable source of energy that mediates the renewal of various nutrients.

There are different classes of biofertilizers based on the type of microorganisms and their application, such as nitrogen fixators, phosphate solubilizing biofertilizers and plant growth-promoting rhizobacteria (PGPR), etc. The application of these biofertilizers has resulted in increased crop productivity in a short period, reduced soil and water pollution, increased soil fertility. It has been recently known that cyanobacteria and microalgae could be utilized as biofertilizers [65, 66, 88].

Microalgae are prokaryotic and eukaryotic microorganisms having the capability to fix carbon through photosynthesis and produce carbohydrates, proteins, and lipids. Structurally, they are either unicellular or multicellular. Microalgae can be prokaryotic (for example, cyanobacteria) or eukaryotic (for example, green algae and diatoms). An estimate shows that there are more than 50,000 microalgae species, but only 30,000 have been identified and studied [74].

The biomass production property of microalgae using solar energy, water, and carbon dioxide (CO_2) has been exploited to fabricate biofuel, animal feed, fertilizers, etc. [45]. Various antioxidants can be produced by microalgae [96] with qualities that can be controlled through clean growth media for microalgae growth, thus making products without any pesticidal and insecticidal properties. The products obtained from microalgae can be purified easily, resulting in more demand for microalgae. Moreover, the growth requirements for microalgae are also minimal as they require less fresh water and can grow in lands that are agriculturally less important or nonproductive.

Microalgae have the potential to act as biofertilizers and biostimulants [18, 34, 82–84], thereby improving the carbon footprint. Microalgae have also been used for wastewater management because of their capability to grow in wastewater that is rich in excess nutrients and accumulates these nutrients that make them an effective and cost-efficient tool for wastewater management.

Microalgae are capable of fixing atmospheric CO_2 using nutrients found in wastewater. Initially, most of the organic matter from wastewater is removed through secondary treatment with the help of microbes. During this process, not all the nitrogen gets fixed, bacterial biomass formation allows fixation of some of the nitrogen and the persisting nitrogen needs to be eliminated as it contributes to toxic algal bloom [26]. For this purpose, various methods are employed, but the use of microalgae proves more beneficial [10, 51]. Hence, microalgae produce biomass with trapped nutrients from wastewater that can be released for plant use [47].

Cyanobacteria are the simplest, autotrophic microalgae found in aquatic habitat (Figure 4.1); and have features, such as adaptation to extreme conditions, short generation time, ability to fix atmospheric nitrogen, making them an effective biofertilizer [1, 91]. Biofertilizers obtained from cyanobacteria can enrich the soil with various essential micronutrients for the growth of crop plants.

Microalgae not just act as biofertilizers, but also have the ability to reduce soil erosion by controlling the discharge of water into the surface of soil and also boost soil fertility [54, 101]. All these different properties of microalgae tend to make them the prime candidate to be used as a sustainable source that can be used for maintaining proper plant growth and soil health.

FIGURE 4.1 Cyanobacteria (blue-green algae) growing in freshwater.

This chapter explores the potential of microalgae as biofertilizers; their production methods and their key roles in improving plant health; plant growth and soil health. The issues on the consequences of using microalgae as biofertilizers or biostimulants are also dealt broadly in this chapter.

4.2 MICROALGAL BIOMASS PRODUCTION

The microalgal growth requirement includes a carbon source and energy source; for example, in the case of autotrophic microalgae, both these sources are different: (i) light and CO_2 as energy and carbon source, respectively; (ii) in case of heterotrophic microalgae, both energy and carbon sources are same that is an organic matter; and (iii) in case of mixotrophic microalgae, the energy source is light but both organic matter and CO_2 are crucial.

4.2.1 *PHOTOTROPHIC CULTURE*

In this type of culture, phototrophic microalgae use sunlight and lead to the assimilation of carbon and inorganic nutrients. This type of culture is beneficial as this involves the assimilation of atmospheric CO_2, but the availability of sunlight provides a major limitation to it. Phototrophic culture produces biomass with less contamination and impurities as compared to heterotrophic culture. Phototrophic culture can be used to produce microalgal biomass at a commercial scale in regions, where there is abundant sunlight that is easily available for the growth of microalgae.

4.2.2 *HETEROTROPHIC CULTURE*

The microalgae growing in the dark can be cultivated using this culture, thereby providing a solution to the limitation of sunlight observed in phototrophic culture. In this culture, microalgae grow by using the available organic substrate as carbon and energy sources. Heterotrophic culture can be contaminated by the organic substrate, and the presence of contaminants contributes to the limitation of this type of culture. Heterotrophic cultivation can be used for wastewater management and biofuel production. Heterotrophic microalgae are quite beneficial as they can use cheap carbon sources (glucose) and wastewater, making biofuel production a cost-efficient method.

4.2.3 *MIXOTROPHIC CULTURE*

Mixotrophic microalgae can use both light energy and organic matter to grow. Therefore, light energy is not a restriction for the performance of this type of microalgae. In this type of culture, improved growth rates are observed as compared to autotrophic or heterotrophic cultures. Mixotrophic production is an important process for the synthesis of biofuel from microalgae due to its ability to use both photosynthesis and organic matter for sustaining algal growth. The organic carbon source is required both in mixotrophic and heterotrophic production, and glucose has been found as the most preferred carbon source for microalgal growth [39]. Microalgae are cultured either in an open system or in a closed system:

- An open system implies that the system is exposed to the environment such as an open tank. The open system requires less money and has an easy setup, but the productivity is low with more contaminants.
- A closed system is not exposed to the environment, such as a photobioreactor. The closed system can be easily optimized and has higher productivity with lesser contaminants, but the setup is costly and complicated.

The type of system to be used depends on the product that is finally required like for biofuel, microalgae can be cultured in both systems. However, for pharmaceutics, microalgae need to be cultivated in a closed system as the contamination should be minimum.

4.2.3.1 OPEN PONDS

Lesser initial investment cost and an easier setup make this system more feasible. Open ponds can be constructed using concrete and kept shallow for proper penetration of sunlight, with a paddle wheel placed inside to allow proper mixing of incoming water. Biomass is produced using aquatic microalgae species and salty water. In open ponds, productivity changes with changes in photoperiod, temperature, and media composition. To increase productivity, a semi-continuous mode is preferred so that nutrients are replenished after a particular time period. The open system has been used for biofuel production and wastewater management. Biomass productivity in an open system is affected by the following factors:

1. **Biological Parameters:** such as salinity, the presence of pathogens, etc. Salinity, water temperature, and the presence of parasites or pathogens need to be estimated to allow the growth of a particular microalgal species.

2. **Surrounding Environmental Factors:** such as rainfall, sunlight, etc. These factors are more critical as compared to the microalgal species used for growth and biomass production.
3. **Type of Microalgal Species:** This decides the design of an open pond. Species having higher photosynthetic efficiencies are selected and based on species preference, the pond is designed in a particular way.

4.2.3.2 PHOTOBIOREACTORS

The use of photobioreactors is more costly because of light illumination, CO_2, feed, culture media availability; and the presence of the circulation system in a closed system requires additional money. However, microalgae cultivation is easy to control without contamination and higher productivity in these photobioreactors. Different designs of photobioreactors have been developed, such as Column photobioreactor; Flat plate photobioreactor; Tubular photobioreactor; and Membrane photobioreactor.

4.3 ROLE OF MICROALGAE IN MAINTAINING THE PLANT HEALTH

Microalgae are economical and renewable energy source, which has been used as biofuel, biofertilizers, biostimulants, and food. Biostimulants are microorganisms or substances that are once applied to plants in a limited amount, lead to enhancement of plant growth, nutrient use efficiency, and forbearance to various stresses. Microalgae have been used as biostimulants because of their capability to synthesize different bioactive substances, such as phytohormones, phenolics, fatty acids, and polysaccharides [11, 73, 76].

Microalgae have not just been observed to have sound effects on the growth of plants, but also provide them with the advantage to grow under various biotic and abiotic stresses due to changing environmental conditions (EC) and increasing global temperature. These stresses ultimately increase the synthesis of ROS (reactive oxygen species), eventually resulting in an oxidative damage [46]. A limited amount of ROS proves beneficial to the plant as it plays various roles, such as signaling.

Plants safeguard themselves from the ROS by synthesizing certain antioxidant molecules and enzymes that together make part of the antioxidant defense system to lower the increased amount of ROS [20, 53, 100].

Microalgae polysaccharides have been found to enhance these ROS scavenging enzymes, thereby leading to the reduction of these ROS. The presence of neutral sugars in polysaccharides allows them to attach to plasma membrane receptors, activating a signaling cascade. Algal polysaccharides act as microbial associated molecular patterns (MAMPs) that, after binding to plasma membrane receptors activate signaling cascade to activate calcium influx, ROS production, phenylalanine ammonia-lyase enzyme, and various ROS scavenging enzyme [15, 33].

It has been observed that the occurrence of various pathogens and parasites on plants leads to decreased growth and productivity. Plants respond to these attacks by employing various defense responses initiated by elicitors from pathogens, for example, chitin. Microalgae polysaccharides also act as elicitors and initiate a series of defense responses. Plant's innate immunity gets triggered by microalgae polysaccharides [15].

Production of pathogenesis related (PR) protein is a major part of the plant defense response triggered by microalgae polysaccharides. These PR proteins prevent the entry and spread of pathogens in plants [24]. Microalgae, for example, cyanobacteria can be used effectively to maintain plant health by acting as a biocontrol agent because of their nematocidal effect [25, 27] to hinder the hatching of nematode eggs and results in immobility and death of juvenile nematodes.

Microalgae have also been found to have antibacterial and antifungal activities [3, 37, 54], where the culture filtrate of microalgae has hydrolytic activity against plant pathogens. Based on the time of application of biostimulants, damage done by harsh EC can be restored to a particular extent, thereby positively favoring plant health [7, 98].

4.4 ROLE OF MICROALGAE IN PLANT GROWTH

The availability of different macronutrients and micronutrients in the soil contributes to increased plant growth and development. Nitrogen, phosphorus, and potassium are some of the essential macronutrients. However, changes in soil conditions (such as soil temperature, soil pH, etc.), and continuous growth of plants have led to the exhaustion of nutrients available in the soil [81, 85, 86]. Therefore, it is important to supplement the soil with by using chemical fertilizers, which have drawbacks such as environmental degradation and soil depletion. Therefore, microalgae can be used as efficient biofertilizers because of the following specific properties:

- The property of microalgae to replenish soil nutrients by producing specific polysaccharides makes microalgae an efficient biofertilizer to prevent soil pollution.
- Biomass produced by microalgae growing in agriculturally nonproductive lands or in wastewater can be further fractionated by microorganisms to produce bioenergy and liquid digestate. This liquid has nutrients that can be easily uptaken by plants; hence its introduction to the soil acts as a biofertilizer. It has also been studied that *Anabaena,* a microalgal strain, can increase soil fertility while decreasing soil density [77]. *Arthrospira* spp. in its marketing form contains organic and inorganic compounds, thus contributing to biofertilizer property of microalgae [65].
- Cyanobacteria can be employed as biofertilizer because of their capability to transform atmospheric nitrogen into a form that is convenient for plant uptake. The inorganic nitrogen in the atmosphere cannot be directly used by plants as it needs to be converted to an organic form, which is available for the plant. This conversion can be done by cyanobacteria [17, 38]. The introduction of filtrate from cyanobacteria and bacteria to *Lupinus termis* has been observed to increase the average seed germination as compared to seeds that were either treated with hormones or were untreated [94].
- Cyanobacteria also contain various bioactive compounds that affect plant growth, such as phytohormones. It has also been studied that algal species isolated from rice plants in Iran could produce phytohormones to improve the plant growth [87]. Non-nitrogen fixing cyanobacteria are indirectly involved in enhancing soil fertility by enriching the soil with potassium and phosphorus [80]. Cyanobacteria also enhance soil biological activity by increasing soil bacteria, nitrogenase, and denitrogenase activities.
- Various elements found in the soil (such as sodium, potassium, iron, copper, etc.), are beneficial for plant growth, but aluminum, cadmium, and silver elements do not have any necessary biological function. Because of their heavier atomic mass and metallic properties, these are known as heavy metals (HMs) contributing to HM stress. Some microbial species have the ability to sequester these HMs, thereby removing toxic HMs from the plant environment.

4.5 ROLE OF MICROALGAE IN MAINTAINING SOIL HEALTH

Algae have ubiquitous nature and can be found almost everywhere in soil, ice, hot springs, areas of hot and cold deserts, snowfields, etc. Microalgae are very simple because of their photosynthetic nature, and their tissues are not categorized into many different organs, which are observed in land plants. Therefore, algae are central-leading group of producers to produce organic compounds and also have a predominant role in the ground stage of the food chain with the following characteristics:

- It aids in the fixation of nitrogen;
- It is also crucial for controlling pests that reduce agricultural harvest;
- It is also crucial for developing a crust that is microbial in nature;
- It is also crucial for maintaining soil fertility;
- It is also crucial for wastewater purification and recirculation of this treated wastewater;
- Microalgae are established in various types of soils that assist in the improvement of soil by changing its characters, for example, appearance, ventilation [29];
- The crop yield in agriculture relies entirely on the fertility of the soil;
- The presence of microalgae reduces soil erosion by regulating the spurt of water into the soil.

Cyanobacteria are a large group of prokaryotes that somewhat resemble algae and higher plants by virtue of being capable of releasing oxygen in the presence of light. These microorganisms are scattered in the world and promote growth and development of plant because they can:

- Contribute to the ecosystems of men by improving the fertility of the soil;
- Generate various biological substances which are bioactive in nature;
- Have a role in HM biosorption [29].

4.5.1 ROLE IN IMPROVEMENT OF SOIL FERTILITY

Some cyanobacteria have the capability to convert nitrogen to ammonia, which is present in the atmosphere, where oxygen is evolved due to photosynthesis in that same cell, which is harmful to this transformation of nitrogen to a more accessible form. Because the evolution of oxygen is a prejudice for the conversion of atmospheric nitrogen to ammonia (a utilizable form of nitrogen), organisms like cyanobacteria have evolved themselves in a way that now they have specialized cells that are free from any oxygen thereby improving the level of nitrogen fixed by these organisms. These specialized oxygen-free compartments or cells are known as heterocysts that are anaerobic or oxygen-free, majorly due to respiration, performing nitrogen fixation that is quite sensitive to the presence of oxygen. Regulation of this nitrogen fixation has been investigated exclusively in the organisms with heterocyst.

4.5.2 ROLE IN NITROGEN FIXATION

It has also been demonstrated that cyanobacteria can act as nitrogen fixating agents in agricultural soils. Many experiments have shown the importance of cyanobacteria in rice fields as a nitrogen fixating agent [28]. The examples of fixing atmospheric nitrogen are *Nitrosomonas* and *Spirillum* that have a significant role in soil fertility improvement [78].

4.5.3 ROLE IN ORGANIC MATTER FORMATION

Microalgae in soil form the major repertoire of organic matter [30]. The decomposition of algae results in the building up of organic matter that ultimately gets blended with the soil and this glue-like substance acts as an agent, which can bind the soil particles and increase the humus content of soil thereby rendering it livable for many other plant species after few years [42]. The collection of humus is crucial for soil moisture-holding capacity [6].

4.5.4 ROLE IN THE PRODUCTION OF EXTRACELLULAR SUBSTANCES

The cyanobacteria can be used as a biofertilizer because they can produce certain bioactive substances to enhance plant growth and development. Researches show that biofertilizers in the form of microorganisms can aid the plant by synthesizing certain growth encouraging regulators, polypeptides, vitamins, amino acids, substances that are antifungal and antibacterial agents that utilize phytopathogen biocontrol and certain exopolysaccharides to modify the physicochemical structure of soil and the activity of exoenzyme.

4.5.5 ROLE IN UPTAKE OF NITROGEN AND PHOSPHORUS

Blue-green algae have some species, which can solubilize phosphate in the soil. In the majority of the aquatic systems, nitrogen, and phosphorus are primary limiting nutrients. Microalgae modify themselves to search their environment for resources by doing changes in structure, storage quantity, or increase capability of the utilization of resources [89]. These adaptations taken by algae require biochemical and physiological changes; algae also excrete substances that improve the accessibility of nutrients.

Algae contain specialized heterocysts for the conversion of atmospheric nitrogen to a form easily uptaken by the plants and thereby fulfill the requirement of soil micro- and macrofauna and flora [4, 31, 99]. Inoculation of blue-green algae can help in 25 to 40% decrease in use of chemical nitrogen fertilizers [48, 49, 58].

The use of microalgae can augment the nitrogen content of the soil, growth hormones production, photosynthesis, biocidal properties and rehabilitation of polluted sites by solubilizing of macro and micronutrients [68]. Microalgae also improve the carbon found in organic compounds present in the soil, boost the soil aggregate stability, enhance the soil microbial activity, and make various types of soil nutrients (macronutrients and micronutrients) accessible to the plants.

4.5.6 ROLE IN MANUFACTURING OF GROWTH HORMONES

Growth hormones are important for the growth of plants, and the exogenous supplementation of phytohormone in agriculture has been used to increase crop yield and also for controlling unwanted plants like weeds [2, 14]. Microalgae can be used as a biofertilizer because they can produce different phytohormones [92], such as cytokinin, jasmonic acid, etc. (Table 4.1). Plants use these as biofertilizers [44]. Microalgae and blue-green algae have intracellular hormones; and some are also known to add on hormones by producing or secreting hormones in the growing medium, where the plant is growing and in the external environment [40, 41, 44]. It has been shown that the impact of hormones presents extracellularly in *Aphanothece* sp. The MBDU 515 was found identical to hormones (such as IBA and IAA) that are artificially developed and are used *in vitro* multiplication of *Arachis hypogaea* and *Moringa oleifera* [19].

TABLE 4.1 Examples of Microalgae to Produce Plant Growth Hormones

Microalgae	Plant Growth Hormones
Anabaena sp.	Cytokinin
Chlorella	Cytokinin
Nannochloropsis	IAA
Scenedesmus	Gibberellins

4.6 ROLE OF MICROALGAE IN BIOREMEDIATION OF SOIL

Microalgae are also used for the bioremediation of soil in many ways. In dry and deserted regions, the soil is mostly saline and soil reclamation is difficult. Many experiments are carried out to check the consequences of the application of salt on plant growth, its metabolism, and the production of the plant and algae [93]. Farmers use hormones (such as gibberellic acid) to enhance the salt tolerance characteristics of plants, but excessive use of growth regulators is non-practical and expensive. It has been proved by research studies that algae play an essential functional role in soil reclamation, enhanced fertility of the soil, and improved plant health. Many microorganisms (including cyanobacteria) are able to focus on ions of metal [81]. Mechanisms followed by cyanobacteria and microalgae for resistance to HMs are:

- Environmental factors;
- Nonspecific protective mechanism of the cell;
- The specific protection mechanism depends on the influence of a metal species being toxic for the cell, in which it is present.

Another intracellular protective mechanism involves biosorption of HMs as the primary mechanism, where there is an exchange of ions in the outer cyanobacterial wall because of the presence of a mucilaginous sheath that behaves as an external vacuole. Properties of these microalgae as metal binders are perhaps because of the presence of a high density of ionic charges, especially carboxyl groups known to be present in the capsular polymer. This cyanobacterium might have more effectiveness in absorption while growing in an environment that is polluted or in the utilization of biomass for the elimination of HMs existing in the soil [21].

By using microorganisms (such as blue-green algae and others), we can overcome the problem of different types of pollutants spreading in the environment. Microorganisms can survive everywhere in different EC. The main process by microorganisms is to degenerate and change pollutants, such as hydrocarbons, oil, HMs, pesticides, dyes, etc.

In recent years research on microorganisms and microalgae in bioremediation is an excellent functional area because microorganisms are renewable and reassuring valuable genetic material to solve environmental constraints (Figure 4.2). Word bioremediation and biodegradation are more interchangeable words.

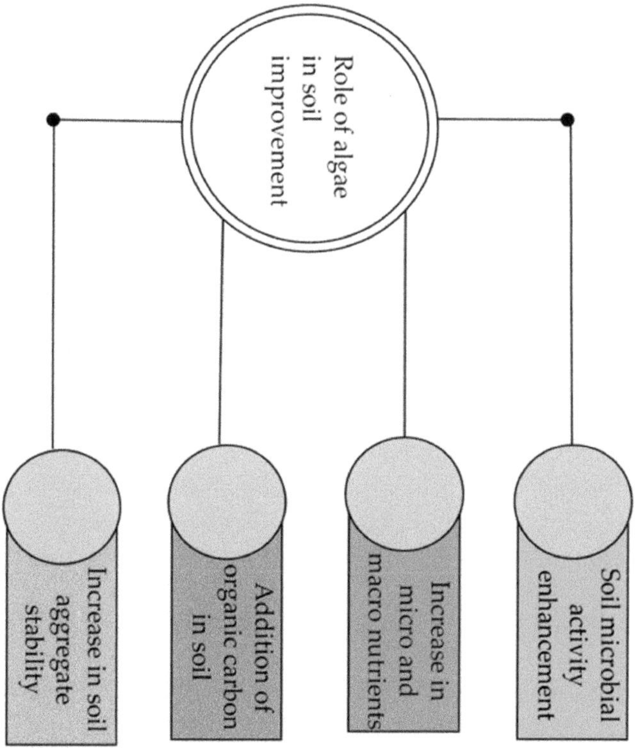

FIGURE 4.2 The role of microalgae in soil improvement.

Currently, microalgae can also be used to clean wastewater because many pollutants present in wastewater are also used by algae as food [79].

HMs that are present in the environment as pollutants are essential for the simple functioning of algae. In marine water, many metals are present; some marine algae have developed many processes to take up these HMs, thereby lowering the amount of HMs.

There are many advantages of using algae for bioremediation over the bacteria, as the microalgae can be used to remove CO_2 and various HMs. The remediation, using plants could potentially be coupled with sequestration of carbon. These have been known as vector suits for the removal of pollutants and have come out as potential low-cost alternative methods for doing physicochemical treatments of the pollutants. However, there are also some advantages and limitations of using microalgae for the bioremediation process:

1. **Advantages:** This method is cheap, non-invasive, natural attenuation (NA), treats water and soil, can be done on-site, emphasizes mass transfer.
2. **Limitations:** Environmental constraints, extended treatment time, monitoring difficulties, require excavation, relatively high capital cost, and operating cost.

The results of the experiment on sewage wastewater concluded that the growth of *chlorella vulgaris* and *Scenedesmus quadricauda* in wastewater negatively affected different pollutants ultimately leading to their reduction [35, 36]. Results show that *Chlorella vulgaris* can be effectively used for removal of excess nitrate and organic matter, thereby reducing chemical oxygen demand (COD). *Scenedesmus quadricauda* has been shown to give excellent results for improvement of BOD (biological oxygen demand) index and the reduction of phosphate content. Phase experiment confirms that microalga may be considered efficient nutrient removers.

4.7 ADVANTAGES OF MICROALGAE AS BIOFERTILIZER

After studying many experiments performed with microalgae, we can say that there are many advantages of microalgae as a biofertilizer. Microalgae show beneficial effects in the agricultural field and hold the relationship between crop plants and algae. Now is the chance to put microalgae in effect as a biofertilizer in place of any other harmful human-made fertilizers. The practice of using algae in the field decreases the number of pollutants present in the soil and plants together. Microalgae are capable of improving soil and plant properties, and microalgae, particularly cyanobacteria are omnipresent everywhere. Hence, their ease of availability makes them a much better tool that can be used in an agricultural field. Microalgae being microscopic, are designated as the first photosynthetic agents that are microbial in nature. Microalgae are quite essential for improving soil fertility and also help in soil reclamation (Figure 4.3). Some of the advantages of microalgae as biofertilizer are:

- Algae produce and release certain bioactive substances that have a positive impact on the growth and yield of crop plants. These bioactive substances include polypeptides, amino acids, plant growth regulators, vitamins, antibacterial or antifungal compounds, which improve soil structure and coenzyme activity;
- Concentrate metal ions that are present in the environment;
- Excretion of organic acid that increases the quantity of phosphorus availability and phosphorus uptake;
- Increased soil matter which is organic in nature;
- It also affects the biological nitrogen fixation system;
- Microorganism improves the wastewater quality;
- Works on the formation of crust.

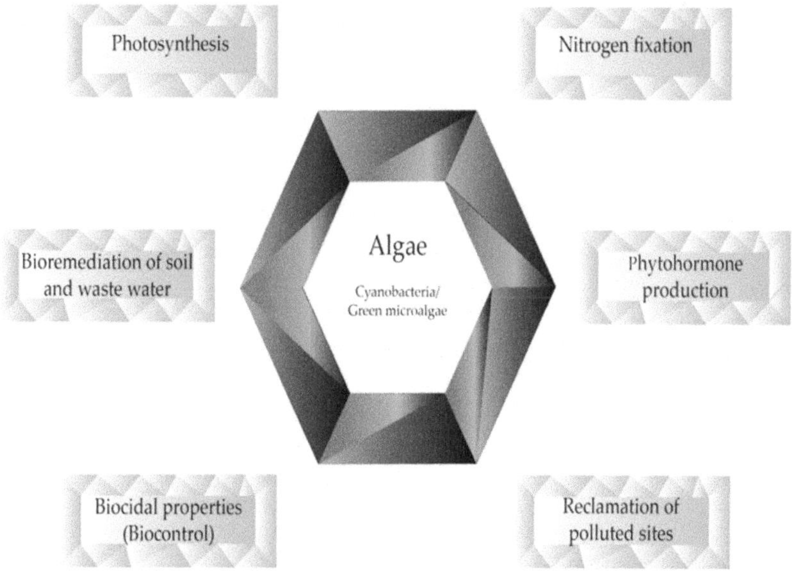

FIGURE 4.3 Advantages of microalgae as biofertilizer.

The incorporation of algal growth for the development of biofertilizer, nutrient recirculation, energy generation, and atomization has many good impacts on plant system, soil, and the environment also (Table 4.2). Microalgae have many functional options in agriculture in modern times, current trends and prospects. Microalgae improve soil fertility and increase the growth promoters in the soil and also fix nitrogen (by cyanobacteria). Microalgae also enhance soil organic matter mineralization and solubilization. They are involved in the production of plant growth hormones, stimulation of plant defense mechanisms, pest, and disease management. They also participate in the suppression of pathogen attack. Microalgae are used in soil conditioners and bioremediation processes (Table 4.2).

Algae are capable of converting solar energy to chemical energy by performing photosynthesis. Microalgae comprise of a number of compounds that are bioactive that can be used for commercial uses. By using biofertilizers which are made up of microalgae, the soil physicochemical properties can be improved. For paddy plants, many cyanobacteria like *Nostoc*, *Anabaena*, *Aulosira* are used to fix nitrogen which is present in the atmosphere. Living cyanobacteria are also known as an active part of cyanobacterial biofertilizer. They are used as a booster for paddy crops which are grown under a plant and low land conditions.

TABLE 4.2 Effect of Some Microalgae on the Growth of Major Crops

Microalgae	Phytohormone	Crop	Effects	References
Anabaena with Trichoderma/ Azotobacter	Indole acetic acid	Rice	Enhance the uptake of micro-macro and micronutrients increased nitrogen availability e in soil change the structure and function of the rhizosphere.	[55–62]
Association of *Anabaena and nostoc species*	Indole 3 propionic acid, indole 3 acetic acid	Rice	Increase the yield of crop and soil fertility	[23, 32, 99]
Association *of cyanobacteria and green algae*	Indole 3 acetic acid	Wheat	Enhance the activity of microbes in soil increase soil organic carbon and growth of plant	[71, 72]

TABLE 4.2 *(Continued)*

Microalgae	Phytohormone	Crop	Effects	References
Association of *cyanobacteria with eubacteria*	Auxin	Wheat	Increase the activity of enzymes in soil	[48, 67]
Calothrix elenkinii	–	Rice	Increase the growth of plant and production of phytohormones, also enhance the production of hydrolytic and defense enzymes.	[64]
Chlorella vulgaris Nannochloropsis	Auxin, IBA	Wheat	Increase the germination rate	[97]
Mesorhizobium, Anabaena-Nostoc association	Indole 3 acetic acid	Wheat	Improve uptake of nutrients both macro and micro	[55–62]
Nostoc species, Calothrix ghosei	Indole 3 acetic acid	Wheat	Enhance the growth of plant and health of grain	[31]

Many rice growers now use cyanobacteria in place of eukaryotic algae as biofertilizers. Cyanobacteria are also used for the before treatment of plant seeds with a cyanobacterial booster to increase the rate of germination of many species. Effects after treatment show that due to hormonal and activities of enzymes, there is an enhancement in plant performance. The use of living algae by making suspension is also hampered by difficulty in biomass conservation. Many studies have been carried out and the results show that the amount of plant phosphorus which is available and released by chemical fertilizers that cannot be assimilated by crop plant is very high in contrast to microalgal based biofertilizers that release a good amount of phosphorus which is easily assimilated by crop and suggesting them to be more sustainable and environment friendly crop fertilizers.

Due to human activities, natural phosphorus from the soil is rapidly decreasing and is also non-renewable, so the demand for phosphorus can be fulfilled by using microalgae as biofertilizer.

4.8 CONSTRAINTS OF USING ALGAE AS BIOFERTILIZER

Microalgae are the emerging tools for realizing eco-friendly sustainable agriculture worldwide. However, many facets of their usage as a promising "no-harm" fertilizer in the agricultural fields are yet to be explored further. Despite the exuberant potential of algae as biofertilizers, several technical challenges are encountered when putting them into practical usage, which need to be surpassed in order to make algae-based processes commercially manifestable [9]. Employment of algae as biofertilizer comes as a challenge on various fronts, namely field-level constraints, production constraints, and marketing constraints [12].

Extreme and drastic fluctuations in the annual and diurnal temperatures of the soil come as trouble for the survivability of the microalgae in the soil. Besides, competition with local microflora for basic nutrients hinders the growth of algae, as well as its benefits to the plant. Additionally, excessive application of fertilizers and pesticides over a decade has now resulted in alteration in pristine soil conditions. Not only has it made the soil more acidic over the years, but it also resulted in the accumulation of harmful salts and toxic ions. These collectively offer a very non-conducive environment for the sustenance of microalgae in the fields. The vision of sustainable use of algae as biofertilizers in the fields is met with challenges on the production front as well. Algal-based bio-fertilizers are prepared as carrier-based inoculants where peat, perlite, charcoal, etc., serve as the carrier medium [43].

For farmers with limited means, these carriers are not easily accessible. Besides, the inadequate availability of carriers in India accounts for the low popularity of bio-fertilizers and their widespread usage amongst the farmers. The production of natural microflora in the soil may, at times, be antagonistic to the artificially added algal strains. Specific removal of these algal strains from the soil is a cumbersome process. Many strains of microalgae require specific agro-climatic conditions and paucity of area-specific strains offers a major constraint in regard to biofertilizer use [43]. It has been suggested that biofertilizers are plant-specific, i.e., what favors one crop might not generate equally favorable results in other crops.

Furthermore, the scarcity of proper infrastructure for the production of biofertilizers is a constant trial met by the biofertilizers industry. Production of the ultimate product without compromising on the quality requires the usage of high-tech instrumentation and well-equipped bioreactors for the growth and propagation of algae. Shortage of well-trained personnel and technicians in the production units and dearth of their hands-on training with intricate techniques and instrumentation handling also serve as a limitation in algal-based bio-fertilizer production [52].

The storage of algal fertilizers is also a serious constraint. The viability period of biofertilizers prepared using carriers such as peat, lignite, etc., is narrow in length. Any interruption in their transportation or unfavorable conditions during their storage compromises their efficacy, leading to reduced interest among the dealers owing to their lower profit margins. When it comes to shelf-life and ease of storage, chemical fertilizers outcompete biofertilizers. In addition to that, farmers are inadequately sensitized towards the adverse effects of chemical fertilizers on plant health, soil health, and eventually environments' health. Besides, biofertilizers are not able to furnish as many major and minor nutrients to the plant as done by inorganic fertilizers. Thus, large quantities of the former are required for field application in order to equate the plants' demand for nourishment [8].

Quite often, farmers refrain from implementing sustainable methods of agricultural practices as the robustness of results in terms of crop yield and 'visible' health, is not as desirable with biofertilizers as they are with the employment of inorganic fertilizers [43]. This keeps the farmer from indulging in the usage of algal-based biofertilizers resulting in the lag between production and purchase logistics. Overall, it does not give any boost to the biofertilizer industry due to disinclination coming from the receiving end.

Thus, lack of awareness amongst farmers, restrictive shelf life, difficulties, and doubts regarding their mode of application, slow course-of-action, and inconsistent responses across all gradients of edaphic and environmental factors, are some major constraints that account for limited commercialization of algae-based biofertilizers, worldwide [12].

4.9 SUMMARY

Microalgae have proved beneficial for the plant's health as they produce specific polysaccharides that can enhance the production of the ROS scavenging enzyme. These enzymes play a significant role in reducing the excess ROS formed because of different stresses acting on plants. By doing so, microalgae play a role in minimizing the ROS that have exceeded their basal level. These polysaccharides are perceived as exogenous substances resulting in a microbe-associated molecular pattern (MAMP) triggered immunity in plants. This type of innate immunity helps in the activation of pathogenesis-related (PR) genes encoding PR proteins. These PR proteins prevent the entry as well as spread of various pathogens in plants. Microalgae act as a biofertilizer as it converts unavailable nutrients to an available form that can be easily uptaken by plants. With this, microalgae promote the overall growth of plants. Cyanobacteria can convert atmospheric nitrogen into a more accessible and usable form which can be easily taken up by the plants. Because plants lack the ability to uptake the inorganic nitrogen in the atmosphere, making it is very important for nitrogen-loving plants to have an association with cyanobacteria. On the other hand, cyanobacteria also contain compounds that act as phytohormones. Microalgae can also protect plants from toxic HM by sequestering them. Hence, microalgae also ameliorate the health of plants growing in the soil saturated with HMs.

Microalgae being a prominent source of renewable energy, yet have challenges at various levels, namely: at the field level, production level, and marketing level. The changing pH and temperature of the soil make it difficult for the microalgal species to grow properly. Their growth is also hindered by the competition with different species of microorganisms growing in the rhizosphere. Also, the regular utilization of chemicals for the past several years has led to an increase in acidity of the soil and accumulation of harmful salts and toxic ions. These together have led to a decline in the growth of microalgal species.

KEYWORDS

- **bioremediation**
- **biostimulant**
- **cyanobacteria**
- **microalgae**
- **photobioreactor**
- **wastewater**

REFERENCES

1. Abdelgani, M. E., & Hassan, I. A., (2006). The use of biofertilizers for increasing food production in Africa. *Africa, Khartoum: International University of Africa Journal Sudan, 3,* 191–201.
2. Aliyu, O. M., Adeigbe, O. O., & Awopetu, J. A., (2011). Foliar application of the exogenous plant hormones at pre-blooming stage improves flowering and fruiting in

cashew (*Anacardium occidentale* L.). *Journal of Crop Science and Biotechnology, 14*(2), 143–150.

3. Alwathnani, H. A., & Perveen, K., (2012). Biological control of fusarium wilt of tomato by antagonist fungi and *Cyanobacteria*. *African Journal of Biotechnology, 11*(5), 1100–1105.
4. Babu, S., Prasanna, R., Bidyarani, N., & Singh, R., (2015). The colonization of inoculated *Cyanobacteria* in wheat plants using biochemical and molecular tools. *Journal of Applied Phycology, 27*(1), 327–338.
5. Bhattacharjee, R., & Dey, U., (2014). Biofertilizer, away towards organic agriculture: A review. *African Journal of Microbiology Research, 8*(24), 2332–2343.
6. Bolyshev, N. N., & Novichkova-Ivanova, L. N., (1978). The algal flora of initial stages of soil formation on blanket loam under the condition of a lysimeter (Moscow & Leningrad). *Botanicheskii Zhurnal, 63*, 690–696.
7. Calvo, P., Nelson, L., & Kloepper, J. W., (2014). Agricultural uses of plant biostimulants. *Plant and Soil, 383*, 3–41.
8. Carvajal-Muñoz, J. S., & Carmona-Garcia, C. E., (2012). Benefits and limitations of bio fertilization in agricultural practices. *Livestock Research for Rural Development, 24*(3), 1–8.
9. Chittora, D., Meena, M., Barupal, T., Swapnil, P., & Sharma, K., (2020). *Cyanobacteria* as a source of biofertilizers for sustainable agriculture. *Biochemistry and Biophysics Reports, 22*, 9. Article ID: 100737.
10. Christenson, L., & Sims, R., (2011). Production and harvesting of microalgae for wastewater treatment, biofuels, and bioproducts. *Biotechnology Advances, 29*(6), 686–702.
11. Cuellar-Bermudez, S. P., & Aguilar-Hernández, I., (2015). Extraction and purification of high-value metabolites from microalgae: essential lipids, astaxanthin and phycobiliproteins. *Microbial Biotechnology, 8*(2), 190–209.
12. Debnath, S., Rawat, D., Mukherjee, A. K., Adhikary, S., & Kundu, R., (2019). Applications and constraints of plant beneficial microorganisms in agriculture. Chapter 3. In: Mirmajlessi, S. M., & Radhakrishnan, R., (eds.), *Biostimulants in Plant Science* (pp. 51–75). Intech Open. doi :10.5772/intechopen.89190.
13. Dias, G. A., Rocha, R. H. C., Araújo, J. L., De Lima, J. F., & Guedes, W. A., (2016). Growth, yield, and postharvest quality in eggplant produced under different foliar fertilizer (*Spirulina platensis*) treatments. *Semina: Ciências Agrárias (Summary: Agricultural Sciences), 37*, 3893–3902.
14. Epp, J. B., Alexander, A. L., & Balko, T. W., (2016). The discovery of arylex active and rinskor active: Two novel auxin herbicides. *Bioorganic & Medicinal Chemistry, 24*(3), 362–371.
15. Farid, R., Mutale-Joan, C., & Redouane, B., (2019). Effect of microalgae polysaccharides on biochemical and metabolomics pathways related to plant defense in *Solanum lycopersicum*. *Applied Biochemistry and Biotechnology, 188*(1), 225–240.
16. Foley, J. A., Ramankutty, N., & Brauman, K. A., (2011). Solutions for a cultivated planet. *Nature, 478*(7369), 337–342.
17. Gantar, M., Kerby, N. W., & Rowell, P., (1993). Colonization of wheat (*Triticumvulgare* L.) by N_2-fixing *Cyanobacteria*, Part III: Role of a hormogonia-promoting factor. *New Phytologist, 124*(3), 505–513.

18. Garcia-Gonzalez, J., & Sommerfeld, M., (2016). Biofertilizer and biostimulant properties of the microalga *Acutodesmus dimorphus*. *Journal of Applied Phycology, 28*(2), 1051–1061.
19. Gayathri, M., Kumar, P. S., Prabha, A. M. L., & Muralitharan, G., (2015). *In vitro* regeneration of *Arachis hypogaea* L. and *Moringa oleifera* Lam. using extracellular phytohormones from *Aphanothece sp*. MBDU 515. *Algal Research, 7*, 100–105.
20. Gill, S. S., & Tuteja, N., (2010). Reactive oxygen species and antioxidant machinery in abiotic stress tolerance in crop plants. *Plant Physiology and Biochemistry, 48*(12), 909–930.
21. Gloaguen, V., Morvan, H., & Hoffmann, L., (1996). Metal accumulation by immobilized *Cyanobacteria*l mats from a thermal spring. *Journal of Environmental Science & Health Part A, 31*(10), 2437–2451.
22. Godfray, H. C. J., Beddington, J. R., & Crute, I. R., (2010). Food security: The challenge of feeding 9 billion people. *Science, 327*(5967), 812–818.
23. Goyal, S. K., Singh, B. V., Nagpal, V., & Marwaha, T. S., (1997). An improved method for production of algal biofertilizer. *Indian Journal of Agricultural Science, 67*, 314–315.
24. Gupta, P., Ravi, I., & Sharma, V., (2013). Induction of β -1, 3-glucanase and chitinase activity in the defense response of *Eruca sativa* plants against the fungal pathogen *Alternaria brassicicola*. *Journal of Plant Interactions, 8*(2), 155–161.
25. Hashem, M., & Abo-Elyousr, K. A., (2011). Management of the root-knot nematode *Meloidogyne incognita* on tomato with combinations of different biocontrol organisms. *Crop Protection, 30*(3), 285–292.
26. Heisler, J., Glibert, P. M., & Burkholder, J. M., (2008). Eutrophication and harmful algal blooms: A scientific consensus. *Harmful Algae, 8*(1), 3–13.
27. Holajjer, P., Kamra, A., Gaur, H. S., & Manjunath, M., (2013). Potential of *Cyanobacteria* for biorational management of plant-parasitic nematodes: Review. *Crop Protection, 53*, 147–151.
28. Hung, T. C., & Chow, T. J., (1988). Comparative studies of some nitrogen-fixing unicellular *Cyanobacteria* isolated from rice fields. *Microbiology, 134*(12), 3089–3097.
29. Ibraheem, I. B., (2007). *Cyanobacteria* as alternative biological conditioners for bioremediation of barren soil. *Egyptian J. of Phycol., 8*(100), 99–116.
30. Ibraheem, I. B., & Abdel-Raouf, N., (2007). Allelopathic activity of some local *Cyanobacteria*l extra-metabolites against some pathogenic bacteria. *Egyptian J. of Phycol., 8*(120), 119–129.
31. Karthikeyan, N., Prasanna, R., Nain, L., & Kaushik, B. D., (2007). Evaluating the potential of plant growth-promoting *Cyanobacteria* as inoculants for wheat. *European Journal of Soil Biology, 43*(1), 23–30.
32. Kaushik, B., (1998). Use of *Cyanobacteria*l biofertilizers in rice cultivation: A technology improvement. *Cyanobacterial Biotechnology, 211*–222.
33. Kemmerling, B., Halter, T., & Mazzotta, S., (2011). Genome-wide survey for *Arabidopsis* leucine-rich repeat receptor kinases implicated in plant immunity. *Frontiers in Plant Science, 2*, 88–93.
34. Khan, S. A., Hussain, M. Z., Prasad, S., & Banerjee, U. C., (2009). Prospects of biodiesel production from microalgae in India. *Renewable and Sustainable Energy Reviews, 13*(9), 2361–2372.
35. Kshirsagar, A. D., & Gunale, V. R., (2011). Pollution status of river Mula (Pune city) Maharashtra, India. *Journal of Ecophysiology and Occupational Health, 11*(1), 81–90.

36. Kshirsagar, A. D., Ahire, M. L., & Gunale, V. R., (2012). Phytoplankton diversity related to pollution from Mula river at Pune City - India. *Terrestrial and Aquatic Environmental Toxicology, 6*(2), 136–142.
37. Kulik, M. M., (1995). The potential for using *Cyanobacteria* (blue-green algae) and algae in the biological control of plant pathogenic bacteria and fungi. *European Journal of Plant Pathology, 101*(6), 585–599.
38. Kumar, N., (2016). Effect of algal bio-fertilizer on the *Vigna radiata*: Critical review. *International Journal of Engineering Research and Applications, 6*(2), 85–94.
39. Kumar, V., Muthuraj, M., Palabhanvi, B., Ghoshal, A. K., & Das, D., (2014). High cell density lipid-rich cultivation of a novel microalgal isolate *Chlorella sorokiniana* FC6 IITG in a single-stage fed-batch mode under mixotrophic condition. *Bioresource Technology, 170*, 115–24.
40. Lefebvre, D. D., Kelly, D., & Budd, K., (2007). Biotransformation of Hg(II) by *Cyanobacteria. Applied and Environmental Microbiology, 73*(1), 243–249.
41. Lu, Y., & Xu, J., (2015). Phytohormones in microalgae: A new opportunity for microalgal biotechnology? *Trends in Plant Science, 20*(5), 273–282.
42. Marathe, K. V., & Chaudhari, P. R., (1975). An example of algae as pioneers in the lithosphere and their role in rock corrosion. *The Journal of Ecology, 63*, 65–70.
43. Mazid, M., & Khan, T. A., (2015). Future of biofertilizers in Indian agriculture: An overview. *International Journal of Agricultural and Food Research, 3*(3), 10–23.
44. Mazur, H., Konop, A., & Synak, R., (2001). Indole-3-acetic acid in the culture medium of two axenic green microalgae. *Journal of Applied Phycology, 13*(1), 35–42.
45. Metting, B., (1990). Microalgae applications in agriculture. *Developments in Industrial Microbiology, 31*, 265–70.
46. Mittler, R., Vanderauwera, S., & Gollery, M., (2004). Reactive oxygen gene network of plants. *Trends in Plant Science, 9*(10), 490–498.
47. Mulbry, W., Westhead, E. K., Pizarro, C., & Sikora, L., (2005). Recycling of manure nutrients: Use of algal biomass from dairy manure treatment as a slow-release fertilizer. *Bioresource Technology, 96*, 451–458.
48. Nain, L., Rana, A., & Joshi, M., (2010). Evaluation of synergistic effects of bacterial and *Cyanobacteria*l strains as biofertilizers for wheat. *Plant and Soil, 331*(1, 2), 217–230.
49. Nisha, R., Kaushik, A., & Kaushik, C. P., (2007). Effect of indigenous *Cyanobacteria*l application on structural stability and productivity of an organically poor semi-arid soil. *Geoderma, 138*(1, 2), 49–56.
50. Odegard, I. Y. R., & Van, D. V. E., (2014). The future of food: Scenarios and the effect on natural resource use in agriculture: 2050. *Ecological Economics, 97*, 51–59.
51. Park, J. B. K., Craggs, R. J., & Shilton, A. N., (2011). Wastewater treatment high rate algal ponds for biofuel production. *Bioresource Technology, 102*(1), 35–42.
52. Pathak, J., Ahmed, H., Singh, P. R., Singh, S. P., Häder, D. P., & Sinha, R. P., (2019). Mechanisms of photoprotection in *Cyanobacteria*. In: Mishra, A. K., Tiwari, D. N., & Rai, A. N., (eds.), *Cyanobacteria* (pp. 145–171). New York: Academic Press.
53. Potters, G., Horemans, N., Bellone, S., Caubergs, R. J., Trost, P., Guisez, Y., & Asard, H., (2004). Dehydroascorbate influences the plant cell cycle through a glutathione independent reduction mechanism. *Plant Physiology, 134*(4), 1479–1487.
54. Prasanna, R., Nain, L., & Tripathi, R., (2008). Evaluation of fungicidal activity of extracellular filtrates of *Cyanobacteria*: Possible role of hydrolytic enzymes. *Journal of Basic Microbiology, 48*(3), 186–194.

55. Prasanna, R., Babu, S., & Bidyarani, N., (2015). Prospecting *Cyanobacteria* fortified composts as plant growth-promoting and biocontrol agents in cotton. *Experimental Agriculture, 51*(1), 42–65.
56. Prasanna, R., Babu, S., Rana, A., & Kabi, S. R., (2013). Evaluating the establishment and agronomic proficiency of *Cyanobacteria*l consortia as organic options in wheat-rice cropping Sequence. *Experimental Agriculture, 49*(3), 416–434.
57. Prasanna, R., Bidyarani, N., & Babu, S., (2015). *Cyanobacteria*l Inoculation elicits plant defense response and enhanced mobilization in maize hybrids. *Cogent Food & Agriculture, 1*(1), 11. Article ID: 998507.
58. Prasanna, R., Kanchan, A., & Kaur, S., (2016). Chrysanthemum growth gains from beneficial microbial interactions and fertility improvements in soil under protected cultivation. *Horticultural Plant Journal, 2*(4), 229–239.
59. Prasanna, R., Kanchan, A., & Ramakrishnan, B., (2016). *Cyanobacteria*-Based bioinoculants influence growth and yields by modulating the microbial communities favorably in the rhizosphere of maize hybrids. *European Journal of Soil Biology, 75*, 15–23.
60. Prasanna, R., Ramakrishnan, B., & Simranjit, K., (2017). Cyanobacterial and rhizobial inoculation modulate the plant physiological attributes and nodule microbial communities of chickpea. *Archives of Microbiology, 199*(9), 1311–1323.
61. Prasanna, R., Sharma, E., & Sharma, P., (2013). Soil fertility and establishment potential of inoculated *cyanobacteria* in rice crop grown under non-flooded conditions. *Paddy and Water Environment, 11*(1), 175–183.
62. Prasanna, R., Triveni, S., & Bidyarani, N., (2014). Evaluating the efficacy of cyanobacterial formulations and biofilm inoculants for leguminous crops. *Archives of Agronomy and Soil Science, 60*(3), 349–366.
63. Prasanna, R., Jaiswal, P., & Kaushik, B. D., (2008). *Cyanobacteria* as potential options for environmental sustainability: Promises and challenges. *Indian Journal of Microbiology, 48*(1), 89.
64. Priya, H., Prasanna, R., & Ramakrishnan, B., (2015). Influence of cyanobacterial inoculation on the culturable microbiome and growth of rice. *Microbiological Research, 171*, 78–89.
65. Priyadarshani, I., & Rath, B., (2012). Commercial and industrial applications of microalgae: Review. *Journal of Algal Biomass Utilization, 3*(4), 89–100.
66. Priyadarshani, I., & Rath, B., (2012). Bioactive compounds from microalgae and *Cyanobacteria*: Utility and applications. *International Journal of Pharmaceutical Sciences and Research, 3*(11), 7. Article ID: 4123.
67. Rana, A., Joshi, M., Prasanna, R., Shivay, Y. S., & Nain, L., (2012). Biofortification of wheat through inoculation of plant growth-promoting rhizobacteria and *Cyanobacteria*. *European Journal of Soil Biology, 50*, 118–126.
68. Rana, A., Kabi, S. R., & Verma, S., (2015). Prospecting plant growth-promoting bacteria and *Cyanobacteria* as options for enrichment of macro and micronutrients in grains in rice-wheat cropping sequence. *Cogent Food & Agriculture, 1*(1), 9. Article ID: 1037379.
69. Ranjan, K., Priya, H., & Ramakrishnan, B., (2016). Cyanobacterial inoculation modifies the rhizosphere microbiome of rice planted to a tropical alluvial soil. *Applied Soil Ecology, 108*, 195–203.
70. Reddy, C. A., & Saravanan, R. S., (2013). Polymicrobial multi-functional approach for enhancement of crop productivity. *Advances in Applied Microbiology, 82*, 53–113.

71. Renuka, N., Prasanna, R., & Sood, A., (2016). Exploring the efficacy of wastewater-grown microalgal biomass as a biofertilizer for wheat. *Environmental Science and Pollution Research, 23*(7), 6608–6620.
72. Renuka, N., Prasanna, R., & Sood, A., (2017). Wastewater grown microalgal biomass as inoculants for improving micronutrient availability in wheat. *Rhizosphere, 1*, 150–159.
73. Renuka, N., Guldhe, A., & Prasanna, R., (2018). Microalgae as multifunctional options in modern agriculture: Current trends, prospects and challenges. *Biotechnology Advances, 36*(4), 1255–1273.
74. Richmond, A., (2004). *Handbook of Microalgal Culture: Biotechnology and Applied Phycology* (P. 289). London: Blackwell Science Ltd.
75. Rodriguez, A. A., & Stella, A. M., (2006). Effects of cyanobacterial extracellular products and gibberellic acid on salinity tolerance in *Oryza sativa* L. *Saline Systems, 2*(1), 1–4.
76. Ronga, D., Setti, L., & Salvarani, C., (2019). Effects of solid and liquid digestate for hydroponic baby leaf lettuce cultivation. *Scientia Horticulturae, 244*, 172–18.
77. Saadatnia, H., & Riahi, H., (2009). *Cyanobacteria* from paddy fields in Iran as a biofertilizer in rice plants. *Plant Soil Environ., 55*(5), 207–212.
78. Saikia, P., & Bordoloi, R. P. M., (1994). Blue-green algal flora from the rice fields of Assan. *Phykos, 33*, 53–57.
79. Samhan, A. F., (2008). *Assessment of the Ability of Microalgae in Removal of Some Industrial Wastewater Pollutants* (p. 161). M.Sc. Thesis; Botany Department, Faculty of Science, Beni-Suef University, Beni-Suef, Egypt.
80. Selvarani, V., (1983). *Studies on the Influence on Nitrogen Fixing and Non-Nitrogen Fixing Blue-Green Algae on the Soil, Growth and Yield of Paddy* (*Oryza sativa-IR 50) M* (p. 133). M. Sc. Thesis; Madurai Kamaraj University, Madurai, India.
81. Shaaban, A. M., Haroun, B. M., & Ibraheem, I. B., (2004). Assessment of the impact of *Microcystis aeruginosa* and *Chlorella vulgaris* in the uptake of some pollutants from culture media. *3rd Int. Conf. Biol. Sci.* (*ICBS) Fac. Sci.*, 28, 29. Tanta University.
82. Shaaban, M. M., (2001). Green microalgae water extract as foliar feeding to wheat plants. *Pakistan Journal of Biological Sciences, 4*, 628–632.
83. Shaaban, M. M., (2001). Nutritional status and growth of maize plants as affected by green microalgae as soil additives. *Journal of Biological Sciences, 6*, 475–479.
84. Shaaban, M. M., El-Fouly, M. M., & Abdel-Maguid, A. A., (2004). Zinc-boron relationship in wheat plants grown under low or high levels of calcium carbonate in the soil. *Pakistan Journal of Biological Sciences, 7*, 633–639.
85. Shaaban, M. M., Hussein, M. M., & El Saady, A. M., (2008). Nutritional status in shoots of barley genotypes as affected by salinity of irrigation water. *American Journal of Plant Physiology, 3*, 89–95.
86. Shaaban, M. M., Loehnertz, O., & El-Fouly, M. M., (2007). Grapevine genotypic tolerance to lime and possibility of chlorosis recovery through micronutrients foliar application. *International Journal of Botany, 3*, 179–187.
87. Shariatmadari, Z., & Riahi, H., (2013). Plant growth-promoting *Cyanobacteria* and their distribution in terrestrial habitats of Iran. *Soil Science and Plant Nutrition, 59*(4), 535–547.
88. Sharma, R., & Khokhar, M. K., (2012). Role of algae and *Cyanobacteria* in sustainable agriculture systems. *Wudpecker Journal of Agricultural Research, 1*(9), 381–388.

89. Singh, N. K., & Dhar, D. W., (2007). Nitrogen and phosphorous scavenging potential in microalgae. *Indian Journal of Biotechnology, 6*, 52–56.
90. Singh, T., & Purohit, S. S., (2011). *Biofertilizers Technology* (p. 320). New Delhi: Agrobios India.
91. Singh, J. S., Kumar, A., Rai, A. N., & Singh, D. P., (2016). *Cyanobacteria*: A precious bio-resource in agriculture, ecosystem, and environmental sustainability. *Frontiers in Microbiology, 7*, 529–535.
92. Stirk, W. A., Ördög, V., Van, S. J., & Jäger, K., (2002). Cytokinin and auxin-like activity in Cyanophyta and microalgae. *Journal of Applied phycology, 14*(3), 215–221.
93. Tang, D., Shi, S., Li, D., Hu, C., & Liu, Y., (2007). Physiological and biochemical responses of *Scytonema javanicum* (Cyanobacterium) to salt stress. *Journal of Arid Environments, 71*(3), 312–320.
94. Tantawy, S. T., & Atef, N. M., (2010). Growth responses of *Lupinus termis* to some plant growth-promoting *Cyanobacteria* and bacteria as biofertilizers. *Journal of Food, Agriculture & Environment, 8*, 1178–1183.
95. Tilman, D., Cassman, K. G., Matson, P. A., Naylor, R., & Polasky, S., (2002). Agricultural sustainability and intensive production practices. *Nature, 418*, 671–678.
96. Tuan, H. M., Lee, C. Y., & Tang, H. C., (1992). Phenolic compounds in food and their effects on health: Volume I - analysis, occurrence, and chemistry; and volume II - antioxidants and cancer prevention. *ACS Symposium Series (USA)*, (507), 992.
97. Uysal, O., Uysal, F. O., & Ekinci, K., (2015). Evaluation of microalgae as microbial fertilizer. *European Journal of Sustainable Development, 4*(2), 77–87.
98. Van, O. M. J., Pepe, O., De Pascale, S., Silletti, S., & Maggio, A., (2017). The role of biostimulants and bioeffectors as alleviators of abiotic stress in crop plants. *Chemical and Biological Technologies in Agriculture, 4*(1), 5–12.
99. Venkataraman, G. S., (1981). Blue-green algae for rice production- a manual for its promotion. Rome: FAO. *Soils Bulletin, 46*, 1–52.
100. You, J., & Chan, Z., (2015). ROS regulation during abiotic stress responses in crop plants. *Frontiers in Plant Science, 6*, 1092.
101. Zhan, J., & Sun, Q., (2012). Diversity of free-living nitrogen-fixing microorganisms in the rhizosphere and non-rhizosphere of pioneer plants growing on wastelands of copper mine tailings. *Microbiological Research, 167*(3), 157–165.

CHAPTER 5

MICROBIAL BIOTECHNOLOGY: POTENTIAL OF FUNGAL ENZYMES IN SOIL BIOREMEDIATION

SUBHOSH CHANDRA MUNI RAMANNA GARI,
SRINIVASULU MANDALA, NAGA RAJU MADDELA,
SURESH YADAV PANYAM, and PALLAVAL VEERA BRAMHACHARI

ABSTRACT

Bioremediation mainly involves the breakdown, abolition, and detoxification of different chemical desecrates, effluents, and harmful substances from the environment by means of comprehensive exploitation of microorganisms. It is an innovative technology in the field of biotechnology for the treatment of a wide variety of contaminants in the area of environmental protection. Among the various bioremediation processes, fungal enzymes are ideal candidates for soil bioremediation. This chapter offers an overview of the variety of bioremediation processes, diversity of fungi, influence of factors, fungal enzymes, contaminants, and pollutants, advantages, disadvantages, and potential of microbial biotechnology in bioremediation of environmental contaminants.

5.1 INTRODUCTION

Bioremediation is a method where waste and harmful compounds are altered into less harmful compounds by microorganisms. The approaches of bioremediation for commercial compounds, based on stimulated soil microorganisms, have been thoroughly studied [83, 96, 125]. However, their performance has been moderately less effective in eradicating constant trace organic contaminants (TrOCs) [124, 133]. Fungi play an important role as degraders and mutualists in many environments, such as, soil and water settings; and are known as fungal remediation or mycoremediation, which has become increasingly popular by using prospective fungi [54]. Fungi are perhaps better for bioremediation than other microbes growing in different conditions [46, 49].

The lignocellulosic substrates and contaminants are degraded by means of several ligninolytic enzymes that are secreted by fungi to breakdown cell walls and more organic compounds. Further, they are utilized in the transformation of waste substances and elimination of contaminants in different ecosystems [15]. Besides, the ability of fungi to breakdown diverse category of pharmaceuticals and personal care products (PPCPs) with unspecific enzymes is well known in the literature [24, 94, 131].

Increasing attention has been focused on the exceptional ability of fungi to breakdown such contaminants using various exo-enzymes and endo-enzymes (namely: peroxidases (POXs) and cytochrome P450), respectively for detoxification and biodegradation [10, 37, 57]. In the case of diffused pollution, *in situ* bioremediation techniques are better adapted for the treatment of large surfaces of contaminated soils [46, 81].

This chapter gives a sketch of fungal enzymes and their potential in soil bioremediation with perspectives on microbial biotechnology.

5.2 DIVERSITY OF FUNGI IN BIOREMEDIATION

Biodiversity refers to the presence of different groups/types of organisms in abundance in any given environment [91, 107]. Fungi can colonize in freshwater and seawater environments; and can grow under stress situation and inseminate spores during air that assists in stability of the environment. Fungi living in effluent treatment plants (ETPs) could be employed for cleaning of wastewater [82].

The prospective fungi, secreting different enzymes exhibiting capability in various environments and as effective organism for remediation have been reported. The remediation of hydrocarbons is dependent on breakdown, and hence fungi are involved in a partial or a complete breakdown of the relevant pollutants [1, 9, 92]. Fundamentally, this mechanism is different from bioaccumulation, which designates the capability of some fungi to accumulate or hyperaccumulate various metals inside the cell from the soil [9]. The different fungi that are utilized for degradation of different pollutants are:

- Extremophilic fungi [34, 106];
- Filamentous fungi [81];
- Marine fungi [19, 22, 29, 65, 67, 114];
- Mutualistic fungi [2, 17, 23, 36, 42, 53, 57, 75, 93, 105, 112];
- White rot fungi [35, 40, 123, 133].

5.3 FACTORS AFFECTING BIOREMEDIATION

The effectiveness of bioremediation is based on chemical characteristics and concentration of contaminants, physicochemical properties of the environment, and their accessibility to microorganisms [39]. These important characteristics include the survival of a microbial species to degrade contaminants, availability of pollutants to microbial species and environmental characteristics (Table 5.1). The major characteristics influencing bioremediation are:

- Accessibility of contaminants;
- Category of pollutants;
- Cultivation substrate vs. co-metabolism;
- Deficiency of minerals;
- Dissemination of minerals;
- Enhancement of potent microbial species;
- Low level of pollutants;
- Mass transfer limits;
- Microbial growth till vital biomass is attained;
- Oxidation/reduction potential;
- Oxygenic *vs.* anoxygenic methods;
- Presence of alternative carbon source;
- Secretion of poisonous metabolites;
- Solubilization in water;
- Structure of chemical compounds;
- Toxic nature of pollutants.

TABLE 5.1 Environmental Characteristics and Optimal Circumstances for Microbial Action in Soil Bioremediation

Environmental Characteristics	Optimal Limits
Heavy metals	Maximum concentration of 2,000 ppm
Minerals	Nitrogen and phosphorus for cultivation of organisms
O_2	Oxygenic, lowest air-fill up pore volume of 10%

TABLE 5.1 *(Continued)*

Environmental Characteristics	Optimal Limits
pH	5.5–8.8
Pollutants	Less-harmful
Soil moisture	25–28%
Soil texture	Little clay
Temperature	15–45 (°C)

5.3.1 *BIOLOGICAL FACTORS*

Competition, antagonism, predation among microorganisms for partial carbon sources, and interactions among microbes like protozoa and bacteriophages are the biotic factors that affect the breakdown of organic substances in the environment. The important biological factors affecting are: mutation, horizontal gene transfer, enzyme action, contact, its own development till significant mycelium, inhabitant dimension and composition is attained [20, 65, 73].

5.3.2 *ENVIRONMENTAL FACTORS*

Microbial cultivation and action are influenced by pH, temperature, water level, soil structure, solubilization in water, minerals, site characteristics, redox potential and O_2 level, lack of trained human resources in this area and elemental availability of contaminants (pollutant intensity, kind, solubilization, chemical structure, and harmfulness). These factors influence the kinetics of breakdown [1, 73].

5.3.3 *ACCESSIBILITY OF NUTRIENTS*

Balancing of nutrients particularly addition of necessary nutrients including nitrogen and phosphorus can influence progress of degradation competence by standardizing bacterium carbon: nitrogen: phosphorus ratio. To live and persist their microbial action, microbes require different minerals, including carbon, nitrogen, and phosphorus, the basic building blocks of life that allow microorganisms to synthesize essential enzymes to degrade pollutants. All of them need N, P, and C. Carbon is the most basic element of living forms, which is needed in higher amount than other minerals [73].

5.3.4 TEMPERATURE

The temperature influences survival of microbes and composition of hydrocarbons [30]. Biological enzymes have optimal temperature and may not have similar metabolic yield for each temperature. Temperature also either speeds up or slows down bioremediation procedure due to its effect on biological characteristics [102]. For every 10°C rise in temperature, it affects biological response rates twice. Above a certain temperature, however, the cells die. Plastic covering can be used to enhance solar warming in late spring, summer, and autumn [73]. It was reported that the rate of degradation of organic pollutants is comparably higher at higher temperatures [43]. Higher rates of degradation of pollutants were noticed in humid soils in contrast to soils from temperate regions [22, 84]. Dimond et al. [34] observed that temperature influences the half-life of pollutants.

5.3.5 CONCENTRATION OF OXYGEN

Biodegradation is performed either in the presence or absence of oxygen. As oxygen is a gaseous, and for a majority of the organisms need oxygen; and degradation of hydrocarbons in most cases can be increased by the presence of oxygen [72].

5.3.6 MOISTURE CONTENT

The soil moisture level leads to unfavorable results in biodegradation mediators. For fungal remediation of polluted soils, the relative moisture normally is beyond 60% [80, 128]. Xiao et al. [128] utilized a relative humidity of 70% compared to 60% in studies by Miligore et al. [80] and 60–70%, according to Singh et al. [104]. Higher relative humidity values of 85–95% were also employed [47]. Seidu et al. [99] showed that relative moisture of 70–80% is favorable for mushroom growth and fruiting.

5.3.7 *PH*

The pH of a substance (acidic, basic, and alkaline character of component) has its own effect on chemical processes and also enhances or reduces the elimination method. The quantification of pH in soil could prove better for organism growth. The breakdown of pollutants in soil is superior with higher pH due to enhanced solubility [129]. Owen et al. [84] explained that higher soil pH showed rapid breakdown of pollutants in contrast to those performed in low pH soil. Studies of Andrea et al. [11] revealed that the breakdown of pollutants is inhibited at pH 4.5 to 4.8.

5.4 FUNGAL ENZYMES IN BIOREMEDIATION

Fungal enzymes (such as cellulolytic enzymes, laccases, POXs, catalases, etc.), have a broad range of applications in biotechnology and managing organic wastes [77]. These enzymes are mainly used for the breakdown of lignocellulosic substrates, such as agricultural residues and forest residues [136]. The hydrolyzed substrates can be used for composting or producing biofuels and also for the transformation of harmful compounds to harmless substances using microorganisms [50]. These fungal enzymes play a key role in soil bioremediation. Currently, much attention has been given to increasing tailor-made enzymes from gene cloning methods and recombinant genes from white-rot fungi, which are efficient means for eco-friendly handling of poisonous substances [44, 96, 109, 126]. The common fungal enzymes employed in bioremediation of various pollutants are listed in Table 5.2.

TABLE 5.2 Potential of Common Fungal Enzymes for Bioremediation of Various Pollutants

Enzyme	Compound	Fungi	References
Catalase	Heavy metals	*Penicillium sp., Rhizopus sp.*	[33]
Hydrolases	Degradation of cellulose	*Aspergillus protuberus*	[132]
	Rice straw degradation	*Penicillium sp., Trichoderma sp., Paecilomyces sp.,* and *Pleurotus ostreatus*	[86]
	Hydrolysis of cellulose	*A. fumigatus* JCM 10253	[97]
	Hydrolysis of cellulose	*Periconia epilithographicola C., Coniochaeta cipronana C.*	[27]
	Degradation of cellulose	*Aspergillus unguis*	[101]

TABLE 5.2 *(Continued)*

Enzyme	Compound	Fungi	References
	Degradation of wheat straw and grass hay	*Curvularia kusanoi* L7, *Hipoxylon sp., Aspergillus fumigatus, Trichoderma sp*	[119]
	Hydrolysis of cellulose	*Trichoderma harzianum, Aspergillus fumigatus, Trichoderma longibrachiatum*	[68]
Laccase	Petroleum hydrocarbons	*White rot fungi, P. ostreatus*	[70]
	PAH	*Trichoderma tomentosum*	[76]
	Toxic dyes like bromophenol blue, methyl violet, and malachite green and other dyes	*Pycnoporus sanguineus, Myceliophthora thermophila, Trametes trogii,* and *Ceriporiopsis subvermispora*	[12]
	Decolorization and detoxification of recalcitrant compounds	*Basidiomycetous, ascomycetous* related organisms	[8]
	Decolorization of humic acid	*Lentinus tigrinus*	[134]
	Phenanthrene	*Ganoderma lucidum*	[3]
	Pyrene (polycyclic aromatic hydrocarbons)		
	PCBS (polychlorinated biphenyls)	*Trametus versicolor, Pleurotus ostreatus*	[14]
	Pentachlorophenol	*Phlebia acanthocystis* TMIC34875, *Phlebia tremellosa* TMIC30511, and *Phlebia aurea* TMIC33908	[127]
	Lignin degradation	*Phanerochaete chrysosporium* (CICC14076), *Lentinula edodes* (CICC14019), and *Trametes versicolor* (CICC50001)	[52]
Lignin peroxidase (LiP)	Phenanthrene	*Ganoderma lucidum*	[3]
	Pyrene (polycyclic aromatic hydrocarbons)		
	PAHs aromatic-containing xenobiotics	*P. chrysosporium*	[14]
	Pentachlorophenol	*Phlebia acanthocystis* TMIC34875, *Phlebia tremellosa* TMIC30511 and *Phlebia aurea* TMIC33908	[127]

TABLE 5.2 *(Continued)*

Enzyme	Compound	Fungi	References
	Lignin degradation	*Phanerochaete chrysosporium* (CICC14076), *Lentinula edodes* (CICC14019), and *Trametes versicolor* (CICC50001)	[52]
Lipase	PAHs	*Drechslera sp., Fusarium sp.,*	[25]
	–	*Lasiodiplodia sp., Mucor sp., Rhizopus sp., Trichoderma sp.*	[16]
Manganese peroxidase (MnP)	Nonylphenol	*Ganoderma lucidum*	[18]
	Triclosan		
	aliphatic and aromatic hydrocarbons	*Penicillium sp. CHY-2*	[48]
	Congo red	*Phanerochaete chrysosporium*	[21]
	Decolorize/detoxify four dyes such as Drimaren Blue CLBR, Drimaren Red K-4B, Drimaren Yellow X-8GN and Disperse Navy Blue HGL	*G. lucidum 00679*	[26]
	Phenanthrene pyrene (Polycyclic aromatic hydrocarbons)	*Ganoderma lucidum*	[3]
	Pentachlorophenol	*Phlebia acanthocystis* TMIC34875, *Phlebia tremellosa* TMIC30511 and *Phlebia aurea* TMIC33908	[127]
	Lignin degradation	*Phanerochaete chrysosporium* (CICC14076), *Lentinula edodes* (CICC14019), and *Trametes versicolor* (CICC50001)	[52]
Peroxidase	Organic pollutants	*White-rot basidiomycetes*	[55]
Versatile peroxidase	Decolorization of humic acid	*Lentinus tigrinus*	[134]

5.4.1 LACCASES

Laccases are capable of catalyzing oxidation of phenolic, polyphenolics, and aromatic compounds [110]. Laccases fit into a huge cluster of proteins named multicopper oxidases, which comprise additional ascorbate oxidases and ceruloplasmin. They carry out lessening of dioxygen to water as oxidizing organic substances with a one-electron redox method. They oxidize a broad array of aromatic substances, mostly phenol and aniline [81]. With a suitable moderator, they show greater oxidation ability ensuing in a number of industrial advantages that involve oxidation of non-phenol lignin materials and degradation of different ecological contaminants [63, 118].

Laccases belonging to blue oxidases utilize "Cu" as cofactor and molecular "O_2" as co-substrate and are copper containing exo-enzymes. Laccases oxidize mainly phenol and non-phenol substances, and their action is 20 times higher in *Trichoderma Versicolor* [78]. The broad range of substrates of their action on different substrates make them perfect catalysts in various potential advantages. These enzymes were widely discovered for their effective bioremediation prospective [122]. The use of laccases is poor and restricted due to their low-life span.

They are phenol oxidases that consist of four copper ions for each molecule, which catalyze reduction of O_2 to H_2O_2 (which is common with oxidation of different phenols). They have quite low redox potential. However, in the presence of redox moderators (such as 2,2'-azino-bis(3-ethylbenzothiazoline-6-sulfonic acid) (ABTS) and 1-hydroxybenzotriazole (HBT)), more organic substances can be oxidized [117].

5.4.2 CATALASES

Primary defense mechanism to reactive oxygen species (ROS) generation in fungi consists of mono-functional catalases and bi-functional POX catalase enzymes. Inhibition of catalase in the presence of pesticide lindane has been reported to manifest in the form of increased ROS generation and hence in ROS-mediated damage resulting in inhibition of growth of *Saccharomyces cerevisiae* [88]. Heavy metals (HMs) (including Pb, Cu, Cd, Zn) were recorded to be among the main causes for ROS stimulation in cells of organisms.

Lin et al. [69] showed that catalase action was used as a device for examination of remediation effectiveness because the catalase action was reduced by enhancing oil level in remediation of oil-polluted soil. Therefore, it gives HM toleration ability to fungi. Hence, fungi secreting this enzyme have been used for remediation of metal-polluted areas. Enhanced catalase action was noticed in fungal strains of *A. niger, Penicillium* sp. and *Rhizopus* sp., when exposed to HMs (such as Pb and Cu at 50 mg/l) [115]. Catalase is an efficient tool for supplying tolerance ability for fungi for cleanup of environmental areas. Therefore, fungal enzymes play an imperative role in remediation of metal-polluted sites.

5.4.3 PEROXIDASES (POXS)

POXs are categorized into lignin peroxidase (LiP), manganese peroxidase (MnP) and versatile peroxidase (VP) based on their origin and action. Among these three enzymes, LiP and MnP are heme POXs and they need hydrogen peroxide (H_2O_2) and manganese for action, which are frequently observed in the breakdown of toxic substances with white-rot fungi and basidiomycetes. In addition, VP enzymes have a wide array of substrate specific, oxidized phenol and non-phenol substances [61].

LiP and MnP have been identified in *Penicillium chrysosporium* due to their higher redox potential [79]. LiP and MnP catalyze oxidation of lignin components with H_2O_2. LiP breaks down non-phenol lignin linkages, while MnP produces Mn^{3+} and serves as a diffusible oxidizer on phenol or non-phenol lignin compounds through lipid peroxidation [58]. The advantage of POXs in soil cleaning was examined in soils polluted with aromatic hydrocarbons and degraded by fungi secreting POXs [28].

5.4.3.1 LIGNIN PEROXIDASES (LIPS)

Lignin peroxidases (LiPs) constitute a class of heme proteins secreted generally with white-rot fungi [130]. Lignin is a composite of ordinary polymer built from random polymerization of phenylpropanoid units. Lignin modifying enzymes (LMEs) have the ability to breakdown lignin with radical reactions [59]. With co-substrate, H_2O_2, and a moderator (such as veratryl alcohol), it breaks down lignin and further phenol containing substances. In this reaction, H_2O_2 gets reduced to H_2O by obtaining an electron from LiP. It plays the main function in biodegradation of lignin, a plant cell-wall component.

It is capable of oxidizing aromatic substances with redox potentials greater than 1.4 V with single electron abstraction, but accurate redox method is still inadequately recognized [87]. They can degrade different aromatic compounds majority of which are usually resistant to degradation with pH 2–5 [100]. When *P. chrysosporium* is demonstrated to produce LiP in media, isozyme forms different basidiomycetes that were isolated with an array ranging from 38 to 43 kDa [41]. They entirely oxidize both methylated lignin form and non-methylated lignin form substances and PAHs [60]. During influential and unspecific catalytic transformation action, it has the ability of straight conversion of up to 90% of lignin structural components [41].

5.4.3.2 MANGANESE PEROXIDASES (MNPS)

Manganese peroxidases (MnPs) are heme class of proteins isolated from lignin-degrading fungi that oxidize Mn^{2+} to Mn^{3+} in a multistep approach. Mn^{2+} motivates MnP secretion and serves as a substrate for MnP. The Mn^{3+} serves as a moderator for oxidation of different phenol containing components. The ensuing Mn^{3+} chelate oxalate is little adequate to disperse into sites unapproachable to enzymes. Whereas, lignin or similar structures (viz.: xenobiotic contaminants covered profound inside soil) were not essentially obtainable to enzymes [113].

They work in comparatively acidic pH 4–7 than LiP [13]. As a heme POX, they share most of the features and mechanisms with LiPs [33]. MnP appears not to occur in the same large gene families, characteristic of other ligninolytic enzymes, such as LAC [116] and was only reported in *Basidiomycota* [49]. MnP from white-rot fungi, *Trametes*, exhibited the capacity to break down azo and indigo colors and PAHs [135].

5.4.3.3 VERSATILE PEROXIDASES (VPS)

Versatile peroxidases (VPs) are capable to openly oxidize Mn^{2+}, methoxybenzenes, phenol aromatic compounds (i.e.: MnP, LiP), and horseradish POX. They have wide substrate activity and affinity to oxidize substrates without manganese contrast to more POXs. It was shown that they are able to oxidize both phenol and non-phenol lignin model dimers [95].

VPs are hybrids among LiPs and MnPs, as they consist of heme portion and oxidize Mn^{2+} to Mn^{3+}, effecting indirect oxidations, and oxidizing phenol and non-phenol containing substances. VPs conjoin mechanisms and substrate ranges between these two enzymes [64, 89]. VPs have thus far only been identified in *Basidiomycetes* [49]. However, little is understood concerning their function in the intervention of xenobiotic conversion;

although, it is identified that their production is influenced by the presence of PAH contaminants [90].

5.4.4 OXIDOREDUCTASES

Oxidoreductases catalyze oxidation and reduction reactions that occur within the cell. Number of ecto-oxidoreductases are produced by fungi to breakdown organic substrates (including laccases, tyrosinases, LiP, MnP, etc.) [49]. These enzymes allowed fungal species to cultivate on refractory substrates, i.e., cellulosic biomass is not effectively broken down by many bacteria. These characteristics of fungi are beneficial than the bacteria in remediation [120].

The well-known laccases were largely secreted by *Basidiomycetes* and *Ascomycetes*; and utilized for removal of color and detoxification of textile industry; remove the color of ink in paper industries, breakdown, and degradation of recalcitrant components in effluents, harmful substances come from coal treatment, and breakdown of soil contaminants [8, 108]. Currently, different POXs are differentiated in microorganisms, particularly from fungi, oxidize contaminants, and transform them into harmless substances by higher redox potentials [7, 31]. The oxidoreductases contribute in humification of different phenol substrates that are secreted from the decay of lignin in soil. In a similar way, oxidoreductases can also degrade toxic xenobiotics (viz.: phenol or anilic substances) by polymerization with other compounds [85].

5.4.5 HYDROLASES

Hydrolases commonly function as biochemical catalysts using water to break a chemical bond, typically resulting in dividing a larger molecule into smaller molecules. In the degradation of lignocellulosic materials, fungi use different hydrolases. The polymer of cellulose is linked jointly by hydrogen bonds to shape glycoproteins and cellulose [71]. Three kinds of cellulolytic enzymes are necessary for the breakdown of cellulose, such as endoglucanase, exoglucanase, and β-glucosidase [98].

Endoglucanases hydrolyze β-1,4-glycosidic links in non-crystalline regions of internal cellulose polymer, which results in a reduction in length of cellulose polymer and an increase in free ends. Exoglucanases breakdown cellulose starting from free ends by splitting off cellobiose. Finally, β-glucosidase degrades cellobiose and release glucose [121]. The lytic polysaccharide mono-oxygenase catalyzes oxidative cleavage of partially depolymerized cellulose and enhances its ease of access to cellulolytic enzymes [51]. Examples of hydrolases are: phosphotriesterases, amidases, proteases, carbohydrases (cellulases and amylases) and depolymerases.

5.4.6 MEMBRANE-BOUND ENZYMES

Various substances of unwanted components and contaminants are chemically degraded by many endo-fungal enzymes, including PAHs and dioxins. Cytochrome P450s have a significant function in the breakdown of contaminants and are involved in total chemical reactions of different medicines viz., anti-inflammatory medicines, fatty acid regulator medicines, antiseizure, and pain-relieving medicines, and diphenyl ether herbicide. The contaminants by mycoremediation go through breakdown in metabolism until their disruption [49]. The total chemical reactions may engage in the accumulation of metabolites at various steps of oxidation in the breakdown procedure.

5.5 CONTAMINANTS AND POLLUTANTS

HMs and mineral oils have been recognized as major soil contaminants, consequently organic substances, such as PAHs and aromatic hydrocarbons [66, 81]. In soils, constant toxic substances, chemicals, salts, radioactive compounds, and HMs are present because of industrial activities and contamination [40, 45]; and have unpleasant results on plant development and animal health. Soil could be contaminated by numerous means, i.e., seepage from a landfill, release of industrial effluents into soil, polluted water that penetrated into soil, breaking of underground tanks, overuse of pesticides and herbicides.

5.5.1 HEAVY METAL (HM) CONTAMINATION OF SOIL

Soil HM contamination mainly refers to the accumulation of HMs, viz., mercury, cadmium, lead, chromium, and other important bio-toxic HMs in soil, ensuing in concentrations that exceed backdrop values. Metals are not recyclable, and through biological amplification, their concentrations can be increased a number of times. Bioremediation represents new approaches for attending difficulty of HM pollution in soil. The majority of HM contaminants in ecology are: As, Sr, Cs, U, Cd, Cr, Cu, Hg, Pb, and Zn. A few metals are minor nutrients and are essential for the survival of plants, for instance, Zn, Cu, Mn, Ni, and Co, whereas the functions of other metals (viz., Cd, Pb, and Hg) are not known [6, 10, 31].

The HMs in polluted soils damage natural ecosystem services and ultimately spoil human healthiness through food chain [38, 56, 111]. Bioremediation decontaminates soil by microbes in place of plants. The approach is most frequently used to detoxify contaminants in soil and groundwater. Higher tolerance and bioremediation capability of fungi to metals like Cd, Cu, and Ni (up to 1,500 mg/L) had gained importance for remediation of these metals from polluted soil and sewage water [4, 46].

5.6 ADVANTAGES OF BIOREMEDIATION

- Easy, less labor-intensive and inexpensive due to their ecologically sound nature.
- Non-intrusive, effectively permitting for continuous site use.
- Novel means of remediation of environment from various pollutants and work as ecologically safe alternatives [103].
- Pollutants are damaged, not easily transported to various ecological media.
- Relative ease of execution [65].
- Reliable [32].

Theoretically, bioremediation is helpful for total degradation of a broad diversity of pollutants. This removes the opportunity of prospective problems linked with treatment and removal of polluted compound.

Bioremediation can frequently be performed on a site, often without any main interruption of regular actions. Bioremediation can prove less expensive than other methods that are employed for the cleanup of dangerous waste [62].

5.7 DISADVANTAGES OF BIOREMEDIATION

- Biological approaches are often very particular. Important site factors required for success include the presence of metabolically capable microbial populations, suitable environmental growth conditions, and appropriate levels of nutrients and contaminants.

- Exploration is required to expand and novel bioremediation approaches that are suitable for sites with various kinds of pollutants that are not consistently isolated in soil. Pollutant may be in the form of solid, liquid, and gas. Bioremediation often takes long time than other treatments, viz., excavation, and elimination of soil or incineration [62].
- It is restricted to those substances that are biodegradable. Not all substances are vulnerable to fast and whole degradation.

5.8 FUTURE PERSPECTIVES

Bioremediation have received a significant focus as various pollutants do not easily breakdown in every environment. Various approaches have been followed successfully to generate various plants and sites that have the capability to propagate in polluted soils of metals and accumulate or tolerate the stress of metals. Usually, biological methods for remediation of pollutants are safe and more advantageous for the environment. Similarly, HMs are absorbed by plants, and their tolerance is based on numerous ecological aspects, and associations among plants and fungi can play an imperative function in successful survival and growth of plants in contaminated soils.

These advantageous effects established by fungi, mutual relationship with pollutant toleration and plant growth-enhancing ability, show their use in remediation of metal-polluted soils might have main approach in near prospect. So, these highlights represent the importance of a consistent link among the research and development for the valuation and management of unindustrialized metal contaminants and the tools, apparatus, and know-how that contribute toward the contentment of these experiments. Future studies will be focused on factors involved in developing *in situ* bioremediation approaches by genetically engineered microorganisms (GEM) and also the applicability and adjustability of these GEMs in every stress circumstance and multiple-heavy-metal-polluted sites.

5.9 SUMMARY

This chapter advocated those fungi as effective candidates with significant economic, natural, and authorized reasons could be employed in remediation of harmful compounds. The bioremediation potential of fungi from extreme environments has been elaborated, which has indicated that HM removal by fungi and nanoparticle synthesis from them was a potential area of research. Exploiting the effect of fungal enzymes is an effective strategy for bioremediation of xenobiotic substances in soil. Implementing a biological strategy to find novel fungal species with better ligninolytic enzymes in situations like rainforests could be an effective path ahead in constant attempts to fight soil contaminants of industrial source in productive and reliable ways.

KEYWORDS

- **bioremediation**
- **environmental factors**
- **fungal enzymes**
- **fungal oxidases**
- **laccases**
- **microorganisms**

REFERENCES

1. Adams, G. O., Fufeyin, P. T., & Okoro, S. E., (2015). Bioremediation, biostimulation and bioaugmentation: A review. *International Journal of Environmental Bioremediation and Biodegradation, 3*(1), 28–39.
2. Agarry, S., & Latinwo, G. K., (2015). Biodegradation of diesel oil in soil and its enhancement by application of bioventing and amendment with brewery waste effluents as biostimulation-bioaugmentation agents. *Journal of Ecological Engineering, 16*(2), 82–91.
3. Agrawal, N., Verma, P., & Shahi, S. K., (2018). Degradation of polycyclic aromatic hydrocarbons (phenanthrene and pyrene) by the ligninolytic fungi *Ganodermalucidum* isolated from the hardwood stump. *Bioresources and Bioprocessing, 5*(1), 1–9.
4. Akhtar, S., Mahmood-ul-Hassan, M., Ahmad, R., Suthor, V., & Yasin, M., (2013). Metal tolerance potential of filamentous fungi isolated from soils irrigated with untreated municipal effluent. *Soil Environment, 32*(1), 55–62.
5. Alberts, B., Bray, D., Hopkin, K., Johnson, A., Lewis, J., Raff, M., Roberts, K., & Walter, P., (2013). *Essential Cell Biology* (4th edn., p. 864). New York: Garland Science/ Taylor & Francis Group.
6. Alloway, B. J., (2013). Sources of heavy metals and metalloids in soils. In: Alloway, B. J., (ed.), *Heavy Metals in Soils: Trace Metals and Metalloids in Soils and their Bioavailability* (pp. 11–50). Netherlands, Dordrecht: Springer.
7. Al-Maqdi, K. A., Hisaindee, S. M., Rauf, M. A., & Ashraf, S. S., (2017). Comparative degradation of a thiazole pollutant by an advanced oxidation process and an enzymatic approach. *Biomolecules, 7*(3), 64–71.

8. Alneyadi, A. H., Rauf, M. A., & Ashraf, S. S., (2018). Oxidoreductases for the remediation of organic pollutants in water: Critical review. *Critical Reviews in Biotechnology, 38*(7), 971–988.
9. Alvarez, A., Saez, J. M., Costa, J. S. D., Colin, V. L., Fuentes, M. S., Cuozzo, S. A., Benimeli, C. S., et al., (2017). Actinobacteria: Current research and perspectives for bioremediation of pesticides and heavy metals. *Chemosphere, 166*, 41–62.
10. Alwan, A. H., Fadil, S. M., & Khadair, S. H., (2013). Bioremediation of the water contaminated by waste of hydrocarbon by use *Ceratophyllaceae* and *Potamogetonaceae* plants. *Journal of Genetic and Environmental Resources Conservation, 1*(2), 106–110.
11. Andrea, M. M., Tomita, R. Y., Luchini, L. C., & Musumeci, M. R., (1994). Laboratory studies on volatilization and mineralization of 14c-p, p'-DDT in soil, release of bound residues and dissipation from solid surfaces. *Journal of Environmental Science and Health, Part B, 29*(1), 133–139.
12. Antosova, Z., Herkommerova, K., Pichova, I., & Sychrova, H., (2018). Efficient secretion of three fungal laccases from *saccharomyces cerevisiae* and their potential for decolorization of textile industry effluent—a comparative study. *Biotechnology Progress, 34*(1), 69–80.
13. Asgher, M., Bhatti, H. N., Ashraf, M., & Legge, R. L., (2008). Recent developments in biodegradation of industrial pollutants by white-rot fungi and their enzyme system. *Biodegradation, 19*(6), 771–783.
14. Baker, P., Tiroumalechetty, A., & Mohan, R., (2019). Fungal enzymes for bioremediation of xenobiotic compounds. Chapter 19. In: Yadav, A. N., Singh, S., Mishra, S., & Gupta, A., (eds.), *Recent Advancement in White Biotechnology through Fungi* (Vol. 3, pp. 463–489). Cham: Springer.
15. Balabanova, L., Slepchenko, L., Son, O., & Tekutyeva, L., (2018). Biotechnology potential of marine fungi degrading plant and algae polymeric substrates. *Frontiers in Microbiology, 9*, 1527–1534.
16. Balaji, V., Arulazhagan, P., & Ebenezer, P., (2014). Enzymatic bioremediation of polyaromatic hydrocarbons by fungal consortia enriched from petroleum-contaminated soil and oilseeds. *Journal of Environmental Biology, 35*(3), 521–529.
17. Bennett, R. M., Cordero, P. R. F., Bautista, G. S., & Dedeles, G. R., (2013). Reduction of hexavalent chromium using fungi and bacteria isolated from contaminated soil and water samples. *Chemistry and Ecology, 29*(4), 320–328.
18. Bilal, M., Asgher, M., Iqbal, H. M., Hu, H., & Zhang, X., (2017). Bio-based degradation of emerging endocrine-disrupting and dye-based pollutants using cross-linked enzyme aggregates. *Environmental Science and Pollution Research, 24*(8), 7035–7041.
19. Bonugli-Santos, R. C., & Dos, S. V. M., (2015). Marine-derived fungi: Diversity of enzymes and biotechnological applications. *Frontiers in Microbiology, 6*, 269–274.
20. Boopathy, R., (2000). Factors limiting bioremediation technologies. *Bioresource Technology, 74*, 63–67.
21. Bosco, F., Mollea, C., & Ruggeri, B., (2017). Decolorization of Congo red by *Phanerochaete chrysosporium*: Role of biosorption and biodegradation. *Environmental Technology, 38*(20), 2581–2588.
22. Boul, H. L., Garnham, M. L., Hucker, D., Baird, D., & Aislabie, J., (1994). Influence of agricultural practices on the levels of DDT and its residues in soil. *Environmental Science and Technology, 28*(8), 1397–1402.

23. Buvaneswari, S., Damodarkumar, S., & Murugesan, S., (2013). Bioremediation studies on sugar-mill effluent by selected fungal species. *International Journal of Current Microbiology and Applied Sciences, 2*(1), 50–58.
24. Cajthaml, T., (2015). Biodegradation of endocrine-disrupting compounds by ligninolytic fungi: Mechanisms involved in the degradation. *Environmental Microbiology, 17*(12), 4822–4834.
25. Chang, Y. T., Lee, J. F., Liu, K. H., Liao, Y. F., & Yang, V., (2016). Immobilization of fungal laccase onto a nonionic surfactant-modified clay material: Application to PAH degradation. *Environmental Science and Pollution Research, 23*(5), 4024–4035.
26. Chowdhary, P., Shukla, G., Raj, G., Ferreira, L. F. R., & Bharagava, R. N., (2018). Microbial manganese peroxidase: Ligninolytic enzyme and its ample opportunities in research. *SN Applied Sciences, 1*(1), 45–52.
27. Coronado-Ruiz, C., Avendano, R., Escudero-Leyva, E., Conejo-Barboza, G., Chaverri, P., & Chavarria, M., (2018). Two new cellulolytic fungal species isolated from a 19th-century art collection. *Scientific Reports, 8*(1), 1–9.
28. D'Annibale, A., Rosetto, F., Leonardi, V., Federici, F., & Petruccioli, M., (2006). Role of autochthonous filamentous fungi in bioremediation of a soil historically contaminated with aromatic hydrocarbons. *Applied Environmental Microbiology, 72*, 28–36.
29. Damare, S., Singh, P., & Raghukumar, S., (2012). Biotechnology of marine fungi. *Progress in Molecular and Subcellular Biology, 53*, 277–297.
30. Das, N., & Chandran, P., (2011). Microbial degradation of petroleum hydrocarbon contaminants: An overview. *Biotechnology Research International, 2011*, 1–13.
31. Delille, D., Duval, A., & Pelletier, E., (2008). Highly efficient pilot biopiles for on-site fertilization treatment of diesel oil-contaminated Sub-Antarctic soil. *Cold Regions Science and Technology, 54*(1), 7–18.
32. Dell'Anno, A., Beolchini, F., & Rocchetti, L., (2012). High bacterial biodiversity increases the degradation performance of hydrocarbons during bioremediation of contaminated harbor marine sediments. *Environmental Pollution, 167*, 85–92.
33. Deshmukh, R., Khardenavis, A. A., & Purohit, H. J., (2016). Diverse metabolic capacities of fungi for bioremediation. *Indian Journal of Microbiology, 56*, 247–264.
34. Dimond, J. B., & Owen, R. B., (1996). Long-term residue of DDT compounds in forest soils in Maine. *Environmental Pollution, 92*(2), 227–230.
35. Dos, S. B. G. C., & Araujo, A. V., (2013). Ligninolytic enzymes from white-rot fungi and application in the removal of synthetic dyes. In: Polizeli, T. M., Rai, M., & De Lourdes, M., (eds.), *Fungal Enzymes* (pp. 258–279). Boca Raton: CRC Press.
36. Duarte, K., Justino, C. I., & Pereira, R., (2013). Removal of the organic content from a bleached kraft pulp mill effluent: Treatment with silica-alginate-fungi biocomposites. *Journal of Environmental Science and Health - Part A, 48*(2), 166–172.
37. Durairaj, P., Malla, S., & Nadarajan, S. P., (2015). Fungal cytochrome P450 monooxygenases of *Fusarium oxysporum* for the synthesis of x-hydroxy fatty acids in engineered *Saccharomyces cerevisiae*. *Microbial Cell Factories, 14*(1), 45–52.
38. Duruibe, J. O., Ogwuegbu, M. O. C., & Egwurugwu, J. N., (2007). Heavy metal pollution and human bio toxic effects. *International Journal of Physical Sciences, 2*(5), 112–118.
39. El Fantroussi, S., & Agathos, S. N., (2005). Is bioaugmentation a feasible strategy for pollutant removal and site remediation? *Current Opinion in Microbiology, 8*(3), 268–275.

40. Emami, S., Pourbabaee, A. A., & Alikhani, H. A., (2012). Bioremediation principles and techniques on petroleum hydrocarbon contaminated soil. *Technical Journal of Engineering and Applied Sciences, 2*(10), 320–323.
41. Falade, A. O., Nwodo, U. U., Iweriebor, B. C., Green, E., Mabinya, L. V., & Okoh, A. I., (2017). Lignin peroxidase functionalities and prospective applications. *Microbiology Open, 6*(1), 1- 14.
42. Fester, T., (2013). Arbuscular mycorrhizal fungi in a wetland constructed for benzene-, methyl tert-butyl ether and ammonia contaminated groundwater bioremediation. *Microbial Biotechnology, 6*(1), 80–84.
43. Foght, J., April, T., Biggar, K., & Aislabie, J., (2001). Bioremediation of DDT-contaminated soils: A review. *Bioremediation Journal, 5*(3), 235–246.
44. Fonseca, M. I., Fariña, J. I., Sanabria, N. I., Villalba, L. L., & Zapata, P. D., (2013). Influence of culture conditions on laccase production, growth, and isoenzymes patterns in native white rot fungi from the Misiones rainforest (Argentina). *BioResources, 8*(2), 2855–2866.
45. Forstner, U., & Wittmann, G. T., (1981). *Metal Pollution in the Aquatic Environment* (2nd edn., p. 488). Berlin Heidelberg: Springer-Verlag.
46. Gomez, F., & Sartaj, M., (2014). Optimization of field scale biopiles for bioremediation of petroleum hydrocarbon contaminated soil at low temperature conditions by response surface methodology (RSM). *International Biodeterioration and Biodegradation, 89*, 103–109.
47. Gorska, E. B., Jankiewicz, U., & Dobrzyński, J. A., (2014). Production of ligninolytic enzymes by cultures of white rot fungi. *Polish Journal of Microbiology, 63*(4), 461–465.
48. Govarthanan, M., Fuzisawa, S., Hosogai, T., & Chang, Y. C., (2017). Biodegradation of aliphatic and aromatic hydrocarbons using the filamentous fungus *Penicillium* sp. CHY-2 and characterization of its manganese peroxidase activity. *RSC Advances, 7*(34), 20716–20723.
49. Harms, H., Schlosser, D., & Wick, L. Y., (2011). Untapped potential: Exploiting fungi in bioremediation of hazardous chemicals. *Nature Reviews Microbiology, 9*(3), 177–192.
50. Hattori, A., Unno, H., Goda, S., Motoyama, K., Yoshimura, T., & Hemmi, H., (2015). *In vivo* formation of the protein disulfide bond that enhances the thermostability of diphosphomevalonate decarboxylase, an intracellular enzyme from the hyperthermophilic archaeon *Sulfolobus solfataricus*. *Journal of Bacteriology, 197*(21), 3463–3471.
51. Horn, S. J., Vaaje-Kolstad, G., Westereng, B., & Eijsink, V., (2012). Novel enzymes for the degradation of cellulose. *Biotechnology for Biofuels, 5*(1), 45–52.
52. Hou, L., Ji, D., Dong, W., Yuan, L., Zhang, F., Li, Y., & Zang, L., (2020). The synergistic action of electro-Fenton and white-rot fungi in the degradation of lignin. *Frontiers in Bioengineering and Biotechnology, 8*, 99–105.
53. Huang, J., Fu, Y., & Liu, Y., (2014). Comparison of alkali-tolerant fungus *Myrothecium* sp. IMER1 and white-rot fungi for decolorization of textile dyes and dye effluents. *Journal of Bioremediation and Biodegradation, 5*(3), 1–5.
54. Igiehon, N. O., & Babalola, O. O., (2017). Biofertilizers and sustainable agriculture: Exploring arbuscular mycorrhizal fungi. *Applied Microbiology and Biotechnology, 101*(12), 4871–4881.
55. Ikehata, K., (2015). Use of fungal laccases and peroxidases for enzymatic treatment of wastewater containing synthetic dyes. Chapter 6. In: Sharma, S. K., (ed.), *Green*

Chemistry for Dyes Removal from Wastewater: Research Trends and Applications (p. 496). New York: John Wiley & Sons.

56. Jaishankar, M., Tseten, T., Anbalagan, N., Mathew, B., & Beeregowda, K., (2014). Toxicity, mechanism and health effects of some heavy metals. *Interdisciplinary Toxicology, 7*(2), 60–72.
57. Jebapriya, G. R., & Gnanadoss, J. J., (2013). Bioremediation of textile dye using white-rot fungi: A review. *International Journal of Current Research and Review, 5*(3), 1–13.
58. Jensen, K. A., Bao, W., Kawai, S., Srebotnik, E., & Hammel, K. E., (1996). Manganese-dependent cleavage of nonphenolic lignin structures by *Ceriporiopsis subvermispora* in the absence of lignin peroxidase. *Applied Environmental Microbiology, 62*(10), 3679–3686.
59. Joy, J., Jose, C., Mathew, P. L., Thomas, S., & Khalaf, M. N., (2015). Biological delignification of biomass. In: Khalaf, M. N., (ed.), *Green Polymers and Environmental Pollution Control* (pp. 271–291). Boca Raton: CRC Press; Taylor & Francis Group.
60. Kadri, T., Rouissi, T., Brar, S. K., Cledon, M., Sarma, S., & Verma, M., (2017). Biodegradation of polycyclic aromatic hydrocarbons (PAHs) by fungal enzymes: A review. *Journal of Environmental Sciences, 51*, 52–74.
61. Karigar, C. S., & Rao, S. S., (2011). Role of microbial enzymes in the bioremediation of pollutants: A review. *Enzyme Research, 2011*, 11–20.
62. Kensa, V. M., (2011). Bioremediation: An overview. *J. of Industrial Pollution Control, 27*(2), 161–168.
63. Khambhaty, Y., Ananth, S., Sreeram, K. J., Rao, J. R., & Nair, B. U., (2015). Dual utility of a novel, copper enhanced laccase from *Trichoderma aureoviridae. International Journal of Biological Macromolecules, 81*, 69–75.
64. Kües, U., (2015). Fungal enzymes for environmental management. *Current Opinion in Biotechnology, 33*, 268–278.
65. Kumar, A., Bisht, B. S., & Joshi, V. D., (2011). Review on bioremediation of polluted environment: A management tool. *International Journal of Environmental Sciences, 1*(6), 1079–1093.
66. Kumar, S., Chaurasia, P., & Kumar, A., (2016). Isolation and characterization of microbial strains from textile industry effluents from Bhilwara - India: Analysis with bioremediation. *Journal of Chemical and Pharmaceutical Research, 8*(4), 143–150.
67. Lee, T. H., Byun, I. G., Kim, Y. O., Hwang, I. S., & Park, T. J., (2006). Monitoring biodegradation of diesel fuel in bioventing processes using in situ respiration rate. *Water Science and Technology, 53*(4, 5), 263–272.
68. Legodi, L. M., La Grange, D., Van, R E. L., & Ncube, I., (2019). Isolation of cellulose-degrading fungi from decaying *Banana pseudostem* and *Strelitzia alba. Enzyme Research*, 1–10.
69. Lin, X., Li, X., Sun, T., Li, P., Zhou, Q., Sun, L., & Hu, X., (2009). Changes in microbial populations and enzyme activities during the bioremediation of oil-contaminated soil. *Bulletin of Environmental Contamination and Toxicology, 83*(4), 542–547.
70. Liu, B., Liu, J., Ju, M., Li, X., & Wang, P., (2017). Bacteria-white-rot fungi joint remediation of petroleum-contaminated soil based on sustained-release of laccase. *RSC Advances, 7*(62), 39075–39081.

71. Lynd, L. R., Weimer, P. J., Van, Z. W. H., & Pretorius, I. S., (2002). Microbial cellulose utilization: Fundamentals and biotechnology. *Microbiology Molecular Biology Reviews, 66*(3), 506–577.
72. Macaulay, B. M., (2015). Understanding the behavior of oil-degrading microorganisms to enhance the microbial remediation of spilled petroleum. *Applied Ecology and Environmental Research, 13*(1), 247–262.
73. Madhavi, G. N., & Mohini, D. D., (2012). Review paper on - parameters affecting bioremediation. *International Journal of Life Science and Pharma Research, 2*(3), 77–80.
74. Malik, Z., & Ahmed, S., (2012). Degradation of petroleum hydrocarbons by oil field isolated bacterial consortium. *African Journal of Biotechnology, 11*(3), 650–658.
75. Mao, J., & Guan, W., (2016). Fungal degradation of polycyclic aromatic hydrocarbons (PAHs) by *Scopulariopsis brevicaulis* and its application in bioremediation of PAH-Contaminated soil. *Acta Agriculturae Scandinavica, Section B—Soil and Plant Science, 66*(5), 399–405.
76. Marchand, C., St-Arnaud, M., Hogland, W., Bell, T. H., & Hijri, M., (2017). Petroleum biodegradation capacity of bacteria and fungi isolated from petroleum-contaminated soil. *International Biodeterioration and Biodegradation, 116*, 48–57.
77. Marco, E., Font, X., Sánchez, A., Gea, T., Gabarrell, X., & Caminal, G., (2013). Co-composting as a management strategy to reuse the white-rot fungus *Trametes Versicolor* after its use in a biotechnological process. *International Journal of Environment and Waste Management, 11*(1), 100–108.
78. Margot, J., Bennati-Granier, C., Maillard, J., Blánquez, P., Barry, D. A., & Holliger, C., (2013). Bacterial versus fungal laccase: Potential for micropollutant degradation. *AMB Express, 3*(1), 1–14.
79. Martínez, A. T., (2002). Molecular biology and structure-function of lignin-degrading heme peroxidases. *Enzyme and Microbial Technology, 30*(4), 425–444.
80. Migliore, L., Fiori, M., Spadoni, A., & Galli, E., (2012). Biodegradation of oxytetracycline by *Pleurotus ostreatus* mycelium: A mycoremediation technique. *Journal of Hazardous Materials, 215*, 227–232.
81. Mougin, C., Boukcim, H., & Jolivalt, C., (2009). Soil bioremediation strategies based on the use of fungal enzymes. In: Singh, A., Kuhad, R. C., & Ward, O. P., (eds.), *Advances in Applied Bioremediation* (pp. 123–149). Berlin, Heidelberg: Springer.
82. Mulligana, C. N., & Yong, R. N., (2004). Natural attenuation of contaminated soils. *Environment International, 30*, 587–601.
83. Ni, S. Q., Wang, Z., Lv, L., Liang, X., Ren, L., & Zhou, Q., (2015). Bioremediation of wastewaters with decabromo diphenyl ether by anaerobic granular sludge. *Colloids and Surfaces B: Biointerfaces, 128*, 522–527.
84. Owen, Jr. R. B., Dimond, J. B., & Getchell, A. S., (1977). DDT: Persistence in northern spodosols. *Journal of Environmental Quality, 6*(4), 359–360.
85. Park, J. W., Park, B. K., & Kim, J. E., (2006). Remediation of soil contaminated with 2, 4-dichlorophenol by treatment of minced shepherd's purse roots. *Archives of Environmental Contamination and Toxicology, 50*(2), 191–195.
86. Pedraza-Zapata, D. C., & Sánchez-Garibello, A. M., (2017). Promising cellulolytic fungi isolates for rice straw degradation. *Journal of Microbiology, 55*(9), 711–719.
87. Piontek, K., Smith, A. T., & Blodig, W., (2001). Lignin peroxidase structure and function. *Biochemical Society Transactions, 29*(2), 111–116.

88. Pita, T., Alves-Pereira, I., & Ferreira, R., (2013). Decline in peroxidase and catalases by lindane may cause an increase in reactive oxygen species in *Saccharomyces cerevisiae*. In: Méndez-Vilas, A., (ed.), *Industrial, Medical and Environmental Applications of Microorganisms: Current Status and Trends* (pp. 83–87). Wageningen: Academic Publishers.
89. Pozdnyakova, N. N., (2012). Involvement of the ligninolytic system of white-rot and litter-decomposing fungi in the degradation of polycyclic aromatic hydrocarbons. *Biotechnology Research International, 2012,* 20–25.
90. Pozdnyakova, N. N., & Balandina, S. A., (2018). Ligninolytic basidiomycetes as promising organisms for the mycoremediation of PAH-contaminated environments. In: *IOP Conference Series: Earth and Environmental Science* (Vol. 107, p. 8). Article ID: 012071.
91. Rana, K. L., Kour, D., & Sheikh, I., (2019). Biodiversity of endophytic fungi from diverse niches and their biotechnological applications. In: Singh, B. P., (ed.), *Advances in Endophytic Fungal Research* (pp. 105–144). Cham: Springer International Publishing.
92. Rastegari, A. A., Yadav, A. N., & Gupta, A., (2019). *Prospects of Renewable Bioprocessing in Future Energy Systems* (p. 518). Cham: Springer International Publishing.
93. Reya, I., Prabha, L. M., & Renitta, E. R., (2013). Equilibrium and kinetic studies on biosorption of Cr (VI) using novel *Aspergillus jegita* isolated from tannery effluent. *Research Journal of Chemistry and Environment, 17*(4), 72–78.
94. Rodarte-Morales, A. I., & Feijoo, G., (2011). Degradation of selected pharmaceutical and personal care products (PPCPs) by white-rot fungi. *World Journal of Microbiology and Biotechnology, 27*(8), 1839–1846.
95. Ruiz-Duenas, F. J., Morales, M., & Pérez-Boada, M., (2007). Manganese oxidation site in *Pleurotus eryngii* versatile peroxidase: A site-directed mutagenesis, kinetic, and crystallographic study. *Biochemistry, 46*(1), 66–77.
96. Sakaki, T., Yamamoto, K., & Ikushiro, S., (2013). Possibility of application of cytochrome P450 to bioremediation of dioxins. *Biotechnology and Applied Biochemistry, 60*(1), 65–70.
97. Saroj, P., Manasa, P., & Narasimhulu, K., (2018). Characterization of thermophilic fungi producing extracellular lignocellulolytic enzymes for lignocellulosic hydrolysis under solid-state fermentation. *Bioresources and Bioprocessing, 5*(1), 31–38.
98. Schwarz, W., (2001). The cellulosome and cellulose degradation by anaerobic bacteria. *Applied Microbiology and Biotechnology, 56*(5, 6), 634–649.
99. Seidu, A., Quainoo, A. K., Addae, G., & Stenchly, K., (2015). Mycoremediation of diesel contaminated soil with oyster mushroom (*Pleurotus ostreatus*) using maize (*Zea mays*) as the test crop. *UDS International Journal of Development, 2*(2), 1–8.
100. Shrivastava, R., Christian, V., & Vyas, B. R. M., (2005). Enzymatic decolorization of sulfon phthalein dyes. *Enzyme and Microbial Technology, 36*(2–3), 333–337.
101. Shruthi, K., Yadav, P. S., Prasad, B. S., & Chandra, M. S., (2018). Cellulase production by *aspergillus unguis* in solid-state fermentation. *Journal of Forestry Research, 30*(1), 205–212.
102. Siddique, T., Okeke, B. C., Arshad, M., & Frankenberger, W. T., (2002). Temperature and pH effects on biodegradation of hexachlorocyclohexane isomers in water and a soil slurry. *Journal of Agricultural and Food Chemistry, 50*(18), 5070–5076.

103. Singh, A., Kumar, V., & Srivastava, J. N., (2013). Assessment of bioremediation of oil and phenol contents in refinery wastewater via bacterial consortium. *Journal of Petroleum and Environmental Biotechnology, 4*(3), 1–4.
104. Singh, M., Srivastava, P. K., Verma, P. C., Kharwar, R. N., Singh, N., & Tripathi, R. D., (2015). Soil fungi for mycoremediation of arsenic pollution in agriculture soils. *Journal of Applied Microbiology, 119*(5), 1278–1290.
105. Singh, R. K., Tripathi, R., Ranjan, A., & Srivastava, A. K., (2020). Fungi as potential candidates for bioremediation. In: Singh, P., Kumar, A., & Borthakur, A., (eds.), *Abatement of Environmental Pollutants* (pp. 177–191). New York: Elsevier Inc.
106. Sinha, A., Singh, V. N., Mehta, B. R., & Khare, S. K., (2011). Synthesis and characterization of monodispersed orthorhombic manganese oxide nanoparticles produced by *Bacillus* sp. cells simultaneous to its bioremediation. *Journal of Hazardous Materials, 192*(2), 620–627.
107. Sogin, M. L., Morrison, H. G., Huber, J. A., Welch, D. M., Huse, S. M., Neal, P. R., Arrieta, J. M., & Herndl, G. J., (2006). Microbial diversity in the deep sea and the underexplored rare biosphere. *Proceedings of the National Academy of Sciences, 103*(32), 12115–12120.
108. Spina, F., Tigini, V., Prigione, V., & Varese, G. C., (2015). Fungal biocatalysts in the textile industry. Chapter 3. In: Gupta, V. K., Mach, R. L., & Sreenivasaprasad, S., (eds.), *Fungal Biomolecules: Sources, Applications and Recent Developments* (pp. 39–50). London: Wiley-Blackwell.
109. Syed, K., Porollo, A., Lam, Y. W., & Grimmett, P., (2013). CYP63A2: A catalytically versatile fungal P450 monooxygenase capable of oxidizing higher molecular weight polycyclic aromatic hydrocarbons, alkylphenols, and alkanes. *Applied Environmental Microbiology ,79*(8), 2692–2702.
110. Tavares, A. P. M., & Coelho, M. A., (2006). Optimization and modeling of laccase production by *Trametes Versicolor* in a bioreactor using statistical experimental design. *Applied Biochemistry and Biotechnology, 134*(3), 233–248.
111. Tchounwou, P. B., Yedjou, C. G., Patlolla, A. K., & Sutton, D. J., (2012). Heavy metal toxicity and the environment. In: Lluch, A., (ed.), *Molecular, Clinical and Environmental Toxicology* (Vol. 101, pp. 133–164). Cham: Springer Basel AG.
112. Tegli, S., Cerboneschi, M., Corsi, M., Bonnanni, M., & Bianchini, R., (2014). Water recycle as a must: Decolorization of textile wastewaters by plant-associated fungi. *Journal of Basic Microbiology, 54*(2), 120–132.
113. Ten, H. R., & Teunissen, P. J., (2001). Oxidative mechanisms involved in lignin degradation by white-rot fungi. *Chemical Reviews, 101*(11), 3397–3414.
114. Thatoi, H., Behera, B. C., & Mishra, R. R., (2013). Ecological role and biotechnological potential of mangrove fungi: A review. *Mycology, 4*(1), 54–71.
115. Thippeswamy, B., & Shivakumar, C. K., (2014). Studies on heavy metals detoxification biomarkers in fungal consortia. *Caribbean Journal of Science and Technology, 2*, 496–502.
116. Torres-Farradá, G., & Manzano, L. A. M., (2017). Diversity of ligninolytic enzymes and their genes in strains of the genus *Ganoderma*: Applicable for biodegradation of xenobiotic compounds. *Frontiers in Microbiology, 8*, 898–905.
117. Tuomela, M., & Hatakka, A., (2011). Oxidative fungal enzymes for bioremediation. In: Young, M., (ed.), *Comprehensive Biotechnology* (2nd edn., pp. 183–196). Madrid: Elsevier.

118. Upadhyay, P., Shrivastava, R., & Agrawal, P. K., (2016). Bioprospecting and biotechnological applications of fungal laccase. *3-Biotech, 6*(1), 15–22.
119. Vazquez, M. A., Cabrera, E. C. V., Aceves, M. A., & Mallol, J. L. F., (2019). Cellulolytic and ligninolytic potential of new strains of fungi for the conversion of fibrous substrates. *Biotechnology Research and Innovation, 3*(1), 177–186.
120. Verma, A., Singh, H., Anwar, S., & Chattopadhyay, A., (2017). Microbial keratinases: Industrial enzymes with waste management potential. *Critical Reviews in Biotechnology, 37*(4), 476–491.
121. Viikari, L., & Alén, R., (2011). Biochemical and chemical conversion of forest biomass. In: Alen, R., (ed.), *Biorefining of Forest Resources* (pp. 225–261). Provo - Finland: Bookwell Oy.
122. Vishwanath, B., Rajesh, B., Janardhan, A., Kumar, A. P., & Narasimha, G., (2014). Fungal laccases and their applications in bioremediation. *Enzyme Research, 2014*, 88–92.
123. Voberkova, S., & Vaverkova, M. D., (2017). Effect of inoculation with white-rot fungi and fungal consortium on the composting efficiency of municipal solid waste. *Waste Management, 61*, 157–164.
124. Wang, S., Yang, Q., Bai, Z., Wang, S., Wang, Y., & Nowak, K. M., (2015). Acclimation of aerobic-activated sludge degrading benzene derivatives and co-metabolic degradation activities of trichloroethylene by benzene derivative-grown aerobic sludge. *Environmental Technology, 36*(1), 115–123.
125. Whiteley, C. G., & Lee, D. J., (2006). Enzyme technology and biological remediation. *Enzyme and Microbial Technology, 38*(3, 4), 291–316.
126. Wong, K. S., Cheung, M. K., Au, C. H., & Kwan, H. S., (2013). A novel *Lentinula edodes* laccase and its comparative enzymology suggest guaiacol-based laccase engineering for bioremediation. *PLoS One, 8*(6), 8. Article ID: e66426.
127. Xiao, P., & Kondo, R., (2020). Biodegradation and biotransformation of pentachlorophenol by wood-decaying white-rot fungus *Phlebia acanthocystis* TMIC34875. *Journal of Wood Science, 66*(1), 1–11.
128. Xiao, X., Chen, H., Si, C., & Wu, L., (2012). Influence of biosurfactant-producing strain *Bacillus subtilis* BS1 on mycoremediation of soils contaminated with phenanthrene. *International Biodeterioration & Biodegradation, 75*, 36–42.
129. Xu, B., Jianying, G., Yongxi, Z., & Haibo, L., (1994). Behaviour of DDT in Chinese tropical soils. *Journal of Environmental Science & Health - Part B, 29*(1), 37–46.
130. Yadav, A. N., Singh, S., Mishra, S., & Gupta, A., (2019). *Recent Advancement in White Biotechnology through Fungi* (Vol. 3, p. 528). Cham: Springer International Publishing.
131. Yadav, A. N., Verma, P., Kumar, V., & Sangwan, P., (2018). Biodiversity of the genus *Penicillium* in different habitats. In: Gupta, V., (ed.), *New and Future Developments in Microbial Biotechnology and Bioengineering* (pp. 3–18). London: Elsevier Inc.
132. Yadav, S. P., Shruthi, K., Siva, P. B. V., & Chandra, M. S., (2017). Isolation and identification of *Aspergillus protuberus* from *Mahanandi forest* sample and investigation of its cellulase production. *Indian Journal of Advances in Chemical Sciences, 5*(1), 8–15.
133. Yang, S., Hai, F. I., Nghiem, L. D., & Price, W. E., (2013). Understanding the factors controlling the removal of trace organic contaminants by white-rot fungi and their lignin modifying enzymes: A critical review. *Bioresource Technology, 141*, 97–108.
134. Zavarzina, A. G., Lisov, A. V., & Leontievsky, A. A., (2018). The role of ligninolytic enzymes laccase and a versatile peroxidase of the white-rot fungus *Lentinus tigrinus*

in biotransformation of soil humic matter: Comparative in vivo study. *Journal of Geophysical Research: Bio-Geosciences, 123*(9), 2727–2742.

135. Zhang, H., Zhang, S., He, F., Qin, X., Zhang, X., & Yang, Y., (2016). Characterization of a manganese peroxidase from white-rot fungus *Trametes* sp. 48424 with strong ability of degrading different types of dyes and polycyclic aromatic hydrocarbons. *Journal of Hazardous Materials, 320*, 265–277.
136. Zhu, J. Y., Chandra, M. S., & Gu, F., (2015). Using sulfite chemistry for robust bioconversion of Douglas-fir forest residue to bioethanol at high titer and lignosulfonate: Pilot-scale evaluation. *Bioresource Technology, 179*, 390–397.

CHAPTER 6

SOIL BIOREMEDIATION STRATEGIES BASED ON FUNGAL ENZYMES

UMAIR RIAZ, GHULAM MURTAZA, SHAZIA IQBAL, and TABINDA ATHAR

ABSTRACT

The modern development in science and technology, industrialization, and urbanization has greatly contaminated the soil with xenobiotics. Although various physical, chemical, and biological methods have been developed for the biotransformation of contaminants, yet the use of fungal mediated enzymes is the most sustainable, socially acceptable, quick, and cost-effective way for the biotransformation of xenobiotics into less harmful or relatively neutral components.

6.1 INTRODUCTION

Rising affluence in the industrialized world, coupled with growth and rapid development in developing countries, have enhanced the large-scale destruction of the climate and environment. Due to these reasons, the management of solid waste is a significant concern to both developed and developing countries. The environment is encumbered with a number of contaminants and recalcitrant compounds (such as polychlorinated dibenzo-dioxins (PCDDs), polyaromatic hydrocarbons (PAHs), polychlorinated dibenzofurans (PCDFs)) and potentially toxic elements (PTEs) (heavy metal (HM)/oids) [76, 77].

Although many traditional, efficient, physicochemical treatments/removal methods of these compounds cannot be applied on a large scale [2], yet the bioremediation, by implementing natural biological processes in particular for polluted land and water, has been documented as eco-friendly and economically successful conversion of harmful, recalcitrant compounds into non-toxic products [29, 61] and it is also made accessible nutrients for crops especially macronutrients. This technique involves adding appropriate microbes in the contaminated environment that conduct numerous chemical and physical reactions as the main metabolism resulting in contaminants being degraded and removed [61].

Because of their vigorous and diverse morphology and metabolic capacity, fungi act as important decomposers in bioremediation. The most effective and renewable route for purifying polluted sites is the toxic organic agents in bioremediation by fungi [29]. Mycoremediation is a method of bioremediation utilized for decontaminating areas by fungi. A growing interest has been in the exclusive ability of fungi to destroy such contaminants by using a variety of intracellular and extracellular enzymes for detoxification and biodegradation, including peroxidases (POXs) and cytochrome P450 [32]. Absorption of fungal biomass has been proposed for the degradation of toxins in addition to the activity of enzymes; e.g.: a Laccases (LAC) as seen during endosulfan transition to endosulfan sulfate and a small volume of endosulfan in the occurrence of white-rot fungi *T. P. Ostreatus* and *T. versicolor* [96].

This chapter focuses on the use and role of fungi and various fungal enzymes in detoxification and biodegradation process.

6.2 FUNGAL ENZYMES

Fungi, being heterotrophic, are natural decomposers and can develop several enzymes. The fungi as a source of biological diversity have a crucial role to play. Enzymes cause complex compounds to degrade into simple sugars. These fungal enzymes are well adapted for a variety of uses. They also regulate soil enzyme function. They are used in well-known agricultural, pulp, and paper, textiles, detergents, milk, feeding stuffs and therapies [53]. It has been important to use different fungal enzymes to eliminate contaminants in the soil polluted by various effluents. White rot fungi are more effective in biodegrading a wide range of organic molecules due to the release of enzymes, which can convert extracellular lignin.

Many enzymes are being used in the biodegradation of lignin, including lignin peroxidase (LiP), manganese peroxidase (MnP), various enzymes that generate H_2O_2 and LAC. By adding carbon sources (such as straw, sawdust, and corn cob) to various soil contaminated sites, biodegradation can be improved. Several environmental assemblages are being deliberated on the *in vitro* biodegradation of single and recalcitrant hydrocarbons, such as PAHs, phenols, halogenated hydrocarbon, persistent organic contaminants (POPs), aromatic hydrocarbon and repro-hydro carbon explosive compounds.

Different mushrooms (such as *Agaricus bisporus, Amanita spissa, Fome fomentarius Phellinus badius, Lactarius piperatus, Suillus grevillei,* and *Tricholoma terreum*) can accumulate potential toxic trace elements (PTEs) and can bioremediate the soil [21].

6.3 POTENTIAL FUNGAL ENZYMES USED IN BIOREMEDIATION

Industrial-intensive fungal enzymes (amylases, xylanases, protease (proteas)), lipases, laccases, POXs, catalases, etc., provide the handling of agricultural waste (such as the organic component of MSW (OFMSW) cells) [59]. The enzyme may be used to hydrolyze polymer substances including cellulose, sugar, xylene, lipid, and protein including waste of fruits, garden, vegetable market, leaf litter, etc., that could be eventually be composted or used in the development of biogas as well as volatile fatty acids [39]. White rot fungi produce one or more types of lignin enzymes, depending on environmental conditions (EC) and species, and can thus be applied to biodegradation studies in degrading lignin in the natural lignocellulosic substrate, but also to the different xenobiotic compounds including the dyes. These enzymes alter the structure of azo-dyes through the degradation of chromophoric compounds that allow phenoxyl radicals to form in reactions [37].

In the extracellular setting of the fungal cell, ligninolytic enzymes secreted by white-red fungi for lignin oxidation are classified into MnP and lignin peroxidase (MnP and LiP) and LAC [40]. LACs and fungal class II White rot basidiomycete peroxidaes are well established in persistent degradation of organic pollutants [47]. These extremophilic fungal enzymes may be useful for remediation under conditions of high salinity and extra-heavy contamination of crude oil, such as waste from oil belt exploration. Significant research is currently focused on the creation of custom enzymes using protein engineering techniques and the recombinant expression of white-red fungi genes, which are successful methods for the environmentally safe disposal of radioactive waste [79, 99]. Fungal enzymes for soil bioremediation are classified into (For more details, reader may refer to Chapter 5 in this book volume):

- Filamentous fungi [3, 10, 65];
- Fungal oxidases [10, 31, 54, 67, 82, 91, 94];
- Laccases [22, 34, 38, 51, 60, 72, 97, 100]; and
- Peroxidases [3, 12, 28, 29, 38, 45, 56, 71, 81, 85, 95].

6.4 EXAMPLES OF XENOBIOTIC BIOTRANSFORMATION MEDIATED BY FUNGAL ENZYMES

Xenobiotics are natural or manufactured organic substances that are inherently harmful to an individual and have adverse environmental effects in terms of toxicity and disease [67]. Agriculture and industry are two main sources of biologically predominant xenobiotics, including insecticides, fertilizers, colorings, poisons, chemicals, and derivatives of hydrocarbons [81]. Most xenobiotics may be altered microbially. To degrade or transform hazardous compounds, bioremediation uses the metabolic potential of biological organisms into less toxic or non-toxic forms [10, 98, 102].

Mycoremediation has achieved extensive attention. Bioremediating fungal enzymes exhibit unique broad ranges of substrates [10]. The fungal strains can tolerate higher pollutant levels. Bacteria may metabolize to use them as energy and carbon source, in certain situations, such as the degradation of PAHs; but they cannot completely mineralize them as fungi can [65]. Laccases (LAC), LiP, MnP, VP, cytochrome P450 monooxygenases (P450) and unspecific peroxygenases (UPO) are enzymes that are studied in this section.

It has been demonstrated that each of these enzymes transforms and remedies toxic xenobiotics into less hazardous or bioactive forms [10]. LAC from *Coriolopsis gallica,* together with proper mediators was able to oxidize the PAHs dibenzothiophene, carbazole, and N-ethylcarbazole [97]. *Trametes versicolor* LACs oxidized the medications, mefenamic acid, and diclofenac in urban wastewater [60]. It has been shown that covalently immobilized *Trametes versicolor* LACs can effectively degrade PAHs, anthracene, phenanthrene, and naphthalene [12]. LACs from *Ceriporiopsis subvermispora, Pycnoporus sanguineus*, *Myceliophthora thermophila,* and *Trametes trogii* successfully metabolized poisonous colors, such as bromophenol brown, malachite green, methyl violet, and other colorants in textiles [6, 24].

LiPs can oxidize both methylated and non-methylated compounds (such as lignin and PAHs) [50]. *Trametes,* a MnP, showed a robust capability for the degradation of azo indigo and PAHs [103].

Peniophora Incarnata related MnP not only showed that anthracene PAH could be broken down, but also this capacity was transferable through yeast heterology, which represents a bioremediation potential. *Ganoderma lucidum* excreted MnP was used as cross-linked enzyme aggregates to effectively destroy triclosan and nonphenol thus disrupting the endocrine [15]. A strong tolerance of *Aspergillus foetidus* in the presence of Pb(II) at the rate of 200 mgL^{-1} was linked with a concomitant increase in Niveaus of anti-oxidant enzymes (such as H_2O_2 and malondialdehyde detoxifying catalase) [62].

6.5 ROLE OF FUNGAL ENZYMES IN XENOBIOTIC BIOTRANSFORMATION IN SOIL

All Phyla of fungi contains potential pollutant degraders, and various species are well known for xenobiotics biotransformation [68]. Fungi cause metabolization of xenobiotics due to secretion of various enzymes, such as LiPs, laccases, MnP, epoxide hydrolase, cytochrome P450. The products include tetraols, dihydrodiol epoxides, quinones, phenols, and trans-dihydrodiols that can be conjugated to make sulfates, xylosides, glucosides, and glucuronides. The toxicity of these resultant fungal metabolites is much less than the original hydrocarbons [20].

Xenobiotics are complex organic contaminants and are widely distributed in both aquatic and terrestrial environments [83]. These PAHs are significant chemical constituents of the petroleum, and their composition consists of merged aromatic rings without any heteroatoms carrying any substituents. Their production is from different natural resources, such as forest fires, and volcanic emissions; and do exist in the oil, coal, and tar deposits. Moreover, these are produced as by-products during the process of fuel-burning [58].

Further, polychlorinated biphenyls (PCBs) also pose significant concerns to the human health and the environment, and are on top of the list among hazardous xenobiotics. *Pleurotus ostreatus* has been proven as significantly effective for the bioremediation of PCBs contaminated soils, and its use could be proven to be cost-effective and prospective strategy [26].

Due to the presence of merged benzene rings, these compounds are categorized as highly toxic due to their carcinogenic and mutagenic potential. Scientific studies have revealed that wood-decaying and litter decaying fungi are highly effective to biodegrade the organo-pollutants [88]. The first scientific study about the significance of ligninolytic fungi to degrade xenobiotics (PAHs) was attributed to the role of white-rot basidiomycete *Phanerochaete chrysosporium* that partially caused degradation of benzo[a] pyrene (BaP) into carbon dioxide (CO_2) [18]. Later on, the potential degradation of PAHs was also reported by using other genera, such as

- *Bjerkandera* [78];
- *Coriolus* [25];
- *Irpex* [8];
- *Nematoloma* [44];
- *Phanerochaete* [16, 17];
- *Pleurotus* [14];
- *Trametes* [27].

Various strains of fungi related to various genera have been characterized for the efficient degradation of different compounds of pesticides in soil [63]. Examples of these genera are:

- *Aspergillus* [80];
- *Fusarium* [80];
- *Lentinus* [66];
- *Mortierella* [9];
- *Penicillium* [69];
- *Phanerochaete* [23];
- *Rhizopus* [80];
- *Trametes* [11];
- *Trichoderma* [80].

G. lucidum produces a significant level of ligninolytic enzymes and is a potent pyrene and phenanthrene degrader [1]. White rot fungus (*Phanerochaete chrysosporium*) is also significantly effective to cause bioremediation of various hazardous compounds in the contaminated soils and is also used as corrective practice to treat different kinds of hazardous wastes [36]. Following species of white-rot fungus are used to bioremediate the contaminated soils with insecticides:

- *Flammulina velutipes* [33];
- *Fusarium oxysporum* [70];
- *Ganoderma austral* [46];

- *Ganoderma lucidum* [55];
- *Marasmiellus chamaecyparidis* [90];
- *Phlebia acanthocystis* [74];
- *Phlebia aurea* [101];
- *Phlebia brevispora* [101];
- *Phlebia lindtneri* [64];
- *Phlebia tremellosa* [101];
- *Pleurotus eryngii* [7];
- *Pleurotus florida* [7];
- *Pleurotus ostreatus* [73];
- *Pleurotus pulmonarius* [42];
- *Pleurotus sajor-caju* [7];
- *Stereum hirsutum* [13];
- *Trametes hirsute* [52];
- *Trametes hirsutus* [86];
- *Trametes versicolor* [13];
- *Trichoderma* [41];
- *Trichoderma viride Helal* [41].

Fusarium graminearum has various important enzymes and fungal Cytochrome P450s; and the P450s are significantly effective for the degradation of xenobiotics [84].

Bjerkandera adusta, *Pleurotus ostreatus*, and *Phanerochaete chrysosporium* are widely used species to accelerate biodegradation of PAHs, as they can produce ligninolytic enzymes (Laccases, MnPs, and LiPs) [92]. The ligninolytic enzymes produced by *L. theobromae* are significantly effective in degrading the BaP. The solitary carbon source of BaP was biodegraded by 53% by the action of L. *theobromae* in 10 days only. Ligninolytic enzymes have two major classes: laccases (LAC), and peroxidases (MnP and LiP).

Among three different ligninolytic enzymes (such as MnP, LAC, and LiP), the role of LAC was most contributing in the biodegradation of BaP [19]. These three enzymes compete for the same substrates and they are extracellularly secreted. They cause catalyzation by the process of oxidation and attack the molecule to cause destabilization of bonds in the molecule. The resultant radicals interact with each other and fungal metabolites that in turn cause the generation of other radicals, and the whole process is continued [4].

6.6 ADVANTAGES OF USING ENZYMES FOR SOIL BIOREMEDIATION

Enzymatic bioremediation of xenobiotics is a socially acceptable, eco-friendly, quick, and easy approach; and it offers various benefits over biological, physical, and chemical contemporary approaches. Stability, activities, and half-life of these can be improved at significant level using enzyme immobilization techniques [82]. The sensitivity of enzymes towards the presence of pollutants concentration in soil is low, therefore, they can efficiently serve the purpose of biotransformation and biodegradation. However, much higher concentration can be toxic for the cells and degradation efficiency may also be reduced.

Enzymes act as biocatalysts, and they facilitate complete and quick breakdown of substrates along with lowering the activation energies. Enzymes can easily contact the contaminants due to smaller size [57]. Enzymes can work well over a wide range of physicochemical gradients, such as temperature, salinity, and pH. Even if these conditions are unfavorable for the microbial cells, fungal enzymes can perform the function of biotransformation [93].

The biotransformation reactions do not generate any toxic products because the xenobiotics are mineralized to the non-toxic substances. Moreover, the process of enzymatic biotransformation can be accelerated by the addition of surfactants or co-solvents [87].

The ecological impact can be greatly controlled under field conditions and in this way, targeted pollutants can be efficiently degraded. The use of fungal enzymes is offering an intrinsic range of advantages; and their functioning and efficiency has even been improved by several folds by using biomolecular engineering, recombinant DNA technology, combination of nanotechnology and enzyme studies by using metabolomics, genomics, proteomics, and transcriptomics [49].

6.7 LIMITATIONS OF USING ENZYMES FOR SOIL BIOREMEDIATION

Although fungous enzymes are efficient and eco-friendly source for the bioremediation of polluted soils, yet their efficient use and functioning need proper assessment and monitoring of enzymatic activities in soil. The external properties of fungal enzymes make the soil tough to comprehend the interaction and mechanism between these interfaces [5]. There is a significant variation in the enzymatic activities under different climatic conditions, ecological zones, and agroecosystems, therefore lack of scientific knowledge is the biggest hurdle in the efficient utilization of fungal enzymes for bioremediation of soil [89].

The physical environment in soil is organized in the form of aggregates, and it holds fluxes of air, water, matter, and numerous biological events at different levels of structures, therefore these complexities must be considered in the experimental studies on fungal enzymes in soil bioremediation [35]. Enzyme based bioremediation techniques also require proper understanding of heterogeneity and complexity of soil, and modeling of thermodynamic status to monitor bioremediation activities. A precise characterization of water interactions in soil structure and thermodynamic state is required to monitor these activities. Therefore, the gap between soil physics and soil pedology should be bridged up to understand the interactions between soil and enzymes for the optimized usage of extracellular fungal enzymes in the bioremediation techniques [75].

6.8 FUTURE PROSPECTS

Global industrialization has led to various challenges and ecotoxic industrial wastes are exerting significant negative consequences to human health by the production of xenobiotics. Fungal enzymes are highly efficient for the detoxification of xenobiotics, and the usage of fungal enzymes has been recognized as an efficient strategy for the biotransformation of pollutants in the soil.

Due to ongoing developments in industrial and agricultural sector, the soils may even be more contaminated, therefore there is urgent need of developing efficient, effective, and cost-effective strategies to cope with the future challenges of soil bioremediation mediated by the fungal enzymes. The use of genetically modified organisms is offering wonderful horizons to develop the capabilities of specific stains by using different genetic engineering approaches. However, there should be a close integration of different ecological, biochemical, molecular, and genetic approaches to ensure sustainable and modern bioremediation mediated by the fungal enzymes.

At present, the use of extracellularly produced enzymes is not a popular approach due to associated high production cost. Moreover, the availability of inadequate information on EC and responsible factors in the vicinity of pollutants and enzymatic active sites is causing restrictions on its wide range of applicability. A good understanding based on scientific evidences should be made by using advanced tools such as metabolomics, genomics, proteomics, and transcriptomics.

Due attention should be given to study the soil enzymatic activities for the degradation of xenobiotics and other contaminants. Measures should be taken for the integration of applied and theoretical models of ecosystem functioning and biomics. The impact of environmental condition on soil organic carbon contents and other elements should be properly studied as all these elements affects soil enzymatic activities. There should be proper testing that either laboratory-based activities are true indicative of filed studies or not. Based on this purpose, nano-probes should be developed to allow the real time in situ observation and measurement.

There should be further elucidation of mechanisms and degradation pathways for the xenobiotic pollutants. Based on the detailed mechanistic approaches, site specific measures should be taken to ensure more cost effective and sustainable use of fungal enzymes for soil bioremediation. In this regard, multidisciplinary scientific approaches by involving chemists, biologists, engineers, and scientists are essential to make effective strategies to control pollutant concentration in the soil.

6.9 SUMMARY

Fungal enzymes are offering excellent solutions to the biotransformation of xenobiotics and degrade the pollutants by turning them into less toxic or harmless components. Use of fungal enzymes is advantageous over physical, chemical, and other biological remedial measures and is providing novel and excellent horizons for bioremediation. However, the scientific studies related to mechanisms and fungal mode of action are not enough, and the application of fungal enzymes in various climatic conditions is still a major challenge. Thus, various fungal species should be tested in different soils, and climatic conditions to understand their mode of action and associated mechanisms. Moreover, there should be strong international communication regarding sharing of knowledge and research activities to offer sustainable and cost-effective solution of soil bioremediation.

KEYWORDS

- **bioremediation**
- **filamentous**
- **fungal enzymes**
- **fungi**
- **white-rot fungi**
- **xenobiotic compounds**

REFERENCES

1. Agrawal, N., Verma, P., & Shahi, S. K., (2018). Degradation of polycyclic aromatic hydrocarbons (phenanthrene and pyrene) by the ligninolytic fungi *Ganoderma lucidum* isolated from the hardwood stump. *Bioresources and Bioprocessing, 5*, 1–9.
2. Akcil, A., Erust, C., Ozdemiroglu, S., Fonti, V., & Beolchini, F., (2015). A review of approaches and techniques used in aquatic contaminated sediments: Metal removal and stabilization by chemical and biotechnological processes. *Journal of Cleaner Production, 86*, 24–36.
3. Akhtar, S., Mahmood-ul-Hassan, M., Ahmad, R., Suthor, V., & Yasin, M., (2013). Metal tolerance potential of filamentous fungi isolated from soils irrigated with untreated municipal effluent. *Soil Environment, 32,* 55–62.
4. Anastasi, A., Coppola, T., Prigione, V., & Varese, G. C., (2009). Pyrene degradation and detoxification in soil by a consortium of basidiomycetes isolated from compost: Role of laccases and peroxidases. *Journal of Hazardous Materials, 165*, 1229–1233.
5. Antonious, G. F., Turley, E. T., & Dawood, M. H., (2020). Monitoring soil enzymes activity before and after animal manure application. *Agriculture, 10*, 166–170.
6. Antosova, Z., Herkommerova, K., Pichova, I., & Sychrova, H., (2018). Efficient secretion of three fungal laccases from *Saccharomyces cerevisiae* and their potential for decolorization of textile industry effluent-a comparative study. *Biotechnology Progress, 34*, 69–80.
7. Arısoy, M., (1998). Biodegradation of chlorinated organic compounds by white-rot fungi. *Bulletin of Environmental Contamination and Toxicology, 60*, 872–876.
8. Baborová, P., Möder, M., Baldrian, P., Cajthamlová, K., & Cajthaml, T., (2006). Purification of a new manganese peroxidase of the white-rot fungus *Irpex lacteus*, and degradation of polycyclic aromatic hydrocarbons by the enzyme. *Research in Microbiology, 157*, 248–253.
9. Badawi, N., Rønhede, S., Olsson, S., Kragelund, B. B., Johnsen, A. H., Jacobsen, O. S., & Aamand, J., (2009). Metabolites of the phenylurea herbicides chlorotoluron, diuron, isoproturon and linuron produced by the soil fungus *Mortierella* sp. *Environmental Pollution, 15*, 2806–2812.
10. Baker, P., Tiroumalechetty, A., & Mohan, R., (2019). Fungal enzymes for bioremediation of xenobiotic compounds. In: Yadav, A., Singh, S., Mishra, S., & Gupta, A., (eds.),

Recent Advancement in White Biotechnology through Fungi. Fungal Biology (pp. 463–489). Cham: Springer.

11. Bastos, A. C., & Magan, N., (2009). *Trametes versicolor*: Potential for atrazine bioremediation in calcareous clay soil, under low water availability conditions. *International Biodeterioration and Biodegradation, 63*, 389–394.
12. Bautista, L. F., Morales, G., & Sanz, R., (2015). Biodegradation of polycyclic aromatic hydrocarbons (PAHs) by laccase from *Trametes versicolor* covalently immobilized on amino-functionalized SBA-15. *Chemosphere, 136,* 273–280.
13. Bending, G. D., Friloux, M., & Walker, A., (2002). Degradation of contrasting pesticides by white-rot fungi and its relationship with lignin aolytic potential. *FEMS Microbiology Letters, 212*, 59–63.
14. Bezalel, L., Hadar, Y., & Cerniglia, C. E., (1997). Enzymatic mechanisms involved in phenanthrene degradation by the white-rot fungus *Pleurotus ostreatus. Applied and Environmental Microbiology, 63*, 2495–2501.
15. Bilal, M., Asgher, M., Iqbal, H. M., Hu, H., & Zhang, X., (2017). Bio-based degradation of emerging endocrine-disrupting and dye-based pollutants using cross-linked enzyme aggregates. *Environmental Science and Pollution Research International, 24*, 7035–7041.
16. Bogan, B. W., & Lamar, R. T., (1996). Polycyclic aromatic hydrocarbon-degrading capabilities of *Phanerochaete laevis* HHB-1625 and its extracellular ligninolytic enzymes. *Applied and Environmental Microbiology, 62*, 1597–1603.
17. Bogan, B. W., & Lamar, R. T., (1995). One-electron oxidation in the degradation of creosote polycyclic aromatic hydrocarbons by *Phanerochaete chrysosporium. Applied and Environmental Microbiology, 5*, 2631–2635.
18. Bumpus, J. A., Tien, M., Wright, D., & Aust, S. D., (1985). Oxidation of persistent environmental pollutants by a white-rot fungus. *Science, 228*, 1434–1436.
19. Cao, H., Wang, C., Liu, H., Jia, W., & Sun, H., (2020). Enzyme activities during benzo [a] pyrene degradation by the fungus *Lasiodiplodia theobromae* isolated from a polluted soil. *Scientific Reports, 10*, 1–10.
20. Cerniglia, C. E., & Sutherland, J. B., (2010). Degradation of polycyclic aromatic hydrocarbons by fungi. In: Timmis, K. N., (ed.), *Handbook of Hydrocarbon and Lipid Microbiology* (pp. 2079–2110). Berlin, Heidelberg: Springer.
21. Chandra, P., (2019). Fungal enzymes for bioremediation of contaminated soil. In: Yadav, A., Singh, S., Mishra, S., & Gupta, A., (eds.), *Recent Advancement in White Biotechnology through Fungi. Fungal Biology* (pp. 189–215). Cham: Springer.
22. Chhaya, U., & Gupte, A., (2013). Possible role of laccase from *Fusarium incarnatum* UC-14 in bioremediation of bisphenol-a using reverse micelles system. *Journal of Hazardous Materials, 254*, 149–156.
23. Chirnside, A. E., Ritter, W. F., & Radosevich, M., (2009). Biodegradation of aged residues of atrazine and alachlor in a mix-load site soil. *Soil Biology and Biochemistry, 41*, 2484–2492.
24. Chmelova, D., & Ondrejovic, M., (2016). Purification and characterization of extracellular laccase produced by *Ceriporiopsis subvermispora* and decolorization of triphenylmethane dyes. *Journal of Basic Microbiology, 56,* 1173–1182.
25. Cho, S. J., Park, S. J., Lim, J. S., Rhee, Y. H., & Shin, K. S., (2002). Oxidation of polycyclic aromatic hydrocarbons by laccase of *Coriolus hirsutus. Biotechnology Letters, 24*, 1337–1340.

26. Chun, S. C., Muthu, M., Hasan, N., Tasneem, S., & Gopal, J., (2019). Mycoremediation of PCBs by *Pleurotus ostreatus*: Possibilities and prospects. *Applied Sciences, 9*, 4185–4190.
27. Collins, P. J., & Dobson, A. D. W., (1996). Oxidation of fluorene and phenanthrene by Mn2+-dependent peroxidase activity in whole cultures of *Trametes versicolor. Biotechnology Letters, 18*, 801–804.
28. Davila-Vazquez, G., Tinoco, R., Pickard, M. A., & Vazquez-Duhalt, R., (2005). Transformation of halogenated pesticides by versatile peroxidase from *Bjerkandera adusta. Enzyme and Microbial Technology, 36,* 223–231.
29. Deshmukh, R., Khardenavis, A. A., & Purohit, H. J., (2016). Diverse metabolic capacities of fungi for bioremediation. *Indian Journal of Microbiology, 56*, 247–264.
30. Dhakar, K., Jain, R., Tamta, S., & Pandey, A., (2014). Prolonged laccase production by a cold and pH tolerant strain of penicillium pinophilum (MCC 1049) isolated from a low temperature environment. *Enzyme Research, 2014*, 10. Article ID online: 120708. doi: 10.1155/2014/120708;
31. Doddapaneni, H., Chakraborty, R., & Yadav, J. S., (2005). Genome-wide structural and evolutionary analysis of the P450 monooxygenase genes (P450ome) in the white-rot fungus *Phanerochaete chrysosporium*: Evidence for gene duplications and extensive gene clustering. *BMC Genomics, 6,* 92–99.
32. Durairaj, P., Malla, S., & Nadarajan, S. P., (2015). Fungal cytochrome P450 monooxygenases of *Fusarium oxysporum* for the synthesis of x-hydroxy fatty acids in engineered *Saccharomyces cerevisiae. Microbial Cell Factories, 14,* 45–51.
33. Fan, B., Zhao, Y., Mo, G., Ma, W., & Wu, J., (2013). Co-remediation of DDT-contaminated soil using white-rot fungi and laccase extract from white-rot fungi. *Journal of Soils and Sediments, 13*, 1232–1245.
34. Fillat, U., Prieto, A., Camararo, S., Martinez, A. T., & Martinez, M. J., (2012). Bio-de-inking of flexographic inks by fungal laccases using synthetic and natural mediators. *Biochemical Engineering Journal, 67,* 97–103.
35. Frąc, M., Hannula, S. E., Bełka, M., & Jędryczka, M., (2018). Fungal biodiversity and their role in soil health. *Frontiers in Microbiology, 9*, 707–714.
36. Fulekar, M. H., Pathak, B., Fulekar, J., & Godambe, T., (2013). Bioremediation of organic pollutants using *Phanerochaete chrysosporium*. In: Goltapeh, E., Danesh, Y., & Varma, A., (eds.), *Fungi as Bioremediators. Soil Biology* (Vol. 32, pp. 135–157). Berlin, Heidelberg: Springer.
37. Gul, U. D., & Donmez, G., (2013). Application of mixed fungal biomass for effective reactive dye removal from textile effluents. *Desalination and Water Treatment, 51,* 3597–3603.
38. Harms, H., Schlosser, D., & Wick, L. Y., (2011). Untapped potential: Exploiting fungi in bioremediation of hazardous chemicals. *Nature Reviews Microbiology, 9,* 177–183.
39. Hattori, T., Hisamori, H., Suzuki, S., Umezawa, T., Yoshimura, T., & Sakai, H., (2015). Rapid copper transfer and precipitation by wood-rotting fungi can affect copper removal from copper sulfate-treated woodblocks during solid-state fungal treatment. *International Biodeterioration and Biodegradation, 97,* 195–201.
40. He, L., Huang, H., Zhang, Z., & Lei, Z., (2015). A review of hydrothermal pretreatment of lignocellulosic biomass for enhanced biogas production. *Current Organic Chemistry, 19,* 437–446.
41. Helal, I. M., & Abo-El-Seoud, M. A., (2015). Fungal biodegradation of pesticide vydate in soil and aquatic system. In: 4th *International Conference on Radiation Sciences and Applications, 2015*, 87–94.

42. Hernandez-Rodriguez, D., Sánchez, J. E., Nieto, M. G., & Márquez-Rocha, F. J., (2006). Degradation of endosulfan during substrate preparation and cultivation of *Pleurotus pulmonarius*. *World Journal of Microbiology and Biotechnology, 22*, 753–760.
43. Hirosue, S., Tazaki, M., Hiratsuka, N., & Yanai, S., (2011). Insight into functional diversity of cytochrome P450 in the white-rot basidiomycete *Phanerochaete chrysosporium*: Involvement of versatile monooxygenase. *Biochemical and Biophysical Research Communications, 407*, 118–123.
44. Hofrichter, M., Scheibner, K., Schneegaß, I., & Fritsche, W., (1998). Enzymatic combustion of aromatic and aliphatic compounds by manganese peroxidase from *Nematoloma frowardii*. *Applied and Environmental Microbiology, 64*, 399–404.
45. Hofrichter, M., & Ullrich, R., (2014). Oxidations catalyzed by fungal peroxygeneases. *Current Opinion in Chemical Biology, 19,* 116–125.
46. Hussaini, S. Z., Shaker, M., & Iqbal, M. A., (2013). Isolation of fungal isolates for degradation of selected pesticides. *Bulletin of Environment, Pharmacology and Life Sciences, 2*, 50–53.
47. Ikehata, K., (2015). Use of fungal laccases and peroxidases for enzymatic treatment of wastewater containing synthetic dyes. Chapter 6. In: Sanjay, K., & Sharma, F. R. S. C., (eds.), *Green Chemical Dyes Removal Wastewater Research Trends Applications* (pp. 203–260). New York: Scrivener Publishing LLC.
48. Iqbal, S., Khan, M. Y., Asghar, H. N., & Akhtar, M. J., (2016). Combined use of phosphate solubilizing bacteria and poultry manure to enhance the growth and yield of mung bean in calcareous soil. *Soil and Environment, 35*, 146–154.
49. Jariyal, M., Yadav, M., Singh, N. K., Yadav, S., Sharma, I., Dahiya, S., & Thanki, A., (2020). Microbial remediation progress and future prospects. Chapter 8. In: *Bioremediation of Pollutants* (pp. 187–214). London: Elsevier.
50. Kadri, T., Rouissi, T., Kaur, B. S., Cledon, M., Sarma, S., & Verma, M., (2017). Biodegradation of polycyclic aromatic hydrocarbons (PAHs) by fungal enzymes: A review. *Journal of Environmental Sciences, 51,* 52–74.
51. Kallio, J. P., Gasparetti, C., Andberg, M., Boer, H., Koivula, A., & Kruus, K., (2011). Crystal structure of an ascomycete fungal laccase from *Thielavia arenaria*-common structural features of ascolaccases. *Federation of European Biochemical Societies Journal, 278*, 2283–2295.
52. Kamei, I., Takagi, K., & Kondo, R., (2011). Degradation of endosulfan and endosulfan sulfate by white-rot fungus *Trametes hirsuta*. *Journal of Wood Science, 57*, 317–322.
53. Kango, N., Jana, U. K., & Choukade, R., (2019). Fungal enzymes: Sources and biotechnological applications. In: Satyanarayana, T., Deshmukh, S., & Deshpande, M., (eds.), *Advancing Frontiers in Mycology & Mycotechnology* (pp. 515–538). Springer, Singapore.
54. Karich, A., Ullrich, R., Scheibner, K., & Hofrichter, M., (2017). Fungal unspecific peroxygenases oxidize the majority of organic EPA priority pollutants. *Frontiers in Microbiology, 8*, 1463–1483.
55. Kaur, H., Kapoor, S., & Kaur, G., (2016). Application of ligninolytic potentials of a white-rot fungus *Ganoderma lucidum* for degradation of lindane. *Environmental Monitoring and Assessment, 188*, 588–596.
56. Kues, U., (2015). Fungal enzymes for environmental management. *Current Opinion in Biotechnology, 33,* 268–278.

57. Kumar, L., & Bharadvaja, N., (2019). Enzymatic bioremediation: A smart tool to fight environmental pollutants. Chapter 6. In: Bhatt, P., (ed.), *Smart Bioremediation Technologies* (pp. 99–118). London: Academic Press.
58. Malhat, F., Loutfy, N., El Menyawi, M. A., & Ahmed, M. T., (2019). Review of contamination by polycyclic aromatic hydrocarbons (PAHs) in Egyptian aquatic environment. *Polycyclic Aromatic Compounds, 4*, 1–9.
59. Marco, E., Font, X., Sa´nchez, A., Gea, T., Gabarrell, X., & Caminal, G., (2013). Co-composting as a management strategy to reuse the white-rot fungus *Trametes versicolor* after its use in a biotechnological process. *International Journal of Environment and Waste Management, 11*, 100–108.
60. Margot, J., Bennati-Granier, C., Maillard, J., Bla´nquez, P., Barry, D. A., & Holliger, C., (2013). Bacterial versus fungal laccase: Potential for micropollutant degradation. *AMB Express, 3*, 63–70.
61. Mishra, A., & Malik, A., (2014). Novel fungal consortium for bioremediation of metals and dyes from mixed waste stream. *Bioresource Technology, 171*, 217–226.
62. Mitra, A., Roy, D., Roy, P., Bor, A. M., Sarkar, & Mitra, A. K., (2014). Sustainability of *Aspergillus spp.* in metal-enriched substrate aiming towards increasing bioremediation potential. *World Journal of Pharmacy and Pharmaceutical Sciences, 3*, 864–878.
63. Mohapatra, D., Rath, S. K., & Mohapatra, P. K., (2018). Bioremediation of insecticides by white-rot fungi and its environmental relevance. In: Prasad, R., (ed.), *Mycoremediation and Environmental Sustainability. Fungal Biology* (pp. 181–212). Cham, Switzerland: Springer.
64. Mori, T., & Kondo, R., (2002). Oxidation of dibenzo-p-dioxin, dibenzofuran, biphenyl, and diphenyl ether by the white-rot fungus *Phlebia lindtneri. Applied Microbiology and Biotechnology, 60*, 200–205.
65. Mougin, C., Boukcim, H., & Jolivalt, C., (2009). Soil bioremediation strategies based on the use of fungal enzymes. In: Singh, A., Kuhad, R. C., & Ward, O. P., (eds.), *Advances in Applied Bioremediation* (pp. 123–149). Berlin/Heidelberg: Springer.
66. Nwachukwu, E. O., & Osuji, J. O., (2007). Bioremedial degradation of some herbicides by indigenous white-rot fungus, *Lentinus subnudus. Journal of Plant Sciences, 2*, 619–624.
67. Olicon-Hernandez, D. R., Gonzalez-Lopez, J., & Aranda, E., (2017). Overview on the biochemical potential of filamentous fungi to degrade pharmaceutical compounds. *Frontiers in Microbiology, 8*, 1792–1799.
68. Olicón-Hernández, D. R., Camacho-Morales, R. L., Pozo, C., González-López, J., & Aranda, E., (2019). Evaluation of diclofenac biodegradation by the ascomycete fungus *Penicillium oxalicum* at flask and bench bioreactor scales. *Science of the Total Environment, 662*, 607–614.
69. Peng, C. L., Gu, P., Li, J., Chen, Q. Y., Feng, C. H., Luo, H. H., & Du, Y. J., (2012). Identification and field bioassay of the sex pheromone of *Trichophysetis cretacea* (Lepidoptera: *Crambidae*). *Journal of Economic Entomology, 105*, 1566–1572.
70. Peter, L., Gajendiran, A., Mani, D., Nagaraj, S., & Abraham, J., (2015). Mineralization of malathion by *Fusarium oxysporum* strain JASA1 isolated from sugarcane fields. *Environmental Progress & Sustainable Energy, 34*, 112–116.
71. Pozdnyakova, N. N., (2012). Involvement of the ligninolytic system of white-rot and litter-decomposing fungi in the degradation of polycyclic aromatic hydrocarbons. *Biotechnology Research International, 2012*, 8. Article ID: 243217.

72. Pozdnyakova, N. N., Balandina, S. A., Dubrovskaya, E. V., Golubev, C. N., & Turkovskaya, O. V., (2018). Ligninolytic basidiomycetes as promising organisms for the mycoremediation of PAH-contaminated environments. *IOP Conference Series Earth and Environmental Science, 107,* 7. Article ID: 012071.
73. Purnomo, A. S., Putra, S. R., Shimizu, K., & Kondo, R., (2014). Biodegradation of heptachlor and heptachlor epoxide-contaminated soils by white-rot fungal. *Environmental Science and Pollution Research, 21,* 11305–11312.
74. Purnomo, A. S., Nawfa, R., Martak, F., Shimizu, K., & Kamei, I., (2017). Biodegradation of aldrin and dieldrin by the white-rot fungus *Pleurotus ostreatus*. *Current Microbiology, 74,* 320–324.
75. Rao, M. A., Scelza, R., & Gianfreda, L., (2014). Soil enzymes. In: Rao, M. A., & Gianfreda, L., (eds.), *Enzymes in Agricultural Sciences* (pp. 10–24). New York: Omics Group eBooks.
76. Riaz, U., Murtaza, G., Saifullah, & Farooq, M., (2018). Comparable effect of commercial composts on chemical properties of sandy clay loam soil and accumulation of trace elements in the soil-plant system. *International Journal of Agriculture and Biology, 20,* 85–92.
77. Riaz, U., Murtaza, G., Saifullah, & Farooq, M., (2018). Influence of various sewage sludges and composts on growth, yield and trace element contents in rice and wheat. *Land Degradation and Development, 29,* 1343–1352.
78. Pointing, S. B., (2001). Feasibility of bioremediation by white-rot fungi. *Applied Microbiology and Biotechnology, 57,* 20–33.
79. Sakaki, T., Yamamoto, K., & Ikushiro, S., (2013). Possibility of application of cytochrome P450 to bioremediation of dioxins. *Biotechnology and Applied Biochemistry, 60,* 65–70.
80. Sene, L., Converti, A., Secchi, G. A., & Simão, R. D., (2010). New aspects on atrazine biodegradation. *Brazilian Archives of Biology and Technology, 53,* 487–496.
81. Sharma, B., Dangi, A. K., & Shukla, P., (2018). Contemporary enzyme-based technologies for bioremediation: A review. *Journal of Environmental Management, 210,* 10–22.
82. Sharma, H. K., Xu, C., & Qin, W., (2019). Biological pretreatment of lignocellulosic biomass for biofuels and bioproducts: An overview. *Waste and Biomass Valorization, 10,* 235–251.
83. Shilla, D. J., & Routh, J., (2018). Distribution, behavior, and sources of polycyclic aromatic hydrocarbon in the water column, sediments, and biota of the Rufiji estuary, Tanzania. *Frontiers in Earth Science, 6,* 70–80.
84. Shin, J., Kim, J. E., Lee, Y. W., & Son, H., (2018). Fungal cytochrome P450s and the P450 complement (CYPome) of *Fusarium graminearum*. *Toxin, 10,* 112–118.
85. Shrivastava, R., Christian, V., & Vyas, B. R. M., (2005). Enzymatic decolorization of sulfonphthalein dyes. *Enzyme and Microbial Technology, 36,* 333–337.
86. Singh, B. K., & Kuhad, R. C., (1999). Biodegradation of lindane (γ-Hexachlorocyclohexane) by the white-rot fungus *Trametes hirsutus*. *Letters in Applied Microbiology, 28,* 238–241.
87. Smitha, M. S., Singh, S., & Singh, R., (2017). Microbial biotransformation: A process for chemical alterations. *Journal of Bacteriology and Mycology: Open Access, 4,* 85–96.
88. Steffen, K. T., Hatakka, A., & Hofrichter, M., (2003). Degradation of benzo[a]Pyrene by the litter-decomposing basidiomycete *Stropharia coronilla*: Role of manganese peroxidase. *Applied and Environmental Microbiology, 69,* 3957–3964.

89. Sudhakaran, M., Ramamoorthy, D., Savitha, V., & Kirubakaran, N., (2019). Soil enzyme activities and their relationship with soil physicochemical properties and oxide minerals in coastal agroecosystem of Puducherry. *Geomicrobiology Journal, 36*, 452–459.
90. Suhara, H., Adachi, A., Kamei, I., & Maekawa, N., (2011). Degradation of chlorinated pesticide DDT by litter-decomposing basidiomycetes. *Biodegradation, 22*, 1075–1086.
91. Syed, K., & Yadav, J. S., (2012). P450 monooxygenases (P450ome) of the model white-rot fungus *Phanerochaete chrysosporium. Critical Reviews in Microbiology, 38,* 339–363.
92. Taha, M., Shahsavari, E., Aburto-Medina, A., Foda, M. F., Clarke, B., Roddick, F., & Ball, A. S., (2018). Bioremediation of biosolids with *Phanerochaete chrysosporium* culture filtrates enhances the degradation of polycyclic aromatic hydrocarbons (PAHs). *Applied Soil Ecology, 124*, 163–70.
93. Tang, J., Zhang, L., Zhang, J., Ren, L., Zhou, Y., Zheng, Y., Luo, L., Yang, Y., Huang, H., & Chen, A., (2020). Physicochemical features, metal availability and enzyme activity in heavy metal-polluted soil remediated by biochar and compost. *Science of the Total Environment, 70*, 8. Article ID: 134751.
94. Thippeswamy, B., Shivakumar, C. K., & Krishnappa, M., (2014). Studies on heavy metals detoxification biomarkers in fungal consortia. *Caribbean Journal of Science and Technology, 2*, 496–502.
95. Torres-Farrada, G., Manzano, L. A. M., Rineau, F., Ledo, A. L. L., Sanchez-Lopez, M. I., Thijs, S., Colpaert, J., et al., (2017). Diversity of ligninolytic enzymes and their genes in strains of the genus *Ganoderma*: Applicable for biodegradation of xenobiotic compounds? *Frontiers in Microbiology, 8,* 898–905.
96. Ulcnik, A., Kralj, C.′ I., & Pohleven, F., (2013). Degradation of lindane and endosulfan by fungi, fungal and bacterial laccases. *World Journal of Microbiology and Biotechnology, 29*, 2239–2247.
97. Vishwanath, B., Rajesh, B., Janardhan, A., Kumar, A. P., & Narasimha, G., (2014). Fungal laccases and their applications in bioremediation. *Enzyme Research, 21*, 110–118.
98. Watanabe, K., (2001). Microorganisms relevant to bioremediation. *Current Opinion in Biotechnology, 12*, 237– 241.
99. Wong, K. S., Cheung, M. K., Au, C. H., & Kwan, H. S., (2013). A novel *Lentinula edodes* laccase and its comparative enzymology suggest guaiacol-based laccase engineering for bioremediation. *PLoS One, 8*, 6. Article ID: e66426.
100. Wu, Y., Teng, Y., Li, Z., Liao, X., & Luo, Y., (2008). Potential role of polycyclic aromatic hydrocarbons (PAHs) oxidation by fungal laccase in the remediation of an aged contaminated soil. *Soil Biology and Biochemistry, 40,* 789–796.
101. Xiao, P., Mori, T., Kamei, I., Kiyota, H., Takagi, K., & Kondo, R., (2011). Novel metabolic pathways of organochlorine pesticides dieldrin and aldrin by the white-rot fungi of the genus *Phlebia. Chemosphere, 85*, 218–224.
102. Yadav, A. N., Mishra, S., Singh, S., & Gupta, A., (2019). *Recent Advancement in White Biotechnology through Fungi: Diversity and Enzymes Perspectives* (Vol. 1, p. 571). Cham: Springer International Publishing.
103. Zhang, H., Zhang, S., He, F., Qin, X., Zhang, X., & Yang, Y., (2016). Characterization of a manganese peroxidase from white-rot fungus *Trametes* sp. 48424 with strong ability of degrading different types of dyes and polycyclic aromatic hydrocarbons. *Journal of Hazardous Materials, 320*, 265–277.

PART III

Recent Trends and Approaches in Bioremediation

CHAPTER 7

ASSESSMENT OF CURRENT SOIL BIOREMEDIATION TECHNOLOGIES

PRITISHA BARIK, TRINATH BISWAL, and JUNAID AHMAD MALIK

ABSTRACT

The environmental degradation has gradually increased in the past few years because of the increase in human activity, energy demand, massive industrialization, unplanned urbanization, and increase in mining activity. These activities have deposited huge wastes in our biosphere resulting in harmful effects on human beings and animals. The remediation of these contaminants from various polluted sites by using microbial community of various kinds in an eco-friendly, cost-effective manner of producing non-toxic or less toxic substances is termed as bioremediation, which can be *ex-situ* and *in-situ* technology on the basis of various factors, such as nature of the contaminants, cost, kinds, and toxicity of the pollutants, site characteristics, chemical composition, etc. The *ex-situ* techniques are comparatively costly as compared to *in situ* method because some extra cost is required for excavation. Examples of modern bioremediation techniques are electrokinetic (EK) process, rhizoremediation, smart bioremediation technology, metagenomics, biofilms, etc.

7.1 INTRODUCTION

The low-cost bioremediation is now a major focus and the interest for many researchers worldwide than the conventional techniques, with the intention of finding out a suitable accepted solution of complete mineralization, neutralization, and reduction of massive contaminants. Now researchers and scientists are trying to develop and design different techniques for bioremediation; however, no single technique of bioremediation is able to fulfill all requirements in a cost-effective and sustainable manner due to the chemical composition, nature, biodegradability, and toxicity of the pollutants [9, 61, 70].

The indigenous microorganisms (called Autochthonous) in the contaminated environment play a key role in solving the problems and challenges linked with bioremediation and biodegradation of the contaminated substances. Here, biological mechanisms can decrease the concentration of the pollutants and convert it into non-toxic state by the process of degradation, mineralization, detoxification, and transformation. In comparison to the physical and chemical methods of remediation, the method of bioremediation has numerous advantages.

The techniques of bioremediation mainly depend on the nature and kind of contaminants, such as agrochemicals, dyes, detergents, polymers, and plastics, organic chemicals, PAHs, chlorinated compounds, toxic HMs, greenhouse gases (GHGs), hazardous nuclear waste, sewage effluents along with different indigenous and non-indigenous wastes. On the basis of the site of application, the techniques applied for bioremediation can be categorized into *ex-situ* and *in-situ* [20, 37, 65].

The application of either *ex-situ* or *in-situ* technology depends on the nature of pollutants, degree of pollution, local environmental state, cost, and some other environmental policies. The success of the bioremediation process also depends on the local environmental conditions (EC), such as percentage of oxygen, temperature, nutrient availability, pH, and some other abiotic factors. Although biodegradation can be applied in diversified fields, yet more focus is given towards hydrocarbons, which is the cause of frequent pollution of groundwater and soil. Recently in addition to *ex-situ* and *in-situ* techniques, some other different modern techniques along with green technology are used for effective bioremediation.

In this chapter, authors have focused on different technologies of bioremediation, their effectiveness, principles, and microbial activities [6, 39, 52, 69].

7.2 PRINCIPLES OF BIOREMEDIATION

The key factor in the process of bioremediation are the microscopic organisms (bacteria), which can survive in any environment and preferably used for the destruction of complex contaminant substances into simple non-contaminant or relatively lower toxic species. The microorganisms take the contaminant substances as their food because these are so small in size that they can comfortably come in interaction with the contaminated waste substances from the environment.

The ability of the microorganisms to degrade the pollutant depends upon three different factors such as The kinds of microorganisms used in the process of bioremediation; the nature and kind of contaminants, which have to be biodegraded; and the geological, climatic factors and chemical state of the contaminant at the waste dumping sites.

The microorganisms exhibit vital role for the destruction of the pollutants. The nature and the type of microorganisms serve as a key role in the process of *in-situ* bioremediation whereas the geological and climatic factor determines which pollutants are most susceptible towards the process of bioremediation in a particular environmental condition.

Most of the microorganisms have the enzymatic capacity to degrade the hydrocarbon of olefin, paraffin, and some other simple aromatic compounds present in the petroleum-based products. The straight-chain paraffin compounds having the carbon atoms in the range of C_{10} to C_{26} and the low molecular mass simple aromatic compounds (e.g.: benzene, xylene, and toluene) are easily degraded by the microbes.

The bioremediation may occur either naturally or by the addition of some suitable microorganisms of one kind or more than one kind. The microorganisms are always metabolizing the chemical substances present in the wastes into CO_2, CH_4, water vapor, some solid biomass and liquid effluents. The contaminants from the wastes may enzymatically be converted into metabolites, which are comparatively less toxic than the original polluting substances present in the wastes.

In *the in-situ* method of bioremediation, the microbial communities cleave the complex waste substances and turn them into non-hazardous and simple forms due to which the sustainability of the life is continued from one generation to another. Therefore, it was concluded that without microorganisms, the whole globe would be covered by waste products and the nutrients required for the existence of life on the surface of the earth is blocked [47, 57].

The microorganisms spontaneously and effectively exist in the biosphere, but they produce some unwanted substances by the process of microbial detoxification, which can impact negatively our environment. The necessary criteria of maintaining consistent bioremediation are the use of advanced technology, which can activate, metabolize, and control the microorganisms effectively at the site of contamination. The criteria depending upon the physiological, genetic, ecological, and biochemical principles, which can be successfully applied in both *ex-situ* and *in-situ* process of bioremediation, are:

- Adaption of metabolic activity of the microbial community;
- Addition of the conservative tracers;
- Addition of the isotopic tracers;
- Addition of the radioactive tracers;
- Formation of stable pattern of isotopic fraction;
- Maintaining the mass balance of reactants, pollutants, and products;
- Maintaining the rates of respiration in the *in-situ* process of bioremediation;
- Molecular biological pointers;
- The field plots have to be replicated;
- The recognition of the intermediate metabolites produced in the landfill sites;
- The tracer for internal conservation;
- Use of modeling technology in order to join transport and the active stoichiometry for the electron acceptors and donors [1, 14].

7.3 CLASSES OF MICROBES USED IN BIOREMEDIATION

The specific bacteria and fungi, which are normally used in the process of bioremediation, are described in this section:

1. **Alcanivorax Borkumensis:** These bacteria can degrade hydrocarbons of both aliphatic and aromatic categories normally present in the fossil fuel and emits CO_2 gas in the atmosphere.
2. **Dechloromonas Aromatic:** These bacteria have the capacity of oxidizing the toxic aromatic compounds, such as benzoate, chlorobenzoate, toluene, and other chlorinated aromatic compounds. Through the coupling reaction, it is also able to reduce oxygen, nitrate or chlorate. It is the only microorganism that has the capability to oxidize benzene in an anaerobic process. This bacterium is specifically applied in the process of *in-situ* bioremediation of the pollutants.
3. **Deinococcus Radiodurans:** This bacterium is resistant to the radiation of extremophile organisms, and is genetically developed for the

bioremediation of the toxic solvents and other HMs and metal ions. This bacterium is much more successful for the biodegradation of Hg and toluene (C_6H_5-CH_3) found in the radioactive waste.

4. **Fungi (Mycoremediation):** These have basically played vital role owing to their contribution towards the cycling of elements through the process of biological degradation and conversion of inorganic and organic materials.
5. **Methylibium Petroleiphilum:** This bacterium is reported to biodegrade the MTBE (methyl tertiary butyl ether) and converts the pollutants in sole carbon and energy source.
6. **Nitrifiers and Denitrifiers:** These bacteria are beneficial for the bioremediation and treatment of industrial wastewater effluents. The removal of undesirable nitrogenous compounds such as ammonia in the form of organic nitrogen, nitrate, and nitrite is essential for proper treatment of industrial wastewater.
7. **Phanerochaete Chrysosporium:** These fungi are interrelated with the degradation of the organic contaminants. These species of fungi have a strong ability to degrade many potential hazardous pollutants, such as dioxins, chlorinated organic compounds, pesticides, herbicides, PAHs, CCL_4, PCBs, insecticides, and many other toxic pollutants through the process of bioremediation [30, 32, 56].
8. **Pseudomonas Putida:** This belongs to gram-negative soil bacterium, which is successful for the bioremediation of toluene and is a common constituent present in the thinner of the paints. It is also able to degrade the naphthalene present in soil.
9. **White Rot Fungi:** These can metabolize a number of vital pollutants in our environment. The most significant feature of these categories of fungi is their corresponding enzymatic functional capability for metabolizing complex structured chemical substances such as protein, lignin, and lipid. These categories of fungi are able to survive in extreme situations of our environment, such as temperature, pH, and moisture content. Although most of the microorganisms need pre-conditioning before being applied in the bioremediation process for their survival within that environment, however this class of the microbes can be directly used without any pre-conditioning because they are capable of disintegrating the toxic materials in extreme situations.

7.3.1 MICROORGANISMS FOR BIOREMEDIATION OF HYDROCARBONS

The most important and toxic hydrocarbon-based pollutants are PAHs and are considered as xenobiotic pollutants, which are produced due to incomplete burning of carbon rich substances. The main anthropogenic activities known to produce the PAHs are burning wood, cigarette smoke and burning of fossil-based fuels.

Nowadays due to the commercialization of the bacteria, bioremediation is considered as the only process, which is effective for the soil polluted with low molecular mass PAHs. However, fungi are comparatively more effective for the degradation of PAHs relative to bacteria due to the following reasons:

- Fungi are able to degrade PAHs, which are normally having high molecular mass and large complex molecules, whereas bacteria are effective towards the degradation of small size molecules [38];
- Fungi are efficiently functioning and active in the non-aqueous environmental medium with less availability of O_2, this proves beneficial for the PAH remediation as they are accumulated at the landfill sites or any contaminated area [3];
- Most of the fungi create their own mechanism for targeting the definite PAHs. In the process of mineralization, fungi are able to produce some extracellular enzymes which can effectively degrade the lignin-based products [3, 38].

7.3.2 MICROORGANISMS FOR BIOREMEDIATION OF HEAVY METALS (HMS)

The toxic metals of different kinds enter in the food chain, food web, and all biogeochemical cycles of our environment at every stage, causing adverse effects on the biological community. Metals are mixed with surface and groundwater due to leaching of mining effluents and solid wastes from the mining sites.

Nonetheless, from the natural environment, the movement of the toxic metals and their ions are due to various geological processes, then mixed with aquatic ecosystem and soil. The metallic pollutants are the largest contributor towards environmental risk because of the property of non-degradability by adopting any of the natural or biological processes. Hence the best way of treatment of the metals from the wastes is separating the metal from landfill wastes and shifting to some other areas, where no negative impact is observed towards the environment. Fungi can interact with the metallic contaminants in many ways thereby decreasing or enhancing the mobility of the metals, cellular uptake, sorption, etc. Fungi, after absorbing the metal, change it chemically by changing the oxidation state of the metal and subsequently store, translocate to different plants and take part in symbiosis [18, 28].

7.3.3 MICROORGANISMS FOR BIOREMEDIATION OF PESTICIDES

The presence of pesticides and its accumulation within the environment is now a major issue because the toxic materials from the pesticides are associated directly with food grains, food materials, and public water supply. There are a number of techniques adopted to onslaught pesticides, but all these processes are costly and not so effective because they need excavation of the polluted soil and moved to some other suitable location for storage and further processing. Nevertheless, the bioremediation technique is treated as a vital alternative for treatment of groundwater and polluted soil without excavation. White rot fungi possess a higher capacity of soil bioremediation containing pesticides, because this kind of fungi has the capability of mineralizing some specific pesticides responsible for contamination of the soil [17].

7.3.4 MICROORGANISMS FOR CH_4 OXIDATION

The traditional method of bioremediation of pollutants is initiated and stimulated by the separation of one or more kinds of microorganisms, which have the capability of disintegrating the targeted contaminants, but this conventional process of separation leads to the biodegradation of only a definite fraction of pollutants in the landfill and the isolated group of microorganisms reveals the kinetics of the degradation of the pollutants, which is different than observed in normal ambient environmental condition. The most trusted microorganisms producing CH_4 are methanotrophs, which are able to metabolize TCE during bioremediation and frequently, CH_4 has to be injected to stimulate the activity of methanotrophs for increasing the effectiveness of bioremediation of TCE [63].

7.3.5 MICROORGANISMS FOR DEGRADATION OF PAHS

PAHs are highly toxic aromatic compounds, which are highly intense towards the public concern because of its persistence within the environment for a longer period and serves as a deadly pollutant for the human health. The PAHs of high molecular mass category can be degraded by using some specific organisms (such as *Burkholderia*, *Mycobacterium*, and *Sphingomonas*), but the mechanism of biodegradation of PAHs by using this kind of bacteria is still not clear. In soil, the bioavailability of PAHs is decreased because of sorption to the biological organic matter, which is considered as a vital factor in controlling the bioremediation of PAHs.

The various kinds of bacteria of the phenanthrene degrading category may be useful in the soil enhancements of various sorptive matrices. The use of surfactants for the improvement of the soil can degrade the aromatic compound hexadecane and phenanthrene and can change the population of the microbial community, which are mainly responsible for the process of bioremediation. However, it is better to use a diversified population of microorganisms for depredating complex organic pollutants for the strategies of bioremediation [43, 44].

7.4 *IN SITU* BIOREMEDIATION TECHNIQUES

The *in-situ* process of bioremediation is efficient for chlorinated substances. The *in-situ* process of bioremediation is extremely effective towards the treatment of groundwater contaminated by NO_3^- and SO_4^{2-}. This process normally is carried out by treating the contaminants present at the sites and is cost-effective relative to the ex-situ process due to non-requirement of excavation. Some of the in-situ processes of bioremediation are enhanced by adopting the processes of bioventing, phytoremediation, and biosparging, and the one without any enhancement called natural attenuation (NA) or more commonly intrinsic bioremediation.

The process of *in situ* bioremediation is effective towards the degradation of chlorinated solvents, HMs, dyes, and paints, as well as the hydrocarbons. The *in-situ* process of bioremediation is successful by maintaining proper environmental parameters, such as availability of nutrients, electron acceptor or O_2, temperature, pH, and moisture. In addition to these, soil porosity also significantly affects the process of *in situ* bioremediation. In addition to many advantages, this process of bioremediation method has limitations, such as more time requirement, less acceptable relative to the conventional processes [55].

7.4.1 NATURAL ATTENUATION (NA)

This process reduces the petroleum-based hydrocarbons in the soil by using natural microorganisms. The site of NA should be properly monitored to estimate the risk factor associated with it and prediction of the endpoint. If the quick removal of the pollutants is necessary, then this process is more significant and the NA process of bioremediation is considered under the following categories:

1. **Biostimulation:** It is the process of mixing of chemicals and nutrients in order to act as indigenous microorganisms.
2. **Bioaugmentation:** It is the immunization with the help of exogenous microbial community and can be used specially for influencing the indigenous biotic community with the occurrence and diffusion of the immunized microorganisms.

Recently many developed technologies in the field of microbial ecology provide us a clear idea about the molecular technologies. For example, the discovery of the degradative genes, metagenomics, and communal fingerprinting are applied for monitoring and analysis of the immunized and indigenous microbial community in the polluted area. Researchers have now initiated advanced novel techniques for estimation and control of NA to enhance its efficiency and consistency [5].

7.4.2 ENHANCED IN SITU BIOREMEDIATION

7.4.2.1 BIOVENTING

Bioventing is a natural *in-situ* method of biodegradation of the pollutants in the soil by supplying O_2 or air to the microorganisms present in the contaminated soil. The process of bioventing requires more oxygen in a controlled flow for the microbial community present in the vadose zone, otherwise less flow of oxygen is needed in all other zones. In the process of bioventing, there is an interaction between the moisture present at the contaminated site and the supplied nutrients to increase the rate of bioremediation with the final objective of attaining the successful conversion of toxic complex form of pollutants to non-toxic or less toxic simple species. This process is popular among all the *in-situ* process of bioremediation, especially of the contaminated sites, where massive petroleum-based products are dumped.

The bioventing design is inspiring process of aeration in the unsaturated zone, whereas in the process of anaerobic bioventing, specially chlorinated compounds are treated in the vadose zone. In the anaerobic bioventing process instead of only O_2 and air, the supply of the mixture of O_2, N_2 along with less proportion of CO_2 and H_2 is more suitable for decreasing the chlorinated vapor along with H_2. The soil having low permeability for oxygen, supply of pure O_2 may be the cause of higher O_2 concentration relative to the supply of air.

However, the process of ozonation may be beneficial for the partial oxidation of waste substances for increasing the process of bioventing. The process of bioventing mainly depends on the increase in the microbial population at the vadose zone for enhancing the rate of biodegradation by supplying moderate concentration of air, extraction of soil vapor, maximizing VOCs, increasing volatilization, and extraction of vapor. The process of soil vapor extraction (SVE) is normally a physical method for remediation of pollutants in soil. However, deceptively, the higher flow rate of the supplied air is the cause of converting the VOCs to the vapor phase and the off-gas treatment method is necessary before it is released to the troposphere. This issue can be noticeably resolved by using the combined effect of the techniques of biotrickling filter and bioventing for decreasing both the pollutants and emission level of the outlet gas which leads to the minimization of the treatment time required in bioventing [21, 72].

7.4.2.1.1 AEROBIC BIOVENTING

In this process of aerobic bioventing, the polluted unsaturated soils having less concentration of oxygen are normally treated by adding O_2 to support bacterial action and to improve the rate of biodegradation. Oxygen is characteristically introduced by injecting air into the subsurface of the pollutants. If the foundations of the building or a similar kind of structures are nearer to the site of contamination, the method of extraction can be used to avoid the accumulation of the pollutants (most probably explosive materials, vapors within the basement of the buildings). In comparison to the method of extraction of soil vapor, the process of bioventing requires a low rate of airflow to increase the removal efficiency of the pollutants.

The aerobic process of bioventing can successfully degrade the pollutants mainly non-chlorinated compounds, VOCs, and petroleum hydrocarbons (PHC) which are mainly present at the capillary fringe or vadose zone. On the basis of field experience, it was found that bioventing is affected by the heterogeneity of the polluted site, impairing the model of the biological activity. Finally, determination of the quantity of removal of the pollutants, supply of O_2, evolution of CO_2 and material balance of all these parameters are always potentially related with the process of bioventing. Not only fuel, but also solvents of non-halogenated class (such as benzene, xylene, acetone, cresol, toluene, resorcinol, and phenol) along with light halogenated solvents (such as chlorobenzene, 1,2-dichloroethane, dichloromethane) and the low molecular mass organic materials like PAHs are also treated successfully by the process of aerobic bioventing. If we require rapid clean up, then extremely intensive *ex-situ* technique may be relatively more suitable [8].

7.4.2.1.2 ANAEROBIC BIOVENTING

In this method of co-metabolic bioventing, many of the hydrocarbons and some lightly chlorinated materials are successfully bioremediated. The microorganisms disintegrate the pollutants directly either by the pathway of anaerobic and co-metabolic or through the process of anaerobic reductive de-chlorination. The latter one is an appropriate biological mechanism, which is able for the successive removal of chlorine from the chlorinated compounds and the microorganisms applied in this path do not gain any energy. The anaerobic co-metabolism method is almost comparable to the aerobic co-metabolism process in the fact that the microbes used accidentally remediate the pollutants, but the method of anaerobic bioventing uses the biological mechanisms for cleaving the targeted contaminants.

Like other methods of bioventing, the anaerobic bioventing takes the similar kind of gas delivery system, but in addition to the mixture of N_2, the electron donor is injected instead of only air to stimulate the reductive anaerobic environment. The added N_2 displaces the O_2 present in the soil and also little quantities of H_2 and CO_2, leading to the creation of reducing environment at the subsurface and facilitates the de-chlorination by the microbial community.

In the process of anaerobic bioventing, the semi-volatile and volatile simple gaseous products are normally produced and finally mixed with the earth's atmosphere. Anaerobic bioventing is an emerging technological process, which is much beneficial in the treatment of highly chlorinated substances, including TCE, RDX (royal demolition explosive/ organic compound-($(O_2NNCH_2)_3$), DDT, PCE (pentachloroethane/tetrachloroethylene), and pesticides [24].

7.4.2.2 BIOSLURPING

This technology is the combination of extraction of soil vapor, vacuum-enhanced pumping and bioventing for achieving the treatment of groundwater and soil by the indirect delivery of O_2 for stimulating the biodegradation of the pollutants. This technology is normally developed and designed for the recovery of LNAPLs (light non-aqueous phase liquids) and remediation of saturated capillary and unsaturated zones. This technology is also adopted for remediating the soils, which are polluted by the semi-volatile and volatile organic materials. "Slurp" is one of the important components of this system, which can be extended up to the layer of free products and able to draw soil gas and free products from the layer, almost similar to drawing liquid from a pot by using straw. The mechanism of pumping is the cause of driving LNAPLs to the soil surface, at which it can be removed from air and water. After the removal of the total soil gases and free products of the system, it may be easily operated similar to the normal bioventing system till the remediation process is finished.

In this method the presence of extreme soil moisture reduces the permeability of air and retards the rate of transfer of O_2, leading to the decrease in microbial activities. This method is not appropriate for remediation of polluted soil having a low value of permeability, but it is cost-effective because of the production of fewer quantities of groundwater during the operation of the system, therefore reduces the cost of treatment, disposal, and storage. Bioslurping is commonly identified as a multi-phase extraction process and is much more efficient for removing the free products, which are normally floating on the water table. For achieving effective bioremediation, bioslurping can be combined with other two remedial methods such as vacuum-enhanced free-product recovery and bioventing [26].

7.4.2.3 BIOSPARGING

This process is comparable to the bioventing technique, where the air is injected inside the subsurface of the soil for promoting the activities of the microbial community in enhancing the removal of the contaminants from the polluted sites. Like the process of bioventing, here also O_2 is injected at the zone of saturation, which leads to the upward motion of the VOCs to the zone of unsaturation to support bioremediation. The efficiency and success of biosparging mainly depend upon two key factors: soil permeability and biodegradability of the pollutants.

All the processes, including the technique of SVE, bioventing, and biosparging are very much similar in operation called IAS (*in situ* air sparging), which depends upon an increase in flow rate to achieve the volatilization of the pollutants, however biosparging stimulates the biodegradation. Biosparging is the widely accepted technique used for the bioremediation of the pollutants present in the aquifers rich in petroleum products, particularly kerosene and diesel.

It has been established that the biosparging of aquifer plume contaminated by BTEX (benzene, toluene, ethylbenzene, and xylene) leads to the change in anaerobic to aerobic environment, which is the cause of the increase in DO, SO_4^{2-}, redox potentials, NO_3^-, and the total cultural heterotrophs with a corresponding decrease in dissolved Fe^{2+}, S^{2-}, CH_4 and total methanogens and anaerobes. Hence the process of biosparging is effective for the bioremediation of BTEX (up to 70%).

Biosparging comprises the introduction of O_2 or air and sometimes the nutrients in the gaseous form, and is injected under pressure into the saturated zone of the contaminated site to induce the aerobic process of biodegradation. In the process of air sparging, the volatile components of the pollutants can be separated from the saturated zone by the process of volatilization and desorption into the stream of air.

There is a specialized technique of biosparging, which normally involves the percolation of organic gases addicted to the saturated zone in order to promote co-metabolic degradation of hazardous chlorinated form of aliphatic hydrocarbons and this process is similar to the co-metabolic bioventing [29, 54].

7.4.2.4 PHYTOREMEDIATION

This process of phytoremediation depends upon the biological, physical, chemical, microbiological, and biochemical interaction of the plants on the contaminated site to diminish the deadly effects of contaminants. There are a number of mechanisms such as extraction, filtration, accumulation, volatilization, stabilization, and degradation adopted in the process of phytoremediation depending upon the nature and kind of pollutants.

The elemental category of pollutants such as radionuclides and HMs are suitably separated by the process of extraction, sequestration, and transformation, whereas the organic contaminants such as chlorinated materials and some other kinds of hydrocarbons are primarily separated by using the mechanisms of stabilization, rhizoremediation, mineralization, volatilization, and degradation along with the interaction of some plants including alfalfa and willow.

The process of removal of contaminants through plants generally involves the ease of acceptance, which may be passive, translocation from the roots of the plants to shoot and is carried out through the xylem flow and accumulation in their shoots. Again, the accumulation and the process of translocation depend upon the separation between neighboring tissues, evaporation rate and transpiration. This process is influenced by the nature, kind of contaminants and categories of plants. It was found that most of the plants grown in the contaminated areas exhibit as good phytoremediators.

The effective phytoremediation principally depends on the optimization of the remediation abilities of the growth of local plants in the contaminated sites by the process of bioaugmentation with exogenous or endogenous plant rhizobacteria, or through biostimulation. The use of PGPR serves a vital role in the process of phytoremediation and likely to increase the production of biomass and enhances the tolerance limit against toxic HMs and some other edaphic conditions of the soil. It was found that the interaction between plants and microorganisms may not be always optimum condition for phytoremediation. In addition to that intervention of human beings is necessary in order to optimize the interaction for increasing the efficiency of the removal of the pollutants.

The combination of organic wastes with the toxins from lubricating oil in the polluted soil, increase the growth of the *Jatropha curcas* and the microbial propagation at the rhizosphere, inspiring the removal of 33% pollutants [4, 42].

7.4.2.5 PERMEABLE REACTIVE BARRIER (PRB)

This method is normally regarded as a suitable physical way for the remediation of the unclean groundwater because of its mechanism and design for the removal of the contaminants. The different mechanisms of biological reaction (such as degradation, sorption, and precipitation) are the various process of removal of the contaminants by adopting PRB method. The different alternative terms, including biological PRB, bio-enhanced PRB, and bioreactive barrier are popularly used for accumulating the aspects of bioremediation or biotechnology. Microorganisms serve a crucial role in increasing the rate of bioremediation rather than simple independent biotechnology.

PRB is a conventional *in situ* method, usually adopted for treatment of contaminated groundwater and soil containing the pollutants like chlorinated substances, PCBs, toxic HMs, and hydrocarbons. In this method, the semi-permanent or permanent reactive medium normally consist of zero-valent iron (ZVI), which is inundated in the route of unhygienic groundwater and contaminated soil.

When the contaminated water passes through the medium under its usual gradient, the pollutants present in it can be trapped easily through initiating a series of chemical reactions resulting in the formation of clean and uncontaminated water. The efficiency of this method mainly depends upon the kinds and nature of the medium used, hydrogeological, biogeochemical reactions, influence of health and environment, cost, and stability. More recently, PRB on coupling with electrokinetics (EK) is used for the effective remediation of many toxic pollutants successfully. This combined system is most successive in the contaminated soil in order to maintain proper environmental circumstances, such as pH, nutrients, moisture, and temperature, which is essential for proper growth of microbial communities and leads to the dispersal of the surfactant biomass across the contaminated soil.

Not only controlling the barrier reactivity is a key factor for evaluating the performance of PRB process, but also preservation of the permeability barrier plays a key role for the effectiveness of PRB and can be attained by sustaining the dispersal of proper size of the particles. The performance of the PRB in long term basis is retarded by the decrease in the reactivity of the medium, presence of ZVI, porosity loss and incapability towards the application of this method to the site polluted with chlorinated compounds, hydrocarbons, and intractable materials, are the major challenges for operating the PRB method.

It was reported that polyhydroxybutyrate (PHB) is a novel polymeric material, which is biodegradable in nature and capable of releasing the nutrients slowly and stimulates the biological activity if used as a medium and leads to the increase in efficiency of the removal of chlorinated materials. By using the iron sulfide (FeS) medium as a barrier, some of the ZVIs can be removed. However, more detailed study is required regarding the reactive medium, design of the system, limitations, percentage of removal of contaminants and advantages for economic and cost point of view [49, 73].

7.4.2.6 BIOAUGMENTATION

It has been found that the rate of biological degradation can be enhanced by the addition of some microorganisms in the sites under prevailing environment, due to an increase in microbial activity, and the process is known as bioaugmentation. This method is much more useful specifically in the sites, where the contaminants present in it are not easily biodegradable in nature or the contaminated site is in extreme condition, such as more concentration of non-biodegradable pollutants or more toxicity of the wastes. For effective bioremediation, the added microorganisms at first are dispersed throughout the polluted site and fight with the indigenous microorganisms already present for the available nutrients. If the dispersal within the site matrix is not uniform, the effect of the added microorganisms becomes localized, and hence they are unable to compete effectively with the microorganisms already present, resulting in a non-effective treatment [51, 62].

7.5 *EX SITU* BIOREMEDIATION TECHNIQUES

Ex-situ bioremediation is a biological process, where the excavated soil is normally kept at a particular level above the treated land surface or transported to some other location and aerated properly in order to upsurge the rate of biodegradation of dangerous organic wastes by using indigenous microbial community. In the aerobic environment, some of the particular class of microorganisms utilizes the organic wastes such as both aliphatic and asphaltic PHC, PAHs, phenols, xylenes, cresols, and some pesticides as a source of carbon (food), energy, and ultimately degraded into CO_2, CH_4 and H_2O.

This method of bioremediation is normally selected on the basis of cost, depth level of the contaminants, nature, and kind of contaminants, degree of pollution, geology, and geographical condition of the landfill site. The measures of performance of ex-situ bioremediation are mainly depend on the addition of extra microorganisms, necessity of the nutrient, existence of organic material in the contaminated soil and O_2 (required for growth and existence of microorganisms). This process of bioremediation normally utilizes extra nutrients and bioreactors for quick breaking of pollutants [59].

7.5.1 EX-SITU BIOREMEDIATION TO DEGRADE CONTAMINANTS

The process of *ex-situ* bioremediation is effective for the biodegradation of a huge variety of hydrocarbons such as kerosene, phenols, cresols, resorcinol, PAHs, diesel, lubricating oil, grease, semi-volatile organic materials, straight-chain normal aliphatic compounds, branched-chain and cyclic aliphatic compounds. The non-chlorinated hydrocarbons having the chain lengths C_6 to C_{14} can be easily treated, whereas the hydrocarbons having the carbon length C_{15} to C_{32} requires longer time periods for degradation. Chlorinated hydrocarbons along with comparatively more complex and toxic materials can be biodegraded, but its detailed analysis and assessment is required [31].

7.5.2 BIOPILES

Biopile-based remediation method is a combined effect of composting and land farming. Basically, the cells are built as composted piles in the aerated environment. This technique normally is applied for the treatment of pollutants of petroleum-based hydrocarbons and is a modified form of land farming process, which is able to minimize the physical losses of the chemicals due to volatilization and leaching. This process is more advantageous for both anaerobic and indigenous microorganisms. The biopile base remediation is generally carried out in a separated location or above the land surface piling through excavating the contaminated soil with the proper adjustment of nutrients, and in some situations, air is injected to create an aeration environment in order to accelerate the activities of the microorganisms leading to the increase in rate of bioremediation. The different stages of this method are irrigation, aeration, leachate collection, nutrient addition and a suitable treatment bed [22]. Although the use of biopile for the remediation of wastes in the landfill is restricted only in pollutants of low molecular mass, yet it can also be used effectively in extreme environmental conditions, i.e., in very cold environments.

7.5.3 WINDROWS

Windrows is a significant ex-situ method of bioremediation and depend on the periodical rotation of the piled contaminated soil for accelerating the remediation by increasing the activity of both indigenous and transient hydrocarbonoclastic bacteria existing in the polluted soil. The intermittent rotation of the contaminated soil with subsequent mixing of water is the cause of creating aeration environment, uniform dispersal of pollutants, increase in nutrient availability, and increase in the activity of microbial community. Controlling and regulating all these factors results in an increase in the rate of bioremediation through the process of assimilation, mineralization, and biotransformation. In comparison to biopiles which exhibits a higher rate of the removal of hydrocarbons, the window treatment method is suitable for the removal of various kinds of hydrocarbons, which mainly depends upon the kind of the soil. However, due to the periodic movement of the windrow process, it cannot be considered as the best method of bioremediation of contaminated soil containing toxic volatile matters. Another disadvantage of this method is the release of more CH_4, which is a potent GHG [26].

7.5.4 BIOREACTOR

The bioreactor refers to a container in which the raw materials are rehabilitated to some specific product materials through series of biological reactions. A bioreactor adopts different modes of operations, such as batch, sequencing batch, fed-batch, continuous, and multistage reactor. The selection of the mode of operation is mainly governed by the economic conditions of the market and capital cost. For maintaining natural process in a bioreactor, we have to initiate and preserve the natural environment to promote the optimum growth of the microorganisms. The soil treatment by a bioreactor may be advantageous relative to any other ex-situ bioremediation methods. The unclean soil samples are fed into the bioreactor in the state of slurry or dry mass or both.

The effectiveness and success of a bioreactor mainly depends on proper controlling of the parameters involved in the process of bioremediation, such as temperature, agitation, pH, and rate of aeration, inoculum, and substrate concentrations. The controlling capacity, manipulating process of parameters, operating ability for a bioreactor indicates the rate of biological reactions by the microbes with an increase in the effectiveness and decrease in bioremediation time. In addition to these parameters, controlled bioaugmentation, addition of nutrients, increasing the bioavailability of the contaminants, the mass transfer in between the microorganisms and pollutants are also considered as some of the significant limiting factors in this method of bioremediation which must have to be established properly for creating the bioreactor more efficient towards bioremediation.

Although this kind of system is used for the remediation of VOCs, BTEX, and other petroleum-based hydrocarbons, yet different kinds of bioreactors are used for removing a wide range of contaminants in the process of bioremediation. Because of good flexible property in the design of the bioreactor, it can help for the optimum biological degradation along with minimal abiotic losses. Since the bioreactor is a closed system, therefore GEMs may be used in case of bioaugmentation and after the functioning of the microorganisms; these may be destroyed and/or reverted back into the landfill or agricultural fields.

The bioremediation by using the bioreactors is not a widely used process because of following reasons [2, 58]:

- Different pollutants respond in different ways towards different bioreactors, hence selecting the most suitable design is very difficult.
- In addition to all the limiting factors, cost is the major factor to make this technique unpopular.
- There are many parameters associated with the bioreactor, which are not appropriately controlled, regulated, and maintained up to the optimum range and cause the delay in the microbial activity, which is the main factor for less effectiveness of the bioreactor towards the process of bioremediation.
- This process is not economical because the contaminated soil and the other hazardous materials to be treated require high capital cost, manpower, and safety measures due to transferring from one site to the site of treatment.

7.5.5 LAND FARMING

It is a land treatment popular bioremediation method, which normally consists of the excavation of polluted soil and its dispersal uniformly on a thin land surface till the improvement of aeration is achieved. It is carried out by indigenous microorganisms as well as physical and chemical methods. The bioremediation of the contaminants is activated in an aerobic environment by plowing or digging the contaminated soil. Some extra minerals and nutrients are added to the soil to support the growth and existence of the indigenous microbial species.

According to EPA (Environmental Protection Agency), before the bioremediation is carried out, the contaminated site must have to be properly prepared by grading and cleaning the soil and connecting the leachate collection with treatment systems, along with all the facilities for the treatment of the vapor produced.

The bioremediation may be increased through the periodical turning of the soil bed and mixing the appropriate nutrients. Due to 0.3 m thickness of the contaminated soil layers, large surface area is required in land farming, therefore it is not an appropriate treatment for a small site, but still, it is the most cost-effective and simplest form of bioremediation. In most of the situations, it is ex situ, but in few cases, it is *in situ*, which depends upon the depth of the pollutant on the site of contamination. In this treatment process, one thing is common that the contaminated soil must have to be excavated, tilled for better results of remediation.

The treatment of the excavated soil carried out on-site is considered as *in-situ* process, and the treatment process carried out off-site is called *ex-situ* method of bioremediation. Cultivation or digging (cause of aeration), adding some extra nutrients such as N, P, and K, and irrigation are the key operational steps, which can stimulate and promote the activities of autochthonous microorganisms leading to the increase in the rate of bioremediation. However, it has been envisioned that the irrigation and digging without the addition of nutrients in the polluted soil with an increase in the biological activity of diesel-degrading and heterotrophic bacteria is the cause of enhancing the rate of bioremediation. The dehydrogenase enzyme and its activity are found to be a good indicator of the biostimulation remediation and usually used as a parameter of biological activity due to land farming.

The approaches of land farming abide the government rules and guidelines and can be applied in any areas and any climatic situation. Finally, we can conclude that the method of land farming is a basic and simple process, easier for designing, implementing, low capital cost and can be applied for the bioremediation of large volumes of the polluted soil having the least necessity of energy and is the most effective popular inexpensive method of bioremediation. But it has some demerits such as the necessity of more operating space, reduction in the activities of microbial community because of adverse environmental situations, extra cost for excavation, less effectiveness of the removal of inorganic pollutant and metal ions [12, 45].

7.5.6 COMPOSTING

Composting is the combination of the polluted soil with any non-hazardous organic matter (such as manure, agricultural wastes, and food wastes) to make the process easygoing and to bring about optimum levels of water and air for growth of microorganisms. The common designs of composting methods are window composting, mechanically agitated composting and static pile composting. In the process of static pile composting, the unused polluted soil is kept in the piles and then aerated by either vacuum pumps or any wind blower. In the mechanically agitated composting, the soil sample is kept in a container made for treatment, where it is aerated by supplying air. In the process of window composting, the sample of contaminated soil is taken in a long pile termed as windows, where the soil is periodically mixed with aeration.

The sample of the polluted soil is added with some compost (bulking agent) to accelerate the growth of microorganisms. Then after mixing the soil with the bulking agent, it is properly covered in order to preserve the soil from erosion and contamination from other sources. The remediation of wastes by composting is a relatively quick process and is regulated by a biological treatment procedure to clean up the organic wastes at a temperature range of 40°–50°C. This thermophilic condition is the basic distinction between biopiles and composting. The disintegration of the bulking agent generates heat, which creates thermophilic environments, and in addition to that, the O_2 content, moisture content, pH, temperature is properly maintained for optimizing the rate of biodegradation [11, 15].

7.6 INTRINSIC AND ENGINEERED BIOREMEDIATION TECHNIQUES

7.6.1 INTRINSIC BIOREMEDIATION

This kind of bioremediation technique (referred to as "cleaning up by nature itself or NA") normally uses the distinctive ability of naturally found microorganisms to biodegrade the pollutants in the soil without any extra assembly for increasing the efficiency of the process. The process of stimulated bioremediation and intrinsic bioremediation are almost similar to each other, however stimulated bioremediation provides a distinctive originality of intrinsic bioremediation. The intrinsic bioremediation proceeds without interference of human activities and proceeds through microbes and nutrients, which can be obtained naturally in the matrix of the contaminated soil.

There are some key factors responsible for evaluating the efficiency of the intrinsic remediation, such as bioavailability of the pollutants, nutrient concentration, the existence of minerals to maintain buffer having the pH of the polluted soil matrix, proper ratio of electron acceptors such as O_2, SO_4^{2-}, Fe^{3+}, and the migration rate of some particular contaminants.

Intrinsic bioremediation is defined by NRC (National Research Council) and EPA as "*biodegradation, dilution, dispersion, volatilization, sorption, radioactive disintegration, biological, and chemical stabilization, oxidation, reduction transformation with demolition of the contaminants.*"

It has been reported that intrinsic bioremediation is not suitable for the remediation of PAHs and the reduced soil is sometimes eco-toxic [34, 48].

7.6.2 ENGINEERED BIOREMEDIATION

In some cases, it is necessary to build an engineered system for the supply of electron acceptors, nutrients, and some other materials that need to accelerate the biodegradation rate, which is commonly known as engineered bioremediation. This mode is needed in the situations, where the rate of immigration of the pollutants is more. The setting of the environmental receptors and other accountability issues need to be optimized to have an improved degradation of the pollutants. The important site characteristics in the case of engineered bioremediation are almost the same as required for intrinsic remediation. Also, it can affect the capability of supplying the chemicals to the ecosystem, including permeability of the subsurface of soils and consistency of the soil subsurface [66].

7.6.3 GENETICALLY ENGINEERED MICROORGANISMS (GEMS)

Genetically engineered microorganisms (GEMs) are the specially prepared microorganisms whose genetic system is previously modified by using special techniques of genetic engineering, which is stimulated either by natural or artificial techniques of genetic exchange in between the microorganisms. This scientific method is a recombination of DNA technology. This technique of using genetically modified microorganisms has been developed for elimination and utilization of hazardous and toxic undesirable wastes under laboratory conditions through the artificially produced GEMs.

Recently, specific highly active enzymes are produced by introducing appropriate genes, which can degrade various pollutants, including hazardous materials. GEMs are potentially active towards the bioremediation of groundwater, soil, and activated sludge and establish an increase in the degradation capacity of a huge range of chemical pollutants. For GEMs, the following four strategies are executed:

- The modification of the activity and specificity of the enzymes;
- The way of regulation, controlling, and construction;
- The development of biological process for controlling, regulating, and monitoring;
- Bioreporter, bioaffinity sensor for evaluating the decrease in toxicity, chemical sensing, and analysis of endpoint.

The vital genes of different bacteria are normally occurring on a single chromosome, whereas particular genes are necessary for the catabolism activity of some rare substrates, which might be established on plasmids. The plasmids are interlinked with catabolic processes; hence GEMs can be applied more efficiently for the purpose of bioremediation [27, 64].

7.7 ELECTROKINETIC (EK) PROCESS OF BIOREMEDIATION

In this method, DC current of low voltage is applied across the polluted soil for separating the pollutants from the dirtied soil, and the process is almost comparable to a battery. Due to the application of electric charge, the charged ions are moved towards the opposite charged electrodes. Each part in the system comprises of an electrode, water, and a pump. The outer cover of the electrode is made with porous ceramic materials, which facilitates the passage of electric current and water through it. The porosity of the soil is filled with water, and it is under pressure, therefore it becomes saturated and cannot move to the neighboring soil. The contaminated ions move through the pores and then separated for treatment.

There is another same kind of process for the movement of pollutants through the contaminated soil by applying electric current which is known as electro-osmosis. In comparison with EKs, the process of electro-osmosis is the movement of the polluted liquid containing various ions. In this process, electric current is applied for remediation of soil. The electric current flows through the pores of the soil containing water between cathode and anode. To raise the efficiency of the process in some situations both of these processes are simultaneously applied [23, 25].

The basic technology of EK bioremediation can be combined with Lasagna Technology (an *in-situ* process), in which the pollutants are separated easily by electro-osmosis through the treatment zones. EK-bioremediation is predominantly applied for the separation of radionuclides and heavy metallic contaminants from the soil of low permeability. It can also be applied for the treatment of complex organic substances along with pesticides and VOCs [25].

7.7.1 MECHANISM AND PROCESS OF EK-BIOREMEDIATION TECHNIQUE

The procedures of this bioremediation technique are carried out at either micro-scale (<10 mm) or macro-scale (>10 mm). The micro-scale involves interactions between the pollutants, microorganisms, and their surrounding subsurface environmental factors. The macro-scale process is carried out in the application for managing the plume-scale and degradation of the pollutants.

7.7.1.1 MICRO-SCALE

In the micro-scale mechanism, the substance or polluted soil is transported by EK for enhancing the rate of bioremediation and prepare bio-accessible pollutants, electron acceptors, nutrients, and electron donors. This process involves the reduction of spatial barriers and inducing more contact between the pollutants and releasing the bonded pollutants from the porous form of the matrix. The local impact towards these facts mainly is governed by the nature of the polluted substances and their unique properties.

The microorganisms as well as the contaminants are dominant towards electrophoresis or electro-osmosis and electromigration. The ionic forms of nutrients are likewise prevailing towards electromigration. The bioremediation is regulated through advection, infiltration, and dispersion. There are mainly two types of mechanisms developed by EK for contact between the polluted soil, when the combined effort of the rate of movement of microbes, pollutants, nutrients are considered.

7.7.1.2 MACRO-SCALE

The macro-scale bioremediation normally applies within a range of technological processes. Each process is restricted to the contaminant and microbe contact or mixing or on issues of hydraulic transport. The EK process can be adopted as an alternative development of the flow fields for enhancing the interaction between the pollutants and microorganisms or bring an obligatory modification to a definite zone, which is restricted under ambient EC.

7.7.2 MODES OF EK-BIOREMEDIATION

1. **Ek-Bioattenuation:** It is restricted to the site areas, where the pollutants are biodegradable in nature and the pollutants, microorganisms, and electron acceptors are properly mixed for adequate remediation at an enhanced rate.
2. **Ek-Biostimulation:** This increases the rate of distribution of nutrients (such as PO_4^{3-}, SO_4^{2-} and $NO3^-$) along with electron donors and acceptors through various materials at a rate more than the rate of diffusion. The application of EK-biostimulation in case of soils causes effective degradation of diesel, PAHs, toluene, xylene, chlorinated hydrocarbons. This process normally occurs in a sequential experiment, where the polluted water is at first reduced by passing through a cathode and then oxidized at the anodic column.
3. **Ek-Bioaugmentation:** This increases the movement of the microorganisms through the soils of low permeability despite of the potential obstruction of the cells because of small pore space in the fine-grained pollutants. The increase in the rate of bioremediation is accredited predominantly due to the movement of microorganisms through the micro-pores present in the soil by the process of electro-osmosis.
4. **Ek-Phytoremediation:** It is more focused on remediation of HMs present in the soil sample, where the applied electric field accumulates the pollutants surrounding the roots of the plants to the increase in bioavailability. The result of the combination of phytoremediation and EK of organic pollutants is although effective to a greater extent, yet it is still not successful up to the desired level, and more research work is needed in this field [50, 67].

7.7.3 LIMITATIONS OF EK-BIOREMEDIATION

- Because of the passage of electric current in aqueous medium, both the metallic electrodes may dissolve due to electrolysis and some corrosive materials are entered into the soil. Hence the electrodes are generally made of chemically inert materials mainly, platinum, graphite, and carbon.

- If the moisture content in the contaminated soil is less than 10%, the efficiency and success of the EK-bioremediation is drastically decreased.
- In the case of unsaturated soil, the introduction of water may possibly wash the pollutants at the contaminated site and accumulated in nearby areas.
- On the implementation of EK mode in thick and non-aqueous liquid phase, the system is fully clogged and not operated.
- The desorption potential and solubility factor of the pollutants limits the overall success of the method.
- The measured electrical conductivity of the polluted soil can change frequently due to the existence of insulating materials or metallic contaminants buried under the soil.
- This method is always more effective in case of clays because the particles of the clay carry negative charges.

7.8 RHIZOFILTRATION

The process of rhizofiltration is the utilization of the roots of the plants to absorb, precipitate, and concentrate the hazardous toxic materials of specific radionuclides and metallic contaminants from the aqueous solutions. This method is very efficient, where the new wetlands are formed and all the polluted water is allowed to pass within the roots of the plants of that area. The Pb, Cr(III), U, As(V) are absorbed strongly through the roots.

The plant roots are able to absorb huge amounts of Pb, Cr from the soil or polluted water, which passes through the root zone of maximum growing area of vegetation. This technology is generally carried out by utilizing suitable plants having healthy root system, where the polluted water is absorbed by the plants. Then after absorbing the pollutants by the roots up to saturation, the plants shift it to another location to remove the contaminants. It is the kind of phytoremediation process, where hydroponically cultured plant roots are used for remediating polluted water by the process of absorption, precipitation, and concentration.

Rhizofiltration and phytoextraction both are similar and follow the same fundamental path of remediation. In the former, once the healthy root systems of the plant species absorb the pollutants present in the soil or water and then are stored in the form of biomass, afterwards transfer the biomass to the stems and leaves. The plant species continue to absorb the pollutants till it is saturated or harvested. The key difference between the phytoextraction and rhizofiltration is that rhizofiltration is applied for the treatment of contaminants in normal aquatic environments, whereas phytoextraction contracts with the remediation of soil.

7.8.1 APPLICATIONS

Some suitable plant species are used in rhizofiltration for removing specific toxic metals including Cu^{2+}, Cd^{2+}, $Cr^{6+,}$ Ni^{2+}, Pb^{2+} and Zn^{2+} from the contaminated water with a long time period for the healthy growth of the root system. The hazardous radioactive materials in the lower level of concentration can also be removed with this method. Rhizofiltration may also be applied for the remediation of both surface and groundwater, effluents from residential and industrial areas, storm waters, agricultural runoffs, acid mine drainage effluents, radionuclides, dilute form of sludge.

7.8.2 DRAWBACKS

- Rhizofiltration is specifically suitable in settings where only a low concentration of the contaminants and high volume of water is available [40, 71].
- The plant species which are effective for translocating HMs to the shoots cannot be used for the method of rhizofiltration due to formation of more plant residues [41, 71].

7.9 SMART BIOREMEDIATION TECHNIQUE

The capability of one kind of microorganisms or group of microorganisms to remediate organic toxic pollutants into harmless, safe materials is considered as smart bioremediation, which has been explored through the biological method of treatment. It is now focusing on the extensive research in the laboratory and belongs to *in situ* process. The microbial community and their enzymes recognize the complex behavior of the metabolites and recruit the apparatus of different stages of degradation. The smart technology of bioremediation includes metagenomics and genomics and appropriately applied for finding the new kind of enzymes from the microbes [35].

7.10 GREEN TECHNOLOGY BIOREMEDIATION

The bioremediation of pollutants by using green technology has a vital significance in the field of biodegradation because it provides some basic concepts and recent information regarding the research output of bioremediation on the basis of green technologies in order to control the pollution load on the environment.

For fighting against the massive environmental complications, this method is considered as a demanding technology in the practical field of sustainable bioremediation. The modern technology of green method of bioremediation includes: the process of genomics, proteomics, and metagenomics. The eco-friendly green technology of bioremediation also includes innovative biochemical and molecular methods of detoxification of contaminants, ecological innovation in bioremediation process, microbial remediation of pollutants, bioleaching, different phytoremediation techniques, novel, and innovative techniques for the remediation of different toxic and hazardous wastes [2].

7.11 AEROBIC BIOREMEDIATION TECHNOLOGIES

The aerobic approach of bioremediation broadly applies for the treatment of a wide range of the oxidized pollutants or toxins, such as chlorinated olefins, chlorinated paraffin (chlorinated ethenes, chloromethanes (CM), and chlorinated ethane), number of chlorinated cyclic hydrocarbons, perchlorate compounds, explosives such as nitrate compounds, RDX, TNT.

In the anaerobic process of bioremediation, the microbial community finally metabolizes and converts the various kinds of complex forms of organic wastes into CH_4 along with less quantities of CO_2 and H_2 gas. In this process, the bacteria consume energy due to degradation of wastes by anaerobic reactions and grow themselves and molecules of the pollutants into non-toxic gaseous species. There are other processes of anaerobic metabolism, which includes a number of methods, such as fermentation, reductive de-chlorination, methanogenesis, denitrification, Fe, and SO_4^{2-} reducing activities.

During the process of anaerobic metabolism, O_2 and electron acceptor are able to replace the hazardous pollutants, including NO_3^-, SO_4^{2-}, CO_2, different oxidized metals, many toxic organic compounds including chlorinated hydrocarbons. H_2 serves as a source for the fermentation of organic wastes.

The chlorinated solvents (such as tetrachloroethene (PCE), 1,1,1-trichloroethane (TCA), carbon tetrachloride (CT), trichloroethene (TCE), dichloroethene (DCE), chloroform (CF), vinyl chloride (VC), chloroethane (CE), CM, dichloroethane (DCA)) and methylene chloride function as electron acceptors; whereas H_2 acts as the electron donor. During the process of anaerobic bioremediation, chloride (Cl^-) ions are periodically released along with the degradation of the chlorinated substances. In this reaction, the electron donor and H_2 is oxidized, whereas the chlorinated compounds and electron acceptors are reduced. The anaerobic method of reductive dechlorination of the comparatively more oxidized form of chlorinated hydrocarbons (TCE and PCE) is generally carried out more effectively compared to the chlorinated hydrocarbons [10, 33].

7.12 METAGENOMICS

The process of metagenomics involves the combined effect of biosynthetic technology and the genomes of soil microflora during the analysis of cloning the metagenome. This process of bioremediation is the low-cost, eco-friendly, and most popular method for the decontamination of polluted environs. Nowadays in the genomic age, the process of metagenomic method has been developed and modified for removal and neutralization of various toxic pollutants. The process of metagenomics is considered as a planned novel method, which analyzes the microorganisms at the genomic level. It provides us the indication regarding the microbe estimation of the "Uncultured Microbiota." Microbes are the vital alternative for separating the hazardous pollutants from our globe. The metagenomics consists of sample collection, screening, and finally finished with the analysis of metagenomic data related to bioremediation. The process of metagenomics can be successfully useful for the bioremediation of polluted water, soil, and air [16, 68].

7.13 ROLE OF BIOFILMS IN BIOREMEDIATION

Biofilm is the accumulation of microorganisms containing various kinds of bacterial species attached with the inert or biological surface and enclosed in an automatically created matrix containing proteins, water, extracellular

DNA, and carbohydrates. It was found that various kinds of microorganisms existing in groups of biofilms have individual capability of metabolic degradation or collective capability of remediation in the metabolic pathway.

The biofilm formed by the bacterial community are modified because of its survival and growth for suitable bioremediation in contest with O_2 and nutrients, and the biofilms possess high tolerance limit in the tough or drastic EC during the various stages of bioremediation. The biofilm-based method of remediation is eco-friendly and cost-effective to eliminate the pollutants from the biosphere. The application of biofilm in the process of bioremediation is able to absorb, degrade, and immobilize different environmental pollutants. The prepared biofilm formed by the bacteria can survive in the indigenous populations surrounding the massive polluted sites, persist, and grows for a longer period of time with proper controlling and management in drastic or tough environmental conditions as well.

The natural biofilms are able to degrade and remove the contaminants from rivers and soil. The biofilm created at the water surface contains bacteria, algae, protozoa, and fungi. The natural biofilms are a vital part of the food web and food chain, where the destructive grazers or protozoa may feed on biofilm material. The use of biofilm is constantly increasing for estimating and monitoring the pollution of the river water and stream due to toxic heavy metallic contaminants [19, 60].

7.14 ROLE OF NANOTECHNOLOGY IN BIOREMEDIATION METHOD

This is a potential method of enhancing the stability, reusability, efficiency, and longevity of the microbes for contact between the microbes and the NPs of magnetic iron. It was observed that due to the attachment of the microbes (enzymes) with the NPs of magnetic iron, the pollutants can be easily separated because of the magnetic field. In the recent years, nano-biotechnology has significantly contributed in the remediation procedure of the impurities present in the soil, air, and water in an eco-friendly manner. The feature of this process depends upon the following three facts:

- Remediation and treatment of the pollutants;
- Detection and sensing; and
- Prevention of the pollution.

The different hazardous materials including HMs, insecticides, herbicides, and pesticides, organic and inorganic toxins from the contaminated environment can be degraded more effectively by the application of suitable nanoparticles (NPs). It is now a promising approach of controlling the pollution load in air, soil, and water either in the direct method or in an integrated process of using NPs. Examples are herbicides such as molinate, atrazine, and pesticides like chlorpyrifos are remediated directly by applying nZVI (Zerovalent nano ions).

The combination of different NPs is more effective for the remediation of organic wastes and also phytoremediation. The normal, cyclic, and branched-chain hydrocarbons, petroleum-based hydrocarbon products, and chlorinated hydrocarbons, which are highly persistent to bioremediation, can be broken down easily by using NPs arbitrated reaction.

7.14.1 REASONS FOR USING NANOMATERIAL IN BIOREMEDIATION

- Shape and size of the non-metallic and metallic nanomaterials are used to remove the pollution load on the environment. Examples are the single metal NPs, carbon-based nanomaterials, bimetallic NPs. Also following forms are used in bioremediation more effectively due to: (i) The NPs are penetrated or diffused inside the zone of the contaminants uniformly and easily at every point, whereas the microparticles never reached at every point. (ii) NPs possess higher efficiency of reactivity to open redox pollutants.
- The nanomaterial exhibits quantum effect; hence less energy of activation is needed for the feasibility of the chemical reaction with the wastes.
- The surface plasmon resonance is considered as additional phenomenon, which is exhibited by the NPs and successfully used for the identification of toxic and hazardous materials.
- When the materials are converted into the nanoscale, the surface area of the material per unit mass increases many times and hence a huge quantity of materials is coming in contact with the surrounding polluted materials and increases the reactivity to a greater extent.
- It was found that oxide-coated Fe NPs forms a weak cover complex with pollutants such as CT and the oxide coating is the cause of drastically increasing the reactivity through the process of electron

transfer and break down the contaminants into CO_2, CH_4, whereas the toxic chemicals like benzoquinone, TCE, and also other chlorinated compounds undergo cleavage into simple chemical species of lower toxicity.

- TiO_2 nanotube NPs can be used for remediating pentachlorophenol through the way of photoelectrocatalytic reaction. Single metal NPs may also be used in the form of biocatalysts for de-chlorination. NPs can also be applied for immobilizing the microbial cells, which can degrade or recover some definite chemicals.

7.14.2 NANOBIOREMEDIATION

The nanobioremediation is the combination of Nano + Bio + Remediation. This means the use of nanotechnology and NPs to increase the action of microbial community for accelerating the method of bioremediation. The process of nanobioremediation possesses a high potential ability to clean up the polluted site in low cost and less time. Iron NPs, nanosorbents, zerovalent NPs of Iron-oxide, bimetallic NPs can effectively activate microbes and increases the remediation of many toxic, hazardous materials, which are normally resistant to remediation in the traditional process. The nanobioremediation is used for the treatment of land, soil, water, and solid wastes and is able to distinguish the pollutants [13, 36, 53].

7.15 SUMMARY

The bioremediation technique has been established as an effective method for remediation of any hazardous materials, such as metallic contaminants, petroleum-based hydrocarbons, normal hydrocarbons, chlorinated compounds, explosives, PAHs, VOCs, complex form of organic materials, pesticides, aromatic nitro-compounds, herbicides, and other contaminants. The unique characteristics of bioremediation are its application in the diversified fields in the state of liquids, solids, and mixture of both in the *ex-situ* and *in situ* methods. The site characterization, proper monitoring and regulating, system design, operation, and implementation factors play an energetic role in the performance of bioremediation. The use of GEMs increases the efficiency or activity of microorganisms and drastically accelerates the rate of bioremediation. The EK method of bioremediation is an advanced method of bioremediation for removing ionic pollutants by using DC current.

KEYWORDS

- **bioremediation techniques**
- **contaminants**
- ***ex-situ***
- ***in-situ***
- **microorganisms**
- **pollutants**

REFERENCES

1. Abatenh, E., Gizaw, B., Tsegaye, Z., & Wassie, M., (2017). Application of microorganisms in bioremediation-review. *Journal of Environmental Microbiology, 1*(1), 2–9.
2. Akomeah, S. A., Fei-Baffoe, B., Belford, E. J. D., & Osei-Fosu, P., (2018). Biostimulation of immobilized microbes in a fixed-bed bioreactor towards hydrocarbon-contaminated water remediation using compost tea and eggshell powder. *American Journal of Environmental Science & Technology, 2*(1), 1–17.
3. Al-Hawash, A. B., Dragh, M. A., Li, S., Alhujaily, A., Abbood, H. A., Zhang, X., & Ma, F., (2018). Principles of microbial degradation of petroleum hydrocarbons in the environment. *Egyptian Journal of Aquatic Research, 44*(2), 71–76.
4. Alikasturi, A. S., Amzar, M. Z., & Kamil, M., (2019). Phytoremediation of copper in mineral, distilled and surface water using *Limnocharis flava* plant. *Materials Today: Proceedings, 19*, 1489–1496.
5. Ameen, F. A., Hamdan, A. M., & El-Naggar, M. Y., (2020). Assessment of the heavy metal bioremediation efficiency of the novel marine lactic acid bacterium, *Lactobacillus plantarum* MF042018. *Scientific Reports, 10*(1), 314–325.
6. Ang, E. L., Zhao, H., & Obbard, J. P., (2005). Recent advances in the bioremediation of persistent organic pollutants via bimolecular engineering. *Enzyme and Microbial Technology, 37*(5), 487–496.
7. Arora, N. K., (2018). Bioremediation: A green approach for restoration of polluted ecosystems. *Environmental Sustainability, 1*, 305–307.
8. Asha, A. J., Singh, S. K., & Mudhoo, A., (2010). A comprehensive overview of elements in bioremediation: A comprehensive overview of elements in bioremediation. *Reviews in Environmental Science and Biotechnology, 9*, 215–288.
9. Azubuike, C. C., Chikere, C. B., & Okpokwasili, G. C., (2016). Bioremediation techniques-classification based on site of application: Principles, advantages, limitations and prospects. *World Journal of Microbiol Biotechnol, 32*, 180–198.
10. Balagurusamy, N., (2005). Anaerobic bioremediation-an untapped potential. *Revista Mexicana de Ingeniería Química* (*Mexican J. of Chemical Engineering*), *4*(3), 273–287.
11. Beškoski, V. P., Cvijovic, G. G., Milic, J., Ilic, M., Miletic, S., Šolevic, T., & Vrvic, M. M., (2011). Ex-situ bioremediation of a soil contaminated by mazut (Heavy residual fuel oil) – A field experiment. *Chemosphere, 83*, 34–40.

12. Brown, D. M., Okoro, S., Gils, J. V., Spanning, R. V., Bonte, M., Hutchings, T., Linden, O., & Egbuche, U., (2017). Comparison of land farming amendments to improve bioremediation of petroleum hydrocarbons in niger delta soils. *Science of the Total Environment, 596*, 284–292.
13. Cecchin, I., Reddy, K. R., Thom, A., Tessaro, E. F., & Schnaid, F., (2017). Nanobioremediation: Integration of nanoparticles and bioremediation for sustainable remediation of chlorinated organic contaminants in soils. *International Biodeterioration & Biodegradation, 119*, 419–428.
14. Chandra, S., Sharma, R., Singh, K., & Sharma, A., (2013). application of bioremediation technology in the environment contaminated with petroleum hydrocarbon. *Ann. Microbial., 63*, 417–431. Online, doi: 10.1007/s13213-012-0543-3.
15. Chen, M., Xu, P., Zeng, G., Yang, C., Huang, D., & Zhang, J., (2015). Bioremediation of soils contaminated with polycyclic aromatic hydrocarbons, petroleum, pesticides, chlorophenols and heavy metals by composting: Applications, microbes and future research needs. *Biotechnology Advances, 33*(6), 745–755.
16. Christgen, B., Yang, Y., Ahammad, S. Z., Li, B., Rodriquez, D. C., & David, W. T. Z., (2015). Metagenomics shows that low-energy anaerobic–aerobic treatment reactors reduce antibiotic resistance gene levels from domestic wastewater. *Environmental Science & Technology, 49*(4), 2577–2584.
17. Diaz, J. M. C., Moreno, L. D., Núñez, R., Nogales, R., & Romero, E., (2014). Enhancing pesticide degradation using indigenous microorganisms isolated under high pesticide load in bioremediation systems with vermicomposts. *Bioresource Technology, 214*, 234–241.
18. Dixit, R., Malaviya, D., Pandiyan, K., Singh, U. B., Sahu, A., Shukla, R., & Paul, D., (2015). Bioremediation of heavy metals from soil and aquatic environment: An overview of principles and criteria of fundamental processes. *Sustainability, 7*(2), 2189–2212.
19. Edwards, S. J., & Kjellerup, B. V., (2013). Applications of biofilms in bioremediation and biotransformation of persistent organic pollutants, pharmaceuticals/personal care products, and heavy metals. *Applied Microbial Biotechnology, 97*, 9909–9921.
20. Elekwachi, C. O., Andresen, J., & Hodgman, T. C., (2014). Global use of bioremediation technologies for decontamination of ecosystems. *Journal of Bioremediation & Bio Degradation, 5*(4), 225–234.
21. Frutos, F. J. G., Escolano, O., García, S., Babín, M., & Fernández, M. D., (2010). Bioventing remediation and ecotoxicity evaluation of phenanthrene-contaminated soil. *Journal of Hazardous Materials, 183*(1–3), 806–813.
22. Germaine, K. J., Liu, J. B., Xuemei, J. K., Culhane, J., Lally, R. D., Kiwanuka, S., David, R., & Dowling, D. N., (2015). Ecopiling: Combined phytoremediation and passive biopiling system for remediating hydrocarbon impacted soils at field scale. *Frontiers in Plant Science, 5*, 756–762.
23. Gill, R. T., Rebottle, M. J., Smith, J. W. N., & Thornton, S. F., (2014). Electrokinetic-enhanced bioremediation of organic contaminants: A review of processes and environmental applications. *Chemosphere, 107*, 31–42.
24. Hashim, M. A., Mukhopadhyay, S., Sahu, J. N., & Sengupta, B., (2011). Remediation technologies for heavy metal contaminated groundwater. *Journal of Environmental Management, 92*, 2355–2388.

25. Hassan, I., Mohamedel-hassan, E., Yanful, E. K., & Yuan, Z. C., (2016). Electrokinetic bioremediation current knowledge and new prospects. *Advances in Microbiology, 6*, 57–72.
26. Iain, M. M., & Gillespie, J. C. P., (2013). Bioremediation, an environmental remediation technology for the bioeconomy. *Trends in Biotechnology, 31*(6), 329–332.
27. Janssen, D. B., & Stuck, G., (2020). Perspectives of genetically engineered microbes for groundwater bioremediation. *Environmental Science Processes & Impacts, 22*, 487–499.
28. Jin, Y., Luan, Y., Ning, Y., & Wang, L., (2018). Effects and mechanisms of microbial remediation of heavy metals in soil: Critical review. *Applied Sciences, 8*, 1336–1343.
29. Johnson, P. C., Richard, L., Johnson, C. L., & Bruce, A. L., (2001). Advances in *in situ* air sparging/biosparging. *Bioremediation Journal, 5*(4), 251–266.
30. Kadri, T., Rouissi, T., Brar, S. K., Cledon, M., Sarma, S., & Verma, M., (2017). Biodegradation of polycyclic aromatic hydrocarbons (PAHs) by fungal enzymes: A review. *Journal of Environmental Sciences, 29*, 52–74.
31. Kathiravan, M. N., Karthick, R., & Muthukumar, K., (2011). *Ex-situ* bioremediation of Cr(VI) contaminated soil by *Bacillus* sp.: Batch and continuous studies. *Chemical Engineering Journal, 169*, 107–115.
32. Kensa, V. M., (2011). Bioremediation: An overview. *Journal of Industrial Pollution Control, 27*(2), 161–168.
33. Kochany, E. L., & Substances, H., (2018). Their microbial interactions and effects on biological transformations of organic pollutants in water and soil: A review. *Chemosphere, 202*, 420–437.
34. Kuhn, K. T., Hamonts, K., Dijk, J. A., Kalka, H., Stichler, W., Springael, D., Dejonghe, W., & Meckenstock, R. U., (2009). Assessment of the intrinsic bioremediation capacity of eutrophic river sediment polluted by discharging chlorinated aliphatic hydrocarbons: A compound-specific isotope approach. *Environmental Science and Technology, 43*(14), 5263–5269.
35. Kumar, L., & Bharadvaja, N., (2019). Enzymatic bioremediation: A smart tool to fight environmental pollutants. Chapter 6. In: Bhat, P., (ed.), *Smart Bioremediation Technologies* (pp. 99–118). London: Academic Press.
36. Kumari, B., & Singh, D. P., (2016). A review on multifaceted application of nanoparticles in the field of bioremediation of petroleum hydrocarbons. *Ecological Engineering, 97*, 98–105.
37. Latha, P. A., & Reddy, S. S., (2013). Review on bioremediation – potential tool for removing environmental pollution. *International Journal of Basic and Applied Chemical Sciences, 3*(3), 21–33.
38. Ławniczak, Ł., Karczewska, M. W., Loibner, A. P., Heipieper, H. J., & Chrzanowski, Ł., (2020). Microbial degradation of hydrocarbons—basic principles for bioremediation. *Molecules, 25*(4), 856–875.
39. Leal, A. J., Rodrigues, E. M., & Leal, P. L., (2017). Changes in the microbial community during bioremediation of gasoline-contaminated soil. *Brazilian Journal of Microbiology, 48*, 342–351.
40. Lee, J. H., (2013). Overview of phytoremediation as a potentially promising technology for environmental pollution control. *Biotechnology and Bioprocess Engineering, 18*, 431–439.

41. Lee, M., & Yang, M., (2010). Rhizofiltration using sunflower (*Helianthus annuus* I.) and bean (*Phaseolus vulgaris*) to remediate uranium-contaminated groundwater. *Journal of Hazardous Materials, 173*, 589–596.
42. Liphadzi, M. S., Kirkham, M. B., & Musi, C. F., (2005). Phytoremediation of soil contaminated with heavy metals: Technology for rehabilitation of the environment. *South African Journal of Botany, 71*(1), 24–37.
43. Liu, S. H., Zeng, G. M., Niu, Q. Y., Liu, Y., Zhou, L., Jiang, L. H., Tan, X., et al., (2017). Bioremediation mechanisms of combined pollution of PAHs and heavy metals by bacteria and fungi: A MINI REVIEW. *Bioresource Technology, 224*, 25–33.
44. Lu, C., Hong, Y., Liu, J., Gao, Y., Ma, Z., Yang, B., Ling, W., & Waigi, M. G., (2019). PAH-degrading bacterial community enriched with contaminated agricultural soil and its utility for microbial bioremediation. *Environmental Pollution, 251*, 773–782.
45. Lukić, B., Panico, A., Huguenot, D., Fabbricino, M., Hullebusch, Eric, D. V., Esposito, G., (2017). A review on the efficiency of land farming integrated with composting as a soil remediation treatment. *Environmental Technology Review, 6*(1), 97–110.
46. Maitra, S., (2018). In situ bioremediation: An overview. *Research Journal of Life Sciences, Bioinformatics, Pharmaceutical and Chemical Sciences, 4*(6), 576–598.
47. Mallavarapu, A., & Mitchison, T., (1999). Regulated actin cytoskeleton assembly at filopodium tips controls their extension and retraction. *Journal of Cell Biology, 146*(5), 1097–1106.
48. Maurer, M., & Rittmann, B. E., (2004). Modeling intrinsic bioremediation for interpret observable biogeochemical footprints of BTEX biodegradation: The need for fermentation and abiotic chemical processes. *Biodegradation, 15*(6), 405–417.
49. Mena, E., Ruiz, C., Villasenor, J., Manuel, A., & Rodrigo, P. C., (2015). Biological permeable reactive barriers coupled with electrokinetic soil flushing for the treatment of diesel-polluted clay soil. *Journal of Hazardous Materials, 283*, 131–139.
50. Moghadam, M. J., Moayedi, H., Sadeghi, M. M., & Hajiannia, A., (2016). Review of combinations of electrokinetic applications. *Environ. Geochem. Health, 38*, 1217–1227.
51. Mrozika, A., & Piotrowska-Seget, Z., (2010). Bioaugmentation as a strategy for cleaning up soils contaminated with aromatic compounds. *Microbiological Research, 165*, 363–375.
52. Ojuederie, O. B., & Babalola, O. O., (2017). Microbial and plant-assisted bioremediation of heavy metal polluted environments: Review. *International Journal of Environmental Research and Public Health, 14*(12), 1504–1521.
53. Patil, S. S., Shedbalkar, U. U., Truskewycz, A., Chopade, B. A., & Ball, A. S., (2016). Nanoparticles for environmental clean-up: A review of potential risks and emerging solutions. *Environmental Technology & Innovation, 5*, 10–21.
54. Paul, B. H., & David, R. L., (2019). Field demonstration of N-nitrosodimethylamine (NDMA) treatment in groundwater using propane biosparging. *Water Research, 164*, 1–10.
55. Perelo, L. W., (2010). Review: *In situ* and bioremediation of organic pollutants in aquatic sediments. *Journal of Hazardous Materials, 177*, 81–89.
56. Perpetuo, E. A., Souza, C. B., & Nascimento, C. A. O., (2011). Engineering bacteria for bioremediation. Chapter 28. In: Carpi, A., (ed.), *Progress in Molecular and Environmental Bioengineering – From Analysis and Modelling to Technology Applications* (pp. 606–631). intechopen.

57. Quintella, C. M., Mata, A. M. T., & Lima, L. C. P., (2019). Overview of bioremediation with technology assessment and emphasis on fungal bioremediation of oil-contaminated soils. *Journal of Environmental Management, 241*, 156–166.
58. Robles-González, I. V., Fava, F., & Poggi-Varaldo, H. M., (2008). Slurry bioreactors for bioremediation of soils and sediments. *Microbial Cell Factories, 7*(5), 1–16.
59. Romeroa, M. B., Monterrosoc, C., Kiddd, P. S., Lu-Chaua, T. A., Gkorezise, P., Vangronsvelde, J., & Casares, J. J., (2019). Modeling the *ex-situ* bioremediation of diesel-contaminated soil in a slurry bioreactor using a hydrocarbon-degrading inoculant. *Journal of Environmental Management, 246*, 480–486.
60. Sharma, A., Jamalia, H., Vaishnavb, A., Giric, B. S., & Srivastava, A. K., (2019). Microbial biofilm: An advanced eco-friendly approach for bioremediation. Chapter 15. In: Yadav, M. K., & Singh, B. P., (eds.), *New and Future Developments in Microbial Biotechnology and Bioengineering: Microbial Biofilms* (pp. 205–219). London: Elsevier.
61. Shishir, T. A., Mahbub, N., & Kamal, N. E., (2019). Review on bioremediation: A tool to resurrect the polluted rivers. *Pollution, 5*(3), 555–568.
62. Sprocati, A. R., Tasso, C. A. F., Sciullo, P. M., Pinto, A. V., Ubaldi, S. C., & Cremisini, C., (2012). Effectiveness of a microbial formula as a bioaugmentation agent: Tailored for bioremediation of diesel oil and heavy metal co-contaminated soil. *Process Biochemistry, 47*(11), 1649–1655.
63. Svenning, M. M., Wartiainen, I., Hestnes, A. G., & Binnerup, S. J., (2003). Isolation of methane-oxidizing bacteria from soil by use of a soil substrate membrane system. *FEMS Microbiology Ecology, 44*(3), 347–354.
64. Urgun-Demirtas, M., Stark, B., & Pagilla, K. K., (2006). Use of genetically engineered microorganisms (GEMs) for the bioremediation of contaminants. *Critical Reviews in Biotechnology, 26*(3), 145–164.
65. Verma, J. P., & Jaiswal, D. K., (2016). Advances in biodegradation and bioremediation of industrial waste. *Front. Microbiology, 6*, 1555–1557.
66. Verma, S., & Kuila, A., (2019). Bioremediation of heavy metals by microbial process. *Environmental Technology & Innovation, 14*, 1–11.
67. Wan, X., Lei, M., & Chen, T., (2020). Review on remediation technologies for arsenic-contaminated soil. *Frontiers of Environmental Science & Engineering, 14*(2), 24–38.
68. Wiseschart, A., Mhuantong, W., Tangphatsornruang, S., Chantasingh, D., & Pootanakit, K., (2019). Shotgun metagenomic sequencing from manao-pee cave, Thailand, reveals insight into the microbial community structure and its metabolic potential. *BMC Microbiol, 19*, 144–150.
69. Wolejko, E. W., Wydro, U., & Łoboda, T., (2016). The ways to increase efficiency of soil bioremediation. *Ecological Chemistry and Engineering, 23*(1), 155–174.
70. Xia, S., Song, Z., Jeyakumar, P., Sabry, M., Rinklebe, S. J., Sik, Y., Bolan, N., & Wang, H., (2019). Review on bioremediation technologies for Cr(VI)-contaminated soils and wastewater. *Critical Reviews in Environmental Science and Technology, 37*(3), 1–52.
71. Yadav, B. K., Siebel, M. A., Johan, J. A., & Bruggen, V., (2011). Rhizofiltration of a heavy metal (Lead) containing wastewater using the wetland plant *Carex pendula. Clean – Soil, Air, Water, 39*(5), 467–474.
72. Yazdi, L. L., Jinb, H. S., Zuoc, Y., & Paul, H., (2014). Enhanced bioremediation of hydrocarbon-contaminated soil using pilot-scale bioelectrochemical systems. *Journal of Hazardous Materials, 274*, 8–15.

73. Yeha, C. H., Linb, C. W., & Wu, C., (2010). permeable reactive barrier for the bioremediation of BTEX-contaminated groundwater: Microbial community distribution and removal efficiencies. *Journal of Hazardous Materials, 178*, (1–3), 74–80.

CHAPTER 8

SUSTAINABLE AGROECOSYSTEMS: RECENT TRENDS AND APPROACHES IN PHYTOREMEDIATION AND RHIZOREMEDIATION

WAJIHA ANUM, UMAIR RIAZ, GHULAM MURTAZA, MUHAMMAD UMAIR RAZA, and ASAD-UR-RAHMAN

ABSTRACT

The exposure to toxic heavy metals (HMs), organic, and inorganic pollutants has created tremendous health risks and environmental menace due to their non-degradable and noxious nature. The phytoextraction, phytosequestration, and phytovolatization have gained popularity because of their effectiveness in reclaiming agroecosystems. Phytoremediation uses green plants that degrade, stabilize, absorb, or accumulate the toxins. Rhizoremediation can clean the environment by enhancing the ability of plant to combat toxins/HMs pollution by efficient synchronic mechanisms among the plants and microbes present in the soil. Both these processes act as natural and sustainable strategies to combat the consequences of degradation without any use of synthetic chemicals.

8.1 INTRODUCTION

The prevailing issues of contaminants (such as heavy metals (HMs) and carbon dioxide (CO_2)) have created challenges both at local and global scale. The non-synthetic amendments are pre-requisites in the current era. Phytoremediation and rhizoremediation are cost-effective and eco-friendly strategies that can utilize plants and microbes for eradicating contaminants from the environment. The basic issues of agriculture are decrease in soil fertility, land degradation, limited use of organic fertilizers, soil erosion, competition for biomass and nutrient mining [3].

Agroecosystems are natural ecosystems that are used for food production and are characterized by many species and their interactions; however, the basic management is done on the productive land patches for farming. As the world is facing food security problems along with many environmental issues, thus the sustainability of the agroecosystems is at the edge of destruction. To overcome the severe threats posed by climate change, it is necessary to implement those practices that do not increase the environmental contamination.

In agroecosystems, slight mismanagement can lead to degradation of the resources (such as soil, water, genetic diversity) and disruption in species interaction. In such cases, the use of green technology is an alternative for maintaining the productivity of any land. The agricultural land utilizes pesticides for avoiding attacks by pests and diseases, losses caused by weeds and diseases; however, the pesticide residue is yet another form of pollution that not only decrease soil fertility but also remain persistent for longer periods [92]. Therefore, phytoremediation of pesticide-polluted soils and the role of microbes for degrading pesticides needs to be understood and implemented accordingly.

The methods previously common for remediation are landfill, acid leaching, thermal treatment that are not suited now because of high cost, decrease soil fertility and destruction of soil structure and contamination types. Therefore, a shift towards phytoremediation is foremostly required. Moreover, the combination of phytoremediation plants along with microorganisms ensures rapid and efficient toxin removal from the environment.

This chapter reviews the concepts of phytoremediation and rhizoremediation, their mode of action, recent trends and practices for enhancing soil health and their suitability in adaptation. The microorganisms, their interaction with plants and conversion of inorganic pollutants into more accessible organic compounds for easy uptake of plants are also addressed.

8.2 PREVAILING ISSUES IN SUSTAINABILITY OF AGROECOSYSTEMS

Sustainable management of agroecosystems is a challenge in the modern agriculture. The unsustainability is due to the over-emphasis on yield production, heavy use of pesticides, use of inappropriate and manipulated methods for crop management and agricultural production [35]. The improved living standards of the increasing population demands the provision of food and fiber on a sustainable basis; thus, the development of agroecosystem is necessary even in less favorable regions [76].

As agroecosystems comprise of genetically depauperate plants grown over large spatial scales, hence it leads to the gradual change in species overtime on a regional scale, which develops a factor in facilitating emergence and propagation of new pathogens, which turn out to be a great threat to the agroecosystems [53]. The soil and water contaminations are the two threats that need to be looked at immediately. Incidentally, modern, and intensive farming can destroy the remaining biota. In this regard, there is a need to shift towards phytoremediation and rhizoremediation for improving the soil structure, soil fertility, removal of contaminants, and overall performance of agroecosystems on a sustainable basis.

8.3 MICROBES FOR SUSTAINABLE AGROECOSYSTEMS

knowing the plant microbiome and its functions, and therefore modifying our management processes to optimize the most important microbiome participants, is probably the most reasonable field for potential research on sustainable agroecosystems [22, 27, 59]. Rhizosphere and spermosphere are important regions in the soil zone. Spermosphere is connected to the exterior layers of seed, which are in contact with the soil and with which microbes can associate before germination [12, 46].

Photosynthetic bacteria are endogenous microorganisms that can harvest energy from the solar system and turn the flux from plant roots, carbon-based material, and toxic gasses into usable substances, such as amino acids, carbohydrates, nucleic acids, and numerous other metabolites [19, 77]. Economically useful microbes use atmospheric nitrogen for improved crop growth, putrefy organic wastes and pesticides, detoxify toxins, mitigate plant disease, increase the supply of plant nutrients, and create bioactive substances, mainly active enzymes, nutrients, and organic acids, which boost soil fertility.

Soil and microorganisms are microbe-friendly ecosystems and are inhabited by a vast array of organisms, including plants, viruses, fungi, bacteria, protozoa, etc. [66, 89]. Microorganisms are essential in the soil surfaces to increase the soil fertility, circulate nutrient materials in the ecosystem, and recycle substances, such as growth hormones, antioxidants, inhibitors, minerals, etc. [23, 77].

Also, native microbial rhizospheric occupants of farming soils are highly affected by the farming activities (e.g.: soil feeding, time, stubble cultivation, burning, etc.), crop genes, and soil types [59]. Different studies have reported that crop exudates can induce variation in soil specificity, such as texture, buffering potential, and the abundance of organic material, influencing the multitude and activity of microbial species. Microbes and the microbial process have complex role in mobilizing and transferring nitrogen, phosphorus, and sulfur and biological residue decay [13]. Changes in the congregation of microbes can help to identify improvements in the physicochemical behavior of the soil and hence offer an early sign of soil development or early warning of soil deterioration [79].

Fostering soil fertility by regulating microbiomes can have ample capacity to endorse the beneficial features of a crop species, thus reducing the usage of artificial pesticides or fertilizers. It has been well established to use microbes as bio-fertilizers and biopesticides in agriculture [42, 59, 90]. It is expected that the use of microbial populations in the agricultural production line will boost the productivity of plants by supplying the crop with nutrients and other growth-promoting compounds not just for a limited period but also for several seasons, as this sustainable method preserves soil quality and fitness [88, 89].

8.4 PHYTOEXTRACTION/PHYTOABSORPTION FOR SUSTAINABLE SOIL

In phytoremediation technique, plants are used for confiscating the pollutants/contaminants through the process of extraction, immobilization, degradation, and containments from soil, water, and air [29]. It is eco-friendly and solar-powered process [77] depending on the ability of plant to sequestrate, translocate, stabilize soil, and water pollutants and accumulate contaminants [61]. As these methods are solely dependent on the plants, therefore do not require any chemical alteration in the environment. Without affecting the soil biota, phytoremediation, phytoextraction, and rhizoremediation are eco-friendly and low-cost input methods [11, 63]. Phytoremediation takes place both for organic and inorganic contaminants (Figure 8.1).

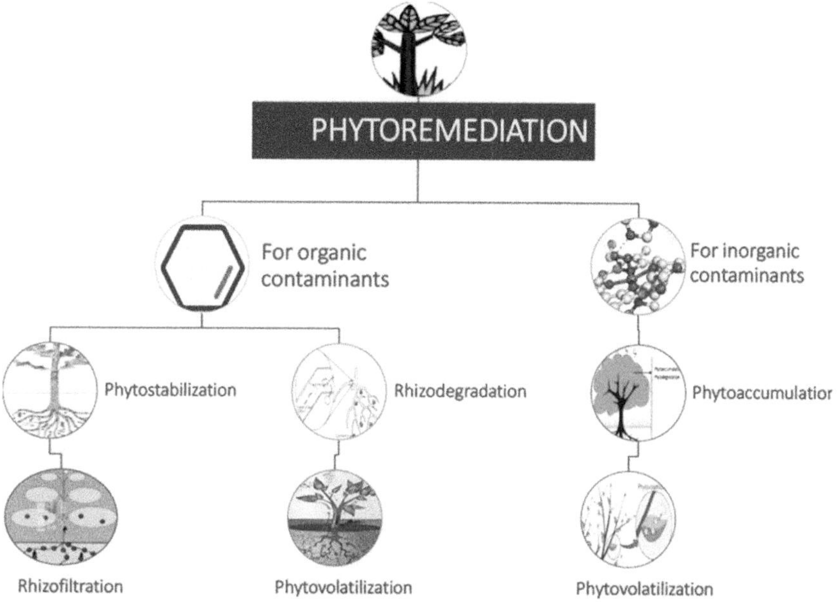

FIGURE 8.1 Brief description of phytoremediation.

Phytoextraction process (also known as phytoabsorption) is the elimination of toxins, such as HMs from the soil/water/environment by up taking and translocating of the toxic substances from soil to the harvestable parts. It is a useful technique for recovering the soils with a high concentration of HMs [12]. It can remove hydrocarbons, chlorinated substances, HMs, explosive, and their components, radionuclides, pesticides, and excess polycyclic aromatic hydrocarbons (PAHs), petroleum hydrocarbons (PHC), solvents like trichloroethylene (TCE), metals, and metalloids, munitions waste like 2,4,6-trinitrotoluene (TNT), salts (NaCl) and radioisotopes [29]. Moreover, the use of advanced chemical technology is also implemented nowadays for the uptake of contaminants and translocating them into different plant parts [43, 86]. It uses the plant species having the higher competence to accumulate HMs, huge biomass, and capability of storing toxins in the above-ground parts.

Phytoextraction process is also known as phytoaccumulation because the contaminants are "accumulated" in the above-ground parts of the plants. Hyper-accumulating plant species have a high potential for accumulating high amounts of metal contaminants in their foliage [25, 66], and they are commonly used in the reclamation of metal-polluted soils [65].

Hyper-accumulating plants have 100 times more ability to absorb metals in comparison to non-accumulating plants [34]. This process is also enhanced by the use of certain chemicals [34], such as ethylene glycol tetraacetic acid (EGTA), sodium dodecyl sulfate (SDS), and ethylene diamine tetra-acetic acid (EDTA). Among them, EDTA is the most studied and used chemical amendment for metal phytoextraction in contaminated soils. New farming skills generated by implementing the technology will be directly applicable to fruits, fiber, and forestry products on areas close to mining operations.

8.4.1 UPTAKE OF POTENTIALLY TOXIC ELEMENTS (PTES)

Alkorta et al. [5] stated that plant factors (such as high tolerance level to Potentially Toxic Element (PTE) content, fast-growing biomass, high root growth, high biomass yield, PTE accumulation in easily harvestable organs, easily manageable crop), genetically stable characteristics and useable biomass for energy production can promise the reclamation of PTE contaminated soils. In the phytoextraction process, the potentially toxic elements (PTEs) are stored mainly into plant leaves, culms, and trunks [33], which are the main sink sites. Belowground parts of the plant, e.g., rhizomes also act as sink sites in some instances [25].

A complex network of homeostatic metabolic mechanism takes place that controls the uptake, distribution, and detoxification of HMs [1, 59]. Yoon et al. [90] explained that plant species with high translocation factor (TF) act as good candidate for translocating metals from soils into the shoots. Phytoextraction process does not alter soil structure or its fertility status [41]. Roots, in some cases, also can control the bioavailability and limit HM mobility in the soil. This process is called phytostabilization. It occurs through sorption, precipitation, and metal valance reduction (complexation) [16].

8.4.2 PLANTS SUITABLE FOR PHYTOEXTRACTION

Plant families (such as *Lamiaceae, Poaceae, Fabaceae, Caryophyllaceae, Asteraceae, Brassicaceae, Cyperaceae, Lamiaceae, Violaceae, Euphorbiaceae, Flacourtiaceae, and Cunouniaceae*) have the majority of the hyper-accumulating plants.

Few species from the *Salicaceae* family (such as willows and poplars) are very important tree species for phytoremediation. Many studies claim that birch, black locust, paulownia, and eucalypt have higher capacity to extract the metal because they have higher bioconcentration for metals. Another factor, which makes them a good candidate in phytoextraction, is their high yielding capacity.

Brassicaceae plants prefer Metals, such as Zn, Cu, and Ni for accumulation. Species with the fast-growing character, high metal tolerance and capability to remove high concentration of HMs and to translocate them to the above-ground plant parts are preferred in phytoextraction [9, 17]. TF (shoot to root ratio) >1 is also an important characteristic of such plants [89]. The disposal of harvested aboveground biomass is one of the basic limitations of phytoextraction; however, it can be decreased by biomass volume reduction through ashing or composting [28].

S. oleracea partakes in the HM accumulating behavior when grown on a contaminated matrix; and owing to this, its edibility has been questioned [78]. A study revealed that *Spinacia oleracea* could efficiently accumulate Cd and Pb in the leaves [17]. Lolium multiflorum and *Brassica juncea* are hyperaccumulator plants. In an experiment, it was revealed that shoots and roots of Indian mustard have the ability to accumulate Zn 4 to 6 times more as compared to control and ryegrass. However, Cu accumulation was higher by ryegrass shoots compared to Indian mustard shoot [55]. *Thlaspi caerulescens* is also a good candidate for absorbing Cd, Cu, Zn, and Pb [20, 44]. Selected plants for effective phytoremediation are listed in Table 8.1.

TABLE 8.1 Plants Used in Phytoremediation

Common Name	Botanical Name	Effective Against Contaminant	References
Alfalfa	*Medicago sativa*	Polycyclic aromatic hydrocarbons	[23]
Alpine pennycress	*Thlaspi caerulescens*	Heavy metals	[76]
Alpine penny grass	*Thlaspi caerulescens*	Cadmium	[72]
Bald cypress	*Taxodium distichum*	Phytodegradation of pesticide Bentazon	[19]
Barley	*Horedeum vulgare*	Arsenic	[30]
Black nightshade	*Solanum nigrum*	Cadmium (hyperaccumulator)	[24]
Black willow	*Salix nigra*	Phytodegradation of pesticide Bentazon	[19]
Bog bulrush	*Scirpus mucronatus*	Lead	[32]
Cherry bark oak	*Quercus falcate*	Phytodegradation of pesticide Bentazon	[19]

TABLE 8.1 *(Continued)*

Common Name	Botanical Name	Effective Against Contaminant	References
Chinese brake	*Pteris vittata* L.	Arsenic hyperaccumulator	[18]
Common reed	*Phragmites australis*	Copper, Lead, zinc	[89]
Corn	*Zea mays*	Cadmium, lead	[57]
Creeping rice paddy mustard	*Arabidopsis halleri*	Zinc (hyperaccumulator)	[94]
Duckweed	*Lemna minor* L.	Cadmium, selenium, and copper	[92]
Field pennycress	*Thlaspi, Brassicaceae*	Zinc (hyperaccumulator)	[85]
Giant cane	*Arundo donax* L.	Arsenic	[10]
High calamine pansy	*Viola calaminaria*	Heavy metals	[81]
Indian grass	*Sorghastrum nutans*	Agrochemical residue detoxifier (pesticides and herbicides)	[44]
Live oak	*Quercus viginiana*	Phytodegradation of pesticide Bentazon	[19]
Maruba-mannengusa	*Sedum alfredii*	Zinc, cadmium	[48]
Milkvetch	*Astragalus*	Selenium	[22]
Mosquito fern	*Azolla caroliniana*	Mercury	[11]
Mustards	*Brassica*	Zinc, copper	[63]
Poplar trees	*Populus*	Cadmium, zinc, copper, lead	[31]
River birch	*Betula nigra*	Phytodegradation of pesticide Bentazon	[19]
Schima	*Schima superba*	Manganese	[88]
Sorghum	*Sorghum bicolor* L.	Cadmium	[41]
Spurge	*Euphorbia cheiradenia*	Heavy metals	[49]
Sunflower	*Helianthus annuus*	Zinc and lead nitrates	[2]
Sweet alyssum	*Alyssum heldreichii*	Nickel	[6]
Tamarind	*Tamarindus Indica*	Chromium	[65]
Tobacco	*Nicotiana tabacum*	Cadmium	[38]
Tulip tree	*Liriodendron tulipifera*	Methyl-Hg	[4]
Water mint	*Mentha aquatica* L.	Arsenic, heavy metals	[6]
Willow	*Salix*	Cadmium, zinc, copper, lead	[80]
Yellow poplar	*Liriodendron tulipifera*	Phytodegradation of pesticide Bentazon	[19]

8.5 PHYTOSEQUESTRATION

Phytosequestration process assimilates atmospheric CO_2 by using land plants. Furthermore, the captured C is fixed in the roots of plants and enters into the soil pool (organic as well as inorganic) and is sequestrated. Dual role is performed by bioenergy crops: firstly, they provide biofuel that balances the greenhouse gas (GHG) emissions and secondly, carbon is sequestrated into the soil through their extensive root systems. Other ways of capturing carbon into the soil are the addition of biochar, wood-burial, and durable plant products. Hence, phytosequestration has remarkable potential for mitigating the effects of GHGs [40].

In terrestrial ecosystems, carbon sequestration occurs due to: phytosequetration (in plant parts) and soil carbon sequestration (in soils). It is a long-term type of storage in a terrestrial ecosystem, where carbon sequestration occurs through photosynthesis and also through live or dead organic matter, which acts as a main sink source [48]. Vegetation and microbial biomass contribute about ~99.9% of carbon biota [40]. Phytosequestration can act as a sink for capturing atmospheric carbon even for centuries, it it remained undisturbed and well managed. Deforestation is one example, which reveals the increase of atmospheric CO_2. Through photosynthesis, the inorganic carbon is converted to organic carbon and stored in plant parts (such as stems, roots, seeds, and other organs). Phytosequestration is thus the most efficient way for storing CO_2 in the soil or in other organic forms.

In agroecosystems, the incorporation of plant species with more efficient photosynthetic capacity (through natural or genetic engineering) can improve the overall functioning of the farm [40]. C4 plants, cyanobacteria, microalgae, and crassulacean acid metabolism (CAM) can fix the carbon; however, C3 plants cannot fix it [58].

8.6 PHYTOVOLATILIZATION

In phytovolatilization, metals/pollutants are absorbed from soil, water, or air by the plants (natural occurring/genetically modified), and then the pollutants are converted into less toxic volatile organic compounds (VOC) that are released into the air [82]. However, this process is only useful for the metals, which can exist as gaseous ones into the environment. The examples of such contaminants/HMs are Se, Hg, and As [70]. Phytovolatilization does not have the problem of disposal on harvesting; however, HMs may recycle through rainfall and redeposit into the oceans and lakes [82].

The advancement in research has made it possible for researchers to identify such plants that have the ability to absorb metal gasses in their elemental. The elemental form is biologically transformed into the gaseous form and released into the air. This whole process is defined as phytovolatilization. However, according to Wilber [85], Se, and Hg are toxic; hence uncertainty is present about its safety in the atmosphere as a result of volatilization. However, phytovolatilization of Se is still attractive in the Se contaminated soils. Recently efforts are being made for incorporating bacterial ion reductase genes in plants for Hg.

Plants have the ability to interact with a variety of organic components; hence it affects the transport of contaminants in the environment. However, it can be reclaimed by phytovolatilization. Organic contaminants may volatilize directly from stems and leaves or indirectly from soil as a result of plant root activities (Figure 8.2). The contaminant fluxes due to volatilization vary from local contaminant spills to global scale (e.g., methane fluxes emitted from ecosystems that reduce the organic carbon) [70, 82].

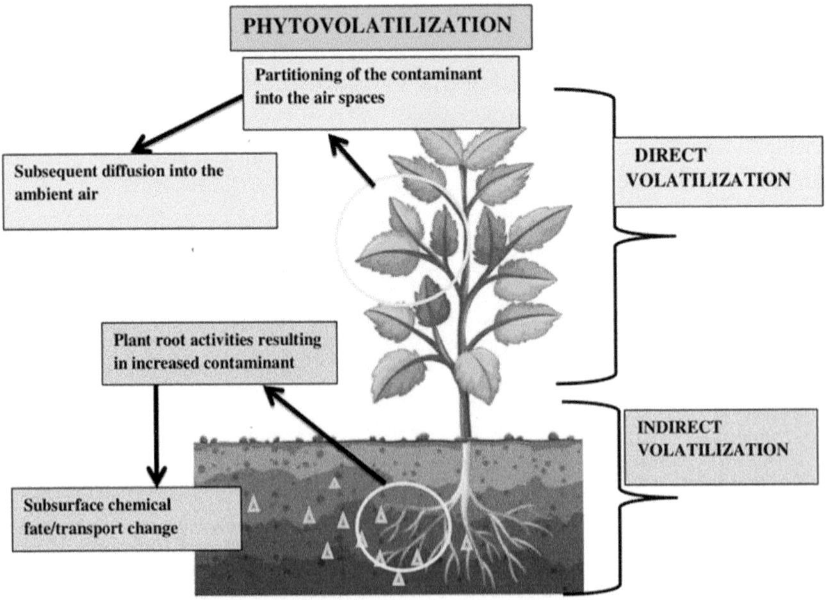

FIGURE 8.2 Process of phytovolatilization.

In a study, it was revealed that leaves of *P. vittata* has the capability to vaporize arsenate, arsenite, and other arsenic compounds. When *P. vittata* was grown under subtropical conditions in the greenhouse, then *P. vittata* could effectively volatilize As because about 90% of the volatilized As was from the contaminated soil. If larger amounts had been released from the contaminated soils, then a secondary As contamination may take place in the surrounding environment [68].

8.7 GENETIC ENGINEERING TO IMPROVE PHYTOREMEDIATION

One of the promising techniques to enhance carbon-capturing/carbon sequestration is the use of genetic engineering that targets the improvement of plants and enhances their ability to absorb more contaminants from a deteriorated agroecosystem. This targets the improvement in photosynthetic ability to capture CO_2 and convert it into plant biomass, to enhance the carbon in cellular carbon pools, and to enhance the allocation of C into the roots, which is then converted or added into the soil organic carbon pools.

8.8 RHIZOREMEDIATION

Rhizoremediation (also called rhizosphere bioremediation, microbe assisted phytoremediation technology and rhizodegradation) is a type of bioremediation of organic contaminants through the aid of rhizospheric microorganisms [37, 73]. This technique is less labor-intensive, lower carbon footprint, minimum environmental damage, preserves biological diversity and less costly as the plants use the energy from the sun [38, 54].

Rhizosphere contains certain types of microorganisms associated with the roots that are capable of catalyzing uptake of metals. Such synchronic connection is vital in augmenting the bioavailability of metal uptake and enhanced root absorption capacity of plants in addition to the non-essential metals [47].

8.8.1 PLANT-MICROBE INTERACTION

Rhizoremediation is reliant on the plant-microbe interaction and their synchrony plays a crucial role for effective contaminant removal. At the soil root interface, the plant-bacteria association degrade the contaminants. When studied at the micro-scale, the most important phenomenon taking place is the bacterial colonization, root exudates, selection, and maintenance of the catalytic genes and inter-kingdom communication.

The bacteria in the rhizosphere can initiate from the infected seed or can be obtained and used from the soil reservoir of contaminated soil during plant growth, and spread through the soil with root growth and development. Through chemotactic movement, it is possible for the bacteria to colonize around the root [77]. Phytoremediation can be more successful if the native plants growing on the polluted soils are optimizing for their remediation potential [7].

8.8.2 MICROBES AS HYDROCARBON DEGRADERS

Plant root exudates certain chemicals that have the ability to stimulate co-metabolism, act as biosurfactants and help in stimulating the initiation of microbial catabolic enzymes [13].

Sesbania cannabina plant was assessed for its ability to bioaugment the rhizodegradation process in polluted petroleum soils when inoculated with free and immobilized bacterial culture. However, bioaugmentation did not work positively for TPH concentration. Maqbool et al. [51] indicated that the concentration of TPH was reduced as 673 mg kg^{-1} and 867 mg kg^{-1} respectively (on 120th day) under the mobile and immobilized bacterial inoculation; however, on the other hand, the concentration was reduced from 2,541 mg kg^{-1} to 679 mg kg^{-1} on the 90th day in non-inoculated soils (control). In *S. cannabina*, rhizosphere (natural plant-microbe interaction) was sufficient enough for degrading the TPH as compared to artificial inoculation of immobilized bacterial culture [51]. In an investigation, *Phragmites australis* was studied on polluted petroleum soils, and the results indicated that its growth performance at the early growth stages was negatively affected [57].

8.8.3 ROOT EXUDATION

The basic driving force in rhizoremediation is root exudation [52, 60]. Enhanced metabolism and microbial growth ensue on the roots present in the rhizosphere [36]. There are four distinct groups of chemicals produced by roots [37}: lysates (released from senescent tissues and roots), secondary plant metabolites, and passive exudates, and mucilage (released from epidermal cells and root tips); moreover, they are also differentiated between water-soluble and insoluble fractions [37].

In soils polluted with TPH, root exudates can degrade the petroleum pollutants [8, 41]. This phenomenon is accelerated by the artificial addition of the plant root exudates [67]. However, for mitigating the TPH soils, certain conditions like growth stage of the plant and prevailing environmental conditions (EC) are crucial to consider first, and secondly biological factors like the point of exudate release, presence of arbuscular mycorrhizal fungi (AMF) and plant type provide the basis for extending of control/reclamation of TPH soils [37].

Rhizospheric microbial population is dependent on the compounds released as root exudates. These compounds include: organic acids, sugars, fatty acids, nucleotides, secondary metabolites, along with inorganic compounds [21]. The plants also release aromatic compounds, such as flavonoids, phenols, terpenoids, and lignin-derived compounds [77].

There are some peculiar characteristics of rhizosphere that make the organic contaminants degradation process appropriate [14]. Similarly, plants produce secondary metabolites similar to the structure of the organic contaminants [72], leading to higher microbial biomass with a more potent degrading activity [89]. The root exudates invite abundant microbes and enhance their activity than those microbes present in the other soils (where there are no exudates) [46, 74, 83].

The rhizosphere has been identified as the hotspot for horizontal gene transfer. Plasmids help microorganisms to degrade the organic contaminants and also help them to adapt to the stressful contaminated environment [69, 81]. Root exudates also help in making the organic contaminants more available (by detaching them) to the micro-organisms [27]. Hence, the importance of root exudates is crucial for initiating as well as carrying out the smooth functioning of the rhizoremediation process.

8.9 SUMMARY

Phytoremediation, along with rhizoremediation, leads to the sustainability of the agroecosystems, which have been destroyed by the extensive use of synthetic chemicals. Microorganisms are essential in the soil surfaces to increase the soil fertility, circulate nutrient materials in the ecosystem. Plants have the potential to communicate with a number of organic elements, thus it influences the ecological distribution of pollutants. Promoting soil fertility by monitoring microflora can have ample capacity to actively support the advantageous properties of different crops and thus reduce the use of artificial chemical fertilizers and pesticides. However, more research is needed for quantifying and improving the relationship between plants and their ability to uptake toxic elements along with the relationship of remedial plants and microbes associated with rhizoremediation.

KEYWORDS

- **agroecosystem**
- **microbes**
- **phytoextraction**
- **phytoremediation**
- **phytovolatization**
- **root exudate**

REFERENCES

1. Abou-Shanab, R. A., Tammam, A. A., El-Aggan, W. H., & Mubarak, M. M., (2017). Phytoremediation potential of wild plants collected from heavy metals contaminated soils. *International Journal of Geology, Agriculture and Environmental Sciences, 5*, 15–19.
2. Adesodun, J. K., Atayese, M. O., Agbaje, T. A., Osadiaye, B. A., Mafe, O. F., & Soretire, A. A., (2010). Phytoremediation potentials of sunflowers (*Tithonia diversifolia* and *Helianthus annuus*) for metals in soils contaminated with zinc and lead nitrates. *Water, Air, and Soil Pollution, 207*(1–4), 195–201.
3. Agegnehu, G., & Amede, T., (2017). Integrated soil fertility and plant nutrient management in tropical agroecosystems: A review. *Pedosphere, 27*(4), 662–680.
4. Alori, E. T., (2015). Phytoremediation using microbial communities: II. Chapter 15. In: Ansari, A. A., Gill, S. S., Gill, R., Lanza, G. R., & Newman, N., (eds.), *Phytoremediation* (pp. 183–190). Cham: Springer.
5. Alkorta, I., Hernández-Allica, J., Becerril, J. M., Amezaga, I., Albizu, I., & Garbisu, C., (2004). Recent findings on the phytoremediation of soils contaminated with environmentally toxic heavy metals and metalloids such as zinc, cadmium, lead, and arsenic. *Reviews in Environmental Science and Biotechnology, 3*(1), 71–90.
6. Anawar, H. M., Garcia-Sanchez, A., Kul-Alam, M. T., & Rahman, M. M., (2008). Phytofiltration of water polluted with arsenic and heavy metals. *International Journal of Environment and Pollution, 33*(2, 3), 292–312.
7. Azubuike, C. C., Chikere, C. B., & Okpokwasili, G. C., (2016). Bioremediation techniques-classification based on site of application: Principles, advantages, limitations and prospects. *World Journal of Microbiology and Biotechnology, 32*(11), 180–188.

8. Baetz, U., & Martinoia, E., (2014). Root exudates: The hidden part of plant defense. *Trends in Plant Science, 19*(2), 90–98.
9. Bais, H. P., Broeckling, C. D., & Vivanco, J. M., (2008). Root exudates modulate plant-microbe interactions in the rhizosphere. In: *Secondary Metabolites in Soil Ecology* (pp. 241–252). Berlin, Heidelberg: Springer.
10. Barbosa, B., Boléo, S., Sidella, S., Costa, J., & Duarte, M. P., (2015). Phytoremediation of heavy metal-contaminated soils using the perennial energy crops *Miscanthus* spp. and *Arundo donax* L. *BioEnergy Research, 8*(4), 1500–1511.
11. Bennicelli, R., **Stępniewska**, Z., Banach, A., Szajnocha, K., & Ostrowski, J., (2004). The ability of *Azolla caroliniana* to remove Hg (II), Cr (III), Cr (VI) from municipal waste water. *Chemosphere, 55*(1), 141–146.
12. Bhat, J. A., Shivaraj, S. M., Singh, P., Navadagi, D. B., Tripathi, D. K., Dash, P. K., Solanke, A. U., et al., (2019). Role of silicon in mitigation of heavy metal stresses in crop plants. *Plants, 8*(3), 71.
13. Blaylock, M. J., & Huang, J., (2000). Phytoextraction of metals. Chapter 7. In: Raskin, I., & Ensley, B. D., (eds.), *Phytoremediation of Toxic Metals: Using Plants to Clean Up the Environment* (pp. 53–70). New York: John Wiley & Sons, Inc.
14. Brennan, M. A., & Shelley, M. L., (1999). Model of the uptake, translocation, and accumulation of lead (Pb) by maize for the purpose of phytoextraction. *Ecological Engineering, 12*(3, 4), 271–297.
15. Brooks, R. R., Chambers, M. F., Nicks, L. J., & Robinson, B. H., (1998). Phytomining. *Trends in Plant Science, 3*(9), 359–362.
16. Chaney, R. L., & Baklanov, I. A., (2017). Phytoremediation and phytomining: Status and promise. Chapter 5. In: Cuypers, A., & Vangronsveld, J., (eds.), *Advances in Botanical Research* (Vol. 83, pp. 189–221). London: Academic Press.
17. Chaturvedi, R., Favas, P. J., Pratas, J., Varun, M., & Paul, M. S., (2019). Metal(loid) induced toxicity and defense mechanisms in *Spinacia oleracea* L.: Ecological hazard and prospects for phytoremediation. *Ecotoxicology and Environmental Safety, 183*, 9. Article ID: 109570.
18. Chen, T., Wei, C., Huang, Z., Huang, Q., Lu, Q., & Fan, Z., (2004). Arsenic hyperaccumulator *Pteris vittata* L. and its arsenic accumulation. *Chinese Science Bulletin, 47*(11), 902–905.
19. Conger, R. M., & Portier, R. J., (1997). Phytoremediation experimentation with the herbicide bentazon. *Remediation Journal, 7*(2), 19–37.
20. Cook, R. L., & Hesterberg, D., (2013). Comparison of trees and grasses for rhizoremediation of petroleum hydrocarbons. *International Journal of Phytoremediation, 15*(9), 844–860.
21. Correa-García, S., Pande, P., Séguin, A., St-Arnaud, M., & Yergeau, E., (2018). Rhizoremediation of petroleum hydrocarbons: Model system for plant microbiome manipulation. *Microbial Biotechnology, 11*(5), 819–832.
22. Davis, A. M., (1986). Selenium uptake in *Astragalus* and *Lupinus* species. *Agronomy Journal, 78*(4), 727–729.
23. Fan, S., Li, P., Gong, Z., Ren, W., & He, N., (2008). Promotion of pyrene degradation in rhizosphere of alfalfa (*Medicago sativa* L.). *Chemosphere, 71*(8), 1593–1598.
24. Ferraz, P., Fidalgo, F., Almeida, A., & Teixeira, J., (2012). Phytostabilization of nickel by the zinc and cadmium hyperaccumulator *Solanum nigrum* L. are metallothioneins involved? *Plant Physiology and Biochemistry, 57*, 254–260.

25. Fiorentino, N., Ventorino, V., Rocco, C., Cenvinzo, V., Agrelli, D., Gioia, L., Mola, I. D., et al., (2017). Giant reed growth and effects on soil biological fertility in assisted phytoremediation of an industrial polluted soil. *Science of the Total Environment, 575*, 1375–1383.
26. Flynn, T. A., (2013). *Evolution of Nickel Hyperaccumulation in Alyssum L* (p. 232). Doctoral Dissertation; University of Oxford, London.
27. Gao, Y., Ren, L., Ling, W., Gong, S., Sun, B., & Zhang, Y., (2010). Desorption of phenanthrene and pyrene in soils by root exudates. *Bioresource Technology, 101*(4), 1159–1165.
28. Garbisu, C., Allica, J. H., Barrutia, O., Alkorta, I., & Becerril, J. M., (2002). Phytoremediation: A technology using green plants to remove contaminants from polluted areas. *Reviews on Environmental Health, 17*(3), 173–188.
29. Gerhardt, K. E., Gerwing, P. D., & Greenberg, B. M., (2017). Opinion: Taking phytoremediation from proven technology to accepted practice. *Plant Science, 256*, 170–185.
30. González, Á., García-Gonzalo, P., Gil-Díaz, M. M., Alonso, J., & Lobo, M. C., (2019). Compost-assisted phytoremediation of As - polluted soil. *Journal of Soils and Sediments, 19*(7), 2971–2983.
31. Guerra, F., Gainza, F., Pérez, R., & Zamudio, F., (2011). Phytoremediation of heavy metals using poplars (*Populus* spp.): Glimpse of the plant responses to copper, cadmium and zinc stress. Chapter 11. In: Golubev, I. A., (ed.), *Handbook of Phytoremediation* (pp. 387–413). New York: Nova Science.
32. Hamzah, A., Sarmani, S. B., & Yatim, N. I., (2015). Phytoremediation of Pb and Hg by using *Scirpus mucronatus* with addition of bacterial inoculums. *Journal of Radioanalytical and Nuclear Chemistry, 304*(1), 151–155.
33. Halim, M., Conte, P., & Piccolo, A., (2003). Potential availability of heavy metals to phytoextraction from contaminated soils induced by exogenous humic substances. *Chemosphere, 52*(1), 265–275.
34. Hasan, M., Uddin, M., Ara-Sharmeen, I., Alharby, H. F., Alzahrani, Y., Hakeem, K. R., & Zhang, L., (2019). Assisting phytoremediation of heavy metals using chemical amendments. *Plants, 8*(9), 295–299.
35. Hill, S. B., (1998). Redesigning agroecosystems for environmental sustainability: A deep systems approach. *Systems Research and Behavioral Science: The Official Journal of the International Federation for Systems Research, 15*(5), 391–402.
36. Hinsinger, P., Gobran, G. R., Gregory, P. J., & Wenzel, W. W., (2005). Rhizosphere geometry and heterogeneity arising from root-mediated physical and chemical processes. *New Phytologist, 168*(2), 293–303.
37. Hussain, I., Puschenreiter, M., Gerhard, S., Schöftner, P., Yousaf, S., Wang, A., Syed, J. H., & Reichenauer, T. G., (2018). Rhizoremediation of petroleum hydrocarbon-contaminated soils: Improvement opportunities and field applications. *Environmental and Experimental Botany, 147*, 202–219.
38. James, C. A., & Strand, S. E., (2009). Phytoremediation of small organic contaminants using transgenic plants. *Current Opinion in Biotechnology, 20*(2), 237–241.
39. Janoušková, M., Vosátka, M., Rossi, L., & Lugon-Moulin, N., (2007). Effects of arbuscular mycorrhizal inoculation on cadmium accumulation by different tobacco (*Nicotiana tabacum* L.) Types. *Applied Soil Ecology, 35*(3), 502–510.

40. Jansson, C., Wullschleger, S. D., Kalluri, U. C., & Tuskan, G. A., (2010). Phytosequestration: Carbon biosequestration by plants and the prospects of genetic engineering. *Bioscience, 60*(9), 685–696.
41. Javaid, S., Zaman, Q., Sultan, K., Riaz, K., Aslam, A., Ehsan, N., Sharif, S., et al., (2020). Heavy metals stress, mechanism and remediation techniques in rice (*Oryza sativa* L.): A review. *Pure and Applied Biology, 9*, 403–426.
42. Jia, W., Lv, S., Feng, J., Li, J., Li, Y., & Li, S., (2016). Morphophysiological characteristic analysis demonstrated the potential of sweet sorghum (*Sorghum bicolor* (L.) in the phytoremediation of cadmium-contaminated soils. *Environmental Science and Pollution Research, 23*(18), 18823–18831.
43. Jiang, M., Liu, S., Li, Y., Li, X., Luo, Z., Song, H., & Chen, Q., (2019). EDTA-facilitated toxic tolerance, absorption and translocation and phytoremediation of lead by dwarf bamboos. *Ecotoxicology and Environmental Safety, 170*, 502–512.
44. Kavamura, V. N., & Esposito, E., (2010). Biotechnological strategies applied to the decontamination of soils polluted with heavy metals. *Biotechnology Advances, 28*(1), 61–69.
45. Keri, L. D., Jason, B., Shaohan, Z., & Joel, R., (2014). Phytoremediation of pesticide wastes in soil. *Zeitschrift fur Naturforschung C* (*A Journal of Biosciences), 61*, 3, 4.
46. Kowalchuk, G. A., Buma, D. S., De Boer, W., Klinkhamer, P. G., & Van, V. J. A., (2002). Effects of above-ground plant species composition and diversity on the diversity of soil-borne microorganisms. *Antonie Van Leeuwenhoek, 81*(1–4), 509.
47. Kumar, S. S., Kadier, A., Malyan, S. K., Ahmad, A., & Bishnoi, N. R., (2017). Phytoremediation and rhizoremediation: Uptake, mobilization, and sequestration of heavy metals by plants. In: Singh, D. P., Singh, H. B., & Prabha, R., (eds.), *Plant-Microbe Interactions in Agro-Ecological Perspectives: Microbial Interactions and Agro-Ecological Impacts* (Vol. 2, pp. 367–394). Singapore: Springer.
48. Lal, R., (2008). Carbon sequestration. *Philosophical Transactions of the Royal Society B: Biological Sciences, 363*(1492), 815–830.
49. Liu, F., Tang, Y., Du, R., Yang, H., Wu, Q., & Qiu, R., (2010). Root foraging for zinc and cadmium requirement in the Zn/Cd hyperaccumulator plant *Sedum alfredii. Plant and Soil, 327*(1, 2), 365–375.
50. Malayeri, B., Chehregani, A., Mohsenzadeh, F., & Golmohammadi, R., (2005). Effect of heavy metals on the development stages of ovule and embryonic sac in *Euphorbia cheiradenia. Pakistan Journal of Biological Sciences, 8*(4), 622–625.
51. Maqbool, F., Wang, Z., Xu, Y., Zhao, J., & Gao, D., (2012). Rhizodegradation of petroleum hydrocarbons by *Sesbania cannabina* in bioaugmented soil with free and immobilized consortium. *Journal of Hazardous Materials, 237*, 262–269.
52. Martin, B. C., George, S. J., Price, C. A., Ryan, M. H., & Tibbett, M., (2014). The role of root exuded low molecular weight organic anions in facilitating petroleum hydrocarbon degradation: Current knowledge and future directions. *Science of the Total Environment, 472*, 642–653.
53. McDonald, B. A., & Stukenbrock, E. H., (2016). Rapid emergence of pathogens in agroecosystems: Global threats to agricultural sustainability and food security. *Philosophical Transactions of the Royal Society B: Biological Sciences, 371*(1709), 9. Article ID: 20160026.

54. McGuinness, M., & Dowling, D., (2009). Plant-associated bacterial degradation of toxic organic compounds in soil. *International Journal of Environmental Research and Public Health, 6*(8), 2226–2247.
55. Mohammed, M. M., Mohammaden, T. F., Eisa, S. S., & Rabie, K. A., (2017). Efficiency of EDTA on Zn and Cu phytoremediation. *Arab Universities Journal of Agricultural Sciences, 25*(2), 403–409.
56. Mojiri, A., (2011). The potential of corn (*Zea mays*) for phytoremediation of soil contaminated with cadmium and lead. *Journal of Biological and Environmental Sciences, 5*(13), 17–22.
57. Nie, M., Wang, Y., Yu, J., Xiao, M., Jiang, L., Yang, J., Fang, C., Chen, J., & Li, B., (2011). Understanding plant-microbe interactions for phytoremediation of petroleum-polluted soil. *PloS One, 6*(3), 8. Article ID: e17961.
58. Nogia, P., Sidhu, G. K., Mehrotra, R., & Mehrotra, S., (2016). Capturing atmospheric carbon: Biological and nonbiological methods. *International Journal of Low-Carbon Technologies, 11*(2), 266–274.
59. Pajević, S., Borišev, M., Nikolić, N., Arsenov, D. D., Orlović, S., & Župunski, M., (2016). Phytoextraction of heavy metals by fast-growing trees: A review. Chapter 2. In: Ansari, A. A., Gill, S. S., Gill, R., Lanza, G. R., & Newman, L., (eds.), *Phytoremediation: Management of Environmental Contaminants* (Vol. 3, pp. 29–64). Cham: Springer.
60. Phillips, L. A., Greer, C. W., Farrell, R. E., & Germida, J. J., (2012). Plant root exudates impact the hydrocarbon degradation potential of a weathered-hydrocarbon contaminated soil. *Applied Soil Ecology, 52*, 56–64.
61. Pilon-Smits, E., (2005). Phytoremediation. *Annu. Rev. Plant Biol., 56*, 15–39.
62. Purakayastha, T. J., Viswanath, T., Bhadraray, S., Chhonkar, P. K., Adhikari, P. P., & Suribabu, K., (2008). Phytoextraction of zinc, copper, nickel, and lead from contaminated soil by different species of *Brassica. International Journal of Phytoremediation, 10*(1), 61–72.
63. Rahman, M. A., & Hasegawa, H., (2011). Aquatic arsenic: Phytoremediation using floating macrophytes. *Chemosphere, 83*(5), 633–646.
64. Rao, S., Jammala, A., Reddy, K. V., & Abburi, K., (2007). Biosorption of hexavalent chromium using tamarind (*Tamarindus indica*) fruit shell - a comparative study. *Electronic Journal of Biotechnology, 10*(3), 358–367.
65. Raskin, I., & Ensley, B. D., (2000). *Phytoremediation of Toxic Metals: Using Plants to Clean up the Environment* (p. 304). New York: John Wiley & Sons Inc.
66. Raskin, I., Smith, R. D., & Salt, D. E., (1997). Phytoremediation of metals: Using plants to remove pollutants from the environment. *Current Opinion in Biotechnology, 8*(2), 221–226.
67. Rentz, J. A., Alvarez, P. J., & Schnoor, J. L., (2005). Benzo [a] pyrene co-metabolism in the presence of plant root extracts and exudates: Implications for phytoremediation. *Environmental Pollution, 136*(3), 477–484.
68. Sakakibara, M., Watanabe, A., Inoue, M., Sano, S., & Kaise, T., (2010). Phytoextraction and phytovolatilization of arsenic from as-contaminated soils by *Pteris vittata. Proceedings of the Annual International Conference on Soils, Sediments, Water and Energy, 12*(1), 26–32.
69. Sentchilo, V., Mayer, A. P., Guy, L., Miyazaki, R., Tringe, S. G., Barry, K., & Van, D. M. J. R., (2013). Community-wide plasmid gene mobilization and selection. *The ISME Journal, 7*(6), 1173–1186.

70. Sheoran, A., & Sheoran, S., (2017). Phytoremediation of heavy metals contaminated soils. *Journal of Plant Development Sciences, 9*(10), 905–915.
71. Simmons, R. W., Chaney, R. L., Angle, J. S., Kruatrachue, M., Klinphoklap, S., Reeves, R. D., & Bellamy, P., (2014). Towards practical cadmium phytoextraction with *Thlaspi caerulescens*. *International Journal of Phytoremediation,* 9. Article ID: 286179.
72. Singer, A. C., Crowley, D. E., & Thompson, I. P., (2003). Secondary plant metabolites in phytoremediation and biotransformation. *Trends in Biotechnology, 21*(3), 123–130.
73. Singh, J. S., Pandey, V. C., & Singh, D. P., (2011). Efficient soil microorganisms: A new dimension for sustainable agriculture and environmental development. *Agriculture, Ecosystems & Environment, 140*(3, 4), 339–353.
74. Smalla, K., Wieland, G., Buchner, A., Zock, A., Parzy, J., Kaiser, S., & Berg, G., (2001). Bulk and rhizosphere soil bacterial communities studied by denaturing gradient gel electrophoresis: Plant-dependent enrichment and seasonal shifts revealed. *Appl. Environ. Microbiol., 67*(10), 4742–4751.
75. Sterckeman, T., Duquène, L., Perriguey, J., & Morel, J. L., (2005). Quantifying the effect of rhizosphere processes on the availability of soil cadmium and zinc. *Plant and Soil, 276*(1–2), 335–345.
76. Stewart, B. A., & Robinson, C. A., (1997). Are agroecosystems sustainable in semiarid regions? In: Sparks, D. L., (ed.), *Advances in Agronomy* (Vol. 60, pp. 191–228). Academic Press.
77. Thijs, S., & Vangronsveld, J., (2015). Rhizoremediation. Chapter 29. In: Lugtenberg, L., (ed.), *Principles of Plant-Microbe Interactions* (pp. 277–286). Springer, Cham.
78. Tiwari, K. K., Singh, N. K., Patel, M. P., Tiwari, M. R., & Rai, U. N., (2011). Metal contamination of soil and translocation in vegetables growing under industrial wastewater irrigated agricultural field of Vadodara - Gujarat, India. *Ecotoxicology and Environmental Safety, 74*(6), 1670–1677.
79. Tlustoš, P., Száková, J., Vysloužilová, M., Pavlíková, D., Weger, J., & Javorská, H., (2007). Variation in the uptake of arsenic, cadmium, lead, and zinc by different species of willows salix spp. grown in contaminated soils. *Open Life Sciences, 2*(2), 254–275.
80. Tonin, C., Vandenkoornhuyse, P., Joner, E. J., Straczek, J., & Leyval, C., (2001). Assessment of arbuscular mycorrhizal fungi diversity in the rhizosphere of *Viola calaminaria* and effect of these fungi on heavy metal uptake by clover. *Mycorrhiza, 10*(4), 161–168.
81. Top, E. M., & Springael, D., (2003). The role of mobile genetic elements in bacterial adaptation to xenobiotic organic compounds. *Current Opinion in Biotechnology, 14*(3), 262–269.
82. United States Environment Protection Agency (USEPA), (2000). *Introduction to Phytoremediation* (p. 25). EPA, 600/R-99/107.
83. Van-Elsas, J. D., & Bailey, M. J., (2002). The ecology of transfer of mobile genetic elements. *FEMS Microbiology Ecology, 42*(2), 187–197.
84. Vogel-Mikuš, K., Drobne, D., & Regvar, M., (2005). Zn, Cd and Pb accumulation and arbuscular mycorrhizal colonization of pennycress *Thlaspi praecox* Wulf. (Brassicaceae) from the vicinity of a lead mine and smelter in Slovenia. *Environmental Pollution, 133*(2), 233–242.
85. Wilber, C. G., (1980). Toxicology of selenium: A review. *Clinical Toxicology, 17*(2), 171–230.

86. Yan, D. Y., & Lo, I. M., (2012). Pyrophosphate coupling with chelant-enhanced soil flushing of field contaminated soils for heavy metal extraction. *Journal of Hazardous Materials, 199*, 51–57.
87. Yang, S. X., Deng, H., & Li, M. S., (2008). Manganese uptake and accumulation in a woody hyperaccumulator, *Schima superba. Plant Soil Environ., 54*(10), 441–446.
88. Ye, Z. H., Baker, A. J. M., Wong, M. H., & Willis, A. J., (1997). Zinc, lead and cadmium tolerance, uptake and accumulation by the common reed, *Phragmites australis* (Cav.) trin. ex Steudel. *Annals of Botany, 80*(3), 363–370.
89. Yergeau, E., Sanschagrin, S., Maynard, C., St-Arnaud, M., & Greer, C. W., (2014). Microbial expression profiles in the rhizosphere of willows depend on soil contamination. *The ISME Journal, 8*(2), 344–358.
90. Yoon, J., Cao, X., Zhou, Q., & Ma, L. Q., (2006). Accumulation of Pb, Cu, and Zn in native plants growing on a contaminated Florida site. *Science of the Total Environment, 368*(2, 3), 456–464.
91. Zayed, A., Gowthaman, S., & Terry, N., (1998). Phytoaccumulation of trace elements by wetland plants. I: Duckweed. *Journal of Environmental Quality, 27*(3), 715–721.
92. Zhang, W., Jiang, F., & Ou, J., (2011). Global pesticide consumption and pollution: With China as a focus. In: *Proceedings of the International Academy of Ecology and Environmental Sciences* (Vol. 1, No. 2, pp. 125–135).
93. Zhao, F. J., Lombi, E., & Breedon, T. M., (2000). Zinc hyperaccumulation and cellular distribution in *Arabidopsis halleri. Plant, Cell & Environment, 23*(5), 507–514.

CHAPTER 9

SUSTAINABLE SOIL HEALTH AND CROP IMPROVEMENT: RECENT TRENDS IN BIOREMEDIATION TOOLS

SURAKSHA CHANOTRA and AZAD GULL

ABSTRACT

Bioremediation technique implies systematic use of biological agents, mostly the derivatives of plant or microbial origin, for effective management and efficient removal of toxicants contaminating various constituents of environment. These pollutants may be in the form of organic wastes, heavy metals (HMs), industrial effluents, polycyclic aromatic hydrocarbons (PAHs), or spillage from the fuel mining and purification units. Accumulation and persistence of such materials for prolonged duration in the environment, in considerably higher amounts results in heavy loss to the quality of environment. Bioremediation methods utilize various microbial agents, such as strains of bacteria, yeast, fungi, and derivatives of plants as remediation agents. A continuous search for the new biological forms is required to regulate increasing pollution and environmental problems hampering the survival and growth of all living organisms. This chapter focusses on recent trends in bioremediation with emphasis on different tools for sustainable soil health and crop improvement.

9.1 INTRODUCTION

Continuous accumulation of heavy metal (HM) ions, such as lead (Pb), mercury (Hg), copper (Cu), gold (Au), iron (Fe) and zinc (Zn), etc., has resulted in genetic and physiological abbreviations in number of living organisms due to their long-term persistence in the environment and accumulation inside their bodies due to ingestion of contaminated food materials [101]. These also have resulted in the overall destruction of mankind, outbreak of pandemics globally, occurrence of many mutations and other irreparable losses to humans, animals, and plant lives on equal extent [32, 78, 101]. These toxicants can be released to the environment more or less accidently during manufacturing, packaging, storage or transportation of metal compounds, extraction, and purification of petroleum oils; and as industrial effluents and agricultural wastes [18, 51].

Prolonged persistence of HM ions or their byproducts may also cause serious threats in the form of oxidative stress, which causes the cell environment to produce free radicals, loss of the ability to eradicate unwanted material or repair cell damage, which results in the formation of reactive oxygen species (ROS) [23, 24, 71].

Therefore, the concept of bioremediation has been introduced as to formulate reclamation measures to save the quality of the environment and for supporting the lives of other biological systems interacting with it. Bioremediation consists of the use of biological agents (mostly the derivatives of plant or microbial origin [43, 79, 105]) for the eradication of hazardous elements from the ecosystem [60, 83]. Being eco-friendly, economically affordable, and practically efficient, bioremediation forms the most effective strategy in the era of pollution [16, 36].

It has been documented that effective use of bioremediation measures may save approximately 50–60% of the cost of the synthetic chemicals [18, 27]. However, the efficiency of bioremediation methods varies with the action rate of microorganisms under different temperature ranges, pH, and available moisture percentage, etc. [57, 104].

This chapter deliberates on the recent trends in bioremediation with emphasis on different tools for sustainable soil health and crop improvement.

9.2 THE RISK OF CHEMICAL FERTILIZERS AND FOOD CONTAMINATION

While the global population continues to increase, hunger, and the demand for food are key issues on the food debate agenda worldwide [41]. To satisfy this need, the introduction of industrial fertilizers to agrarian lands to improve food production continues to grow [41, 66]. Chemical fertilizers are used to increase plant growth and to improve tissue safety of plants. It is better achieved at the optimal levels of nutrients, where plant growth and production are maximized [32].

Chemical fertilizer utilization is popular in vegetables than in any other food crop. Health issues caused by chemical fertilizers were reported higher after vegetable foods were eaten than the other food crops around the world [41]. Diarrhea and stomach ache has become the most severe forms of food-contaminated disease [65, 66].

Farmers and consumers of agricultural crops cultivated with synthetic fertilizers testify to loss of flavor, shortened shelf-life and the depletion of adaptive seed potential [71, 82]. Many experiments have mentioned that crops cultivated using artificial fertilizers are less safe than the sustainably sourced or developed crops. Chemical fertilizer nitrites can also lessen the quality of foods, decrease the integration of protein, fat, and beta-carotene, decompose the B-group of vitamins, and reduce the vitamin A content [41, 81].

9.3 HISTORY OF BIOREMEDIATION

Since the 1800s, farmers have used bioremediation tools to increase crop productivity and to improve the soil health. One of the most suitable examples are the cultivation of leguminous crops to improve the nitrogen fixation capacity of the soil, which in turn helps to gain increased crop yield. *Rhizobium* spp. and free-living bacteria (such as *Azotobacter* or *Rhizobacteria*) have been reported to impart significant benefits to the soil and in turn allow symbiotic association with root zones for completing their life cycles and producing their progenies. Thus, it acts as a biological agent for cleaning the environment and contributes significantly to soil health [45, 98].

9.4 BIOREMEDIATION TOOLS FOR SUSTAINABLE FUTURE

The process of bioremediation aims at the eradication of hazardous pollutants from the environment, and it mainly involves the application of bio-agents in the form of plant and microbial derivatives [16]. The pollutants include HM ions, pesticidal residues, synthetic wastes, etc., which remain persistent in the environment for longer period of time and posing deleterious effect on environmental health. It works on the principle of: (i) interrelated mechanisms by changing the pH value of the target pollutant to make it soluble to remove the pollutant easily by bioremediation tools; (ii) the redox reaction; and (iii) enhanced absorption of hazardous pollutants.

To maintain the quality of the environment by checking the spread of pollution any further, the existing pollutants are degraded to less toxic or non-toxic levels [46, 47, 65], for ensuring optimum survival, growth, and development of other flora and fauna constituting important section of the ecosystem for a clean environment [7].

Modern and allied fields of microbiology, molecular biology, biotechnology, nanotechnology, analytical chemistry, and environmental sciences have proved the efficiency of bioremediation tools for sustaining the environment [102]. These tools mostly emphasize the utilization of bacterial strains and directives of plant species for biodegradation process [44, 51, 87, 94]. Bioremediation is based on three basic steps:

1. **Bioaugmentation:** This involves the addition of archaea or bacterial cultures to speed up the process of degradation of pollutants.
2. **Biodegradation:** of pollutants by process of biostimulation which involves the addition of nutrients and artificial supply of oxygen in order to enhance the process of degradation so as to make the environment a safer place for survival of beneficial organisms.
3. **Complete Natural Attenuation (NA):** which is based on active participation of microorganisms without human interference.

9.4.1 *IN-SITU BIOREMEDIATION*

Bioremediation of any contaminant depends on toxicity level of the pollutant, level of contamination and target site under treatment. Therefore, the majority of the techniques applied under the bioremediation process have been categorized into *in-situ* and *ex-situ* techniques (Table 9.1). In-situ techniques of bioremediation involve the application of remedial agents for the treatment of soil and underground water resources with low level of contamination [41].

In-situ bioremediation is eco-friendly, considerably affordable, and easy to adopt technology with a promising potential to purify the target site. In this context, the use of mercury (Hg) resistant bacteria has been reported as an emerging approach towards bioremediation [12–15]. Attempts have been made to raise genetically modified organisms comprising of Hg resistant bacterial strains for treatment of organic and metal toxicants [87, 108].

In-situ tools comprises of bioventing, bioattenuation, bioreactors [100], biostimulation, etc. [43, 63, 64].

TABLE 9.1 Categorization of Bioremediation Techniques

Specificity For	Advantages	Limitations
***In Situ* Bioremediation: Examples (Bioaugmentation and Bioventing)**		
• bioremediation of soil and groundwater sites contaminated with aromatic compounds, e.g.: catechol, phenol, and cresols, etc. • Contaminants polluting the soil surface and groundwater with PAHs and metals.	• Cost-efficient. • Eco-friendly. • Action takes place in a natural way.	• Effected by environmental fluctuations. • Time consuming. • Difficult to record performance details on a daily basis.
***Ex Situ* Bioremediation: Examples (Biostimulation and Rhizofiltration)**		
• Degradation of metals ions. • contaminating soil and groundwater. • Metal contaminants. • polluting water bodies.	• Cost-efficient. • Can be applied directly to the target site.	• Requires sophisticated working conditions. • Skill requirements by the operator. • Difficult to conduct mass transfer action.

9.4.2 EX-SITU BIOREMEDIATION

The *ex-situ* techniques are used for the treatment of soil and groundwater at some specific target sites, which have been reported to be replaced from the original site by means of soil excavation and pumping of water, etc. In this direction, attempts have been made to identify suitable plants and microbial derivatives for the elimination of contaminants from soil and groundwater. *Ex-situ* biodegradation with the use of *Zoogloea, Lepthothrix, Sphingomonas, Novosphingobium,* and *Comamonadaceae* have been effective for the treatment of polycyclic aromatic hydrocarbons (PAHs) and reported to yield 95% of naphthalene and methylnaphthalene and 90% of total PAHs elimination in the target site under treatment [55]. *Ex-situ* tools comprise of rhizofiltration, bioleaching, and phytoremediation, etc.

9.5 FACTORS AFFECTING BIOREMEDIATION

Bioremediation is a complex process involving synchronized action of physical (biotic and abiotic), chemical, biological (action of microorganisms) and environmental factors (edaphic factors) favoring the biodegradation of toxic substances to make them non-toxic and to eliminate them the polluted environment [29, 48]. Important factors involved in bioremediation process are:

- Application of natural agents is the prime focus;
- Availability of optimum instruments, equipment, and means required for introduction of bioremediation tools;
- Prevalence of favorable edaphic factors (including light, temperature, water, air, photoperiod, and air current, etc.), and environmental factors for accelerating the microbial activity of the biological tools [103];
- Specificity of microbial agents against the target pollutant;
- Sufficient availability of biological agents for bioremediation of the target contaminants.

9.6 BIOREMEDIATION METHODS FOR CROP IMPROVEMENT

9.6.1 *BIOATTENUATION*

Bioattenuation is an efficient bioremediation tool to remove the contaminants to a greater extent without any considerable loss to the environment. Recently, special attention has been paid to enhancing the biodegradative potential of microorganisms by transfer of packaged catabolic genes from one or more donor strains to indigenous microflora existing in target areas under contamination zones [80].

Bioattenuation agents serve to reclaim the target toxicants by altering their chemical composition and thereby disturbing its metabolic processes resulting in biodegradation of the toxic substances, such as recalcitrant or xenobiotics. It includes physical, biological, and chemical reactions to make treatment more effective by reducing the toxicity level of the contaminant to low or non-toxic levels either by complete elimination or by decreasing volume/concentration of the contaminants below the economic injury level (EIL).

PAHs form one of the extensively studied group of petroleum hydrocarbons (PHC) and are mostly employed for the treatment by bioattenuation measures [2, 37, 53, 67, 96, 97, 107]. Thus, polluting environment need to be considered under the bioremediation process with utmost importance.

PAHs contamination have been reported to be increased by many folds in soil and water resources, particularly in areas adjoining the industrial units in the last century [40, 58]. Application *Cycloclasticus, Pseudomonas, Pseudo alteromonas, Halomonas, Marinomonas,* and *Dietzia* have been most efficient measures for bioattenuation of PAHs as it helps in mineralization of the hydrocarbons [33].

Moreover, addition of salt as nutrient supplement for bioattenuation of PAH mixtures comprised of varying contents of fluorine, phenanthrene, waste generated in daily household activities and wastewater from industrial plants, had resulted in the elimination of fluorine and phenanthrene up to 99% compared to 98% of pyrene [109]. Application of lignocellulosic substrate as a bio-stimulant has been effective for bioremediation of total petroleum hydrocarbons (TPHs) to a greater extent [68].

9.6.2 *BIOSTIMULATION*

Biostimulation involves the addition of nutrient materials to the site of contamination to accelerate the growth and activity of natural chemical degraders, including a vast range of microorganisms and derivatives of microbial agents. It involves the addition of inorganic compounds (such as macro-nutrients including nitrogen (N), phosphorous (P) and potassium (K) and micro-nutrients including magnesium (Mg), sulfur (S), iron (Fe), chlorine (Cl), zinc (Zn), manganese (Mn), copper (Cu), sodium (Na), etc.). All these nutrients, when applied for biostimulation, help to gain soil fertility, thus contributing to maintaining ecological balance by supporting the lives of microflora and fauna.

Biostimulation have been reported to be effective against PAHs and mode of action involves the acceleration of degradation activities of the microorganisms. Efficiency of biostimulation process against PAHs depends on the concentration of nutrients added for reclamation. For example, addition of nitrogen dominant macro-nutrient with 67 to 87% (high level) and 12 to 32% (low level) formulations helps in phenanthrene elimination from the site of contamination [61]. Application of biostimulation tool with organic compost accelerated elimination of PAHs in experimental soil site [96]. Also, biodegradation of phenanthrene, anthracene, and benzopyrene with vermicompost and organic manure, has posed a transient effect in the successful elimination of PAHs during the prior first month [9]. Polyacrylamide, an active flocculant, was used in wastewater treatment [39].

9.6.3 *BIOAUGMENTATION*

Bioaugmentation involves the application of suitable species of microorganisms for accelerating the process of bioremediation to eliminate pollution-causing agents. The selection of microorganisms to be applied for bioremediation depends on physical, chemical, and biological factors and should be capable of eating away the contaminants, specifically the aromatic compounds of catechol, phenol, and cresols, etc. For example, *Pseudoonasputida* sp. (*P. putida* OR45a and *P. putida* KB3) and *Acinetobacter* sp. and *Comamonas testosterone* 12 have been identified as effective bioaugmentation agents [76, 84]. Bacteria *Acinetobacter* sp. and *Comamonas testosterone* were effective against 4-flouroaniline and 3-chloroanilne for bioremediation of water resources [19, 106].

Results of bacterial application over PAHs showed acceleration in mineralization rate by 36% after 150 days of treatment for pyrene and 5% after 70 days of treatment for benzo[a]pyrene (BaP), respectively [30]. *Scopulariopsis brevicaulis* exhibited similar results for bioaugmentation of phenanthrene by 60%, fluoranthene by 62%, pyrene by 64% and benzo (a) pyrene by 82% after 30 days of treatment [72]. Application of bacterial consortium comprising of 12 indigenous strains possessing different catabolic potentials was effective for treatment of soil contaminated with metals like lead (Pb) and zinc (Zn) and diesel oil contaminants [99].

9.6.4 *PHYTOREMEDIATION*

Phytoremediation is an eco-friendly tool for purifying the contaminated soil, air, and water resources and aims at the total biodegradation of organic pollutants and HMs mainly with the help of plant derivatives and rhizospheric microorganisms [8, 32, 57]. The efficacy of the remediation process depends on the toxicity level of pollutants, level of pollution, catabolic potential of derivatives/agents applied and schedule of treatment [93, 113]. Increased doses of contaminant may inhibit the growth and development of the plant, and the phenomenon is known as hormesis [23].

On the basis of phytoremedial properties, about 500 taxa of kingdom Plantae have been categorized as hyperaccumulators [1, 28, 56, 59, 82, 88] for treatment of metal compounds and important ones include *Brassicaceae, Caryophyllaceae, Violaceae, Fabaceae, Euphorbiaceae, Lamiaceae, Asteraceae, Cyperaceae, Poaceae, Cunouniaceae*, and *Flacourtiaceae* [82].

The actual site of phytoremediation varies from plant to plant, and the site of accumulation for HMs in the root zone with slight transport to aboveground parts [65, 69, 73]; and on the other hand, effective movement and accelerated transport is from root zone to above-ground parts [4, 55].

Distribution of HMs for different parts of the plant (*Helianthus annuus)* has been observed to vary greatly for different metals and organs, e.g., accumulation of 59% lead and 47% zinc and cadmium in the leaves. Whereas, accumulation of Pb and Zn and Cd was 1.0% and 79% in seeds of *Helianthus annuus*, respectively [11]. *Brassica napus, Brassica juncea* and *Brassica rapa* depicted growth at a comparatively faster rate, thus accelerating the phytoremediation process [35].

Application of fungi as an efficient biosorbent has been employed for the treatment of industrial effluents and sewage contaminating the groundwater [3, 91, 92]. In addition, the introduction of several strains of lactic acid bacteria (LAB) could eliminate HMs from the target site to help in the revitalization of biodiversity and ecological balance in the area under treatment [10, 38, 82]. The biodegradation efficiency of LAB is relatively higher, and it also increases the rate of degradation of metals when applied at recommended dosage [6, 49].

Phytoremediation technique involves series of reactions, which work in a subsequent manner resulting in eradication of pollutants aiming at the formation of clean environment [85]. General processes include: immobilization of pollutants, removal of residual contaminants, and acceleration of biodegradation process. Non-volatile compounds undergo process of phytostabilization to degrade toxic compounds [94].

9.7 LIMITATIONS OF BIOREMEDIATION

On the other hand, the effectiveness of bioremediation techniques can be altered considerably by a large number of factors and sometimes the cost of inputs required makes the adoption of technique less preferred [74, 110], and these act as limitations for realizing the full potential of the methodology as indicated below:

- Addition of nutrients and artificial oxygen supply is compulsory while treating with bacteria or fungi;
- *Ex-situ* bioremediation demands sophisticated working space;
- Identification of target contaminant is costly, time consuming and hectic procedure;
- *In-situ* bioremediation is dependent on the depth of the soil, which in turn affects the efficacy of the process;
- Maintenance of optimum environmental conditions (EC) becomes difficult with increase in area under treatment;
- Nature of the organism applied for degradation of a particular toxicant should be specific so as to meet the target of bioremediation, which otherwise serves as a limiting factor towards success of the process;
- Requirement of particular environment type for imitation of bioremediation process;
- Requires artificial induction of microorganisms for accelerating the rate of reaction;
- Target organisms must be below extreme levels with minimum span of spread in the target area under treatment which otherwise be difficult to handle;
- Time consuming series of reactions.

9.8 RECENT TRENDS IN BIOREMEDIATION FOR CROP IMPROVEMENT

- Fungi had been reported as potential bioremediation agents for biodegradation of aromatic amines (AA). Fungi possess advantageous metabolic capabilities that can be utilized for biodegradation of various classes of AA, but more research is needed [52].
- Mammalian P450 cytochrome genes exhibit a high degree of herbicide resistance to crop plants. It can be used as an efficient bioremediation technique in herbicide rotation system to impart delayed herbicide resistance in weeds, which reduces the chances of herbicide resistance in weeds, thereby checks their further growth and spread. For example, using Mammalian P450 had been used to suppress overexpression of maize glutathione S-transferase I (GSTI) gene responsible occurrence of herbicide resistance in weeds [31, 62, 75, 77].
- Minimum inhibitory concentration (MIC) can be employed for assessing resistance in isolates of bacteria, yeast, and fungi, etc. MIC

method offers accessibility over a large number of microbes, and results can be obtained in a short time span [5].

- More recently a bacterial strain namely *Lactobacillus plantarum* MF042018 was reported to be thermo-resistant in nature; and exhibited considerably maximum uptake capacity of Cd^{2+} and Pb^{2+} at pH 2.0 and a temperature of 22°C, up to 500 and 100 ppm, to both nickel and chromium after 1 h of treatment with multiple antibiotic resistance (MAR) index above 0.5 [10].
- *Podospora anserina*, having arylamine *N*-acetyltransferase 2 (NAT2) enzyme, exhibits detoxification potential and can be utilized for bioremediation of highly toxic pesticide residue 3,4-dichloroaniline (3,4-DCA) for reclamation of contaminated soil sites [75].
- *Pseudomonas* sps., *Burkholderia* sps., *Rhodococcus*sps., *Herbaspirillum* sps., *Sphingomonas* sps., *Sphingobium* sps., *Novosphingobium* sps., *Methylobacterium populi, Dehalospirillum multivorans, Bacillus* sps., *Acinetobacter* sps., *Nocardioides* sps. are most extensively studied strains of bacteria used as efficient bioremediation agents [25, 26, 31].
- Soil bacteria and filamentous fungi have been reported to exhibit major detoxification pathway for degradation of AA and its potentially was reported to be dependent on the activity of *N*-acetyltransferase14 (NATs14) [75, 89].
- The bioremediation properties of *Aspergillus niger* and *Mycobacterium chlorophenolicum* are pH-dependent; and were suitable for biodegradation of pentachlorophenol (PCP) compounds [5, 20, 21].

9.9 SIGNIFICANCE OF BIOREMEDIATION IN CROP IMPROVEMENT

Bioremediation technique is an efficient, clean, cheap, and promising technology for achieving the target of sustainable agriculture development as it ensures the maintenance of environment quality when compared to conventional methods, which fails to cure problems of residual effects [25, 31]. With the indiscriminate use of chemicals (such as fertilizers and pesticides) for improving the agricultural crop production, serious issues of pollution, resistance in pests, disturbance in ecological balance, and many more have cropped up to hamper all living organisms to a greater extent. The residues of such chemicals and fertilizers get percolated into deeper layers of the soil, thereby polluting soil/water/air and in the tissues of the plants growing in the harboring environment [68]. The majority of the toxic compounds, such as dichlorodiphenyltrichloroethane (DDT), dieldrin, PAHs, polychlorinated biphenyls (PCBs), benzene hexachloride (BHC) and propachlor, etc., have higher degrees of persistence, thereby polluting the environment [57, 66]. Bioremediation is a sustainable remediation technique, in which microorganisms interact chemically and physically with the toxic compounds and cause alterations in their structural, physiological, and functional properties, making them less toxic or non-toxic or complete degradation. Therefore, bioremediation is a total organic technique applied for purifying the soil sites contaminated with HMs, industrial effluents, oil spills, and other soil wastes; and involves active participation of derivatives of plants and microbes for biodegradation of toxicants [34]. Significance of bioremediation tools in sustainable crop improvement can be realized by following considerations:

- Biodegradation of organic matter in the form of composting helps in improvement of soil fertility level which can be utilized by the crop grown there.
- Phytoremediation technologies can be used to raise genetically modified organisms with improved growth and yield parameters and high resistance to pests and diseases.
- bioremediation helps in maintenance of gene flow among the interacting species thereby preserving existing biodiversity.
- The true costs of benefits of phytoremediation with biotech plants must be determined.
- Political will and funding are required, both to pursue basic research into phytoremediation, and to implement novel strategies.
- bioremediation can be viewed as green techniques with a wider scope of environment safety with removal of contaminants. These climatic alterations may also affect the crop and soil production capacity; therefore, more research needed to be done on this aspect.

9.10 SUMMARY

In the modern era of technology and rapid development, pollution has emerged as the most serious issue hampering the survival and normal development of almost all living organisms, including microflora and fauna. In this concept, the concept of bioremediation has been introduced with a sole objective to clean up the surrounding environment and to restore its viability that can be cherished by all living organisms. Bioremediation is an emerging powerful tool designed with a variety of specifications to meet the target of complete eradication of pollutants from the environment; and it involves the application of plants and microbial derivatives having potential to digest away the persistent contaminants such as organic compounds, metals, and residues of industrial plants associated with extraction, purification, and transportation of fuels. The effectiveness of the process depends on the physical state of the toxicants, level of contamination, biological, and catabolic activities of the agents, chemical nature of the complex formed by toxicants and bioremediation interaction and environmental factors (such as availability of light, temperature, air current, and moisture), etc., time, and skill requirements, identification of specific strains for treating particular toxicant, etc. Therefore, the introduction of bioremediation tools is important for achieving the target of sustainable development with wider scope of environment safety and improving crop production; and it can be viewed as an emerging technique with promising results in the near future.

KEYWORDS

- **biodegradation**
- **bioremediation**
- **minimum inhibitory concentration**
- **pentachloro phenol**
- **remediation**
- **toxicants**

REFERENCES

1. Abbaszadeh-Dahaji, P., Omidvari, M., & Ghorbanpour, M., (2016). Increasing phytoremediation efficiency of heavy metal-contaminated soil using PGPR for sustainable agriculture. Chapter 9. In: Choudhary, D. K., Varma, A., & Tuteja, N., (eds.), *Plant-Microbe Interaction: An Approach to Sustainable Agriculture* (pp. 187–204). Singapore: Springer Nature.
2. Abdul, A. S., Gibson, T. L., & Rai, D. N., (1990). Use of humic acid solution to remove organic contaminants from hydrogeologic systems. *Environmental Science and Technology, 24*, 328–333.
3. Abigail, M., Samuel, M., & Chidambaram, R., (2014). Hexavalent chromium biosorption studies using *Penicillium griseofulvum* MSR1; a novel isolate from tannery effluent site:

Box-Behnken optimization, equilibrium, kinetics and thermodynamic studies. *J. Taiwan Inst. Chem. Eng., 49*, 156–164.

4. Adesodun, J. K., Atayese, M. O., Agbaje, T., Osadiaye, B. A., Mafe, O., & Soretire, A. A., (2010). Phytoremediation potentials of sunflowers (*Tithonia diversifolia* and *Helianthus annuus*) for metals in soils contaminated with zinc and lead nitrates. *Water, Air and Soil Pollution, 207*, 195–201.
5. Agarwal, M., Rathore, S., & Chauhan, A., (2020). A rapid and high throughput mic determination method to screen uranium resistant microorganisms. *Methods Protoc., 3*, 21. Online, doi: 10.3390/mps3010021.
6. Ahmed, S., Islam, M., Ferdousi, J., & Iqbal, T., (2017). Probiotic *Lactobacillus* Sp. with bioremediation potential of toxic heavy metals. *Bangladesh J. Microbiol, 34*, 43–46.
7. Akash, S., Vinay, M. P., Neelesh, B., & Navneet, (2018). Bioremediation of environmental pollutants. Chapter 5. In: Pathak, V. M., & Navneet, (eds.), *Handbook of Research on Microbial Tools for Environmental Waste Management* (pp. 80–105). New York - USA: IGI Global.
8. Ali, H., Khan, E., & Sajad, M. A., (2013). Phytoremediation of heavy metals: Concepts and applications. *Chemosphere, 91*, 869–881.
9. Alvarez-Bernal, D., García-Díaz, E. L., Contreras-Ramos, S. M., & Dendooven, L., (2006). Dissipation of polycyclic aromatic hydrocarbons from soil added with manure or vermicompost. *Chemosphere, 65*(9), 1642–1651. Online, doi: 10.1016/j.chemosphere.2006.02.028.
10. Ameen, F. A., Hamdan, A. M., & Moustafa, Y., (2020). Assessment of the heavy metal bioremediation efficiency of the novel marine lactic acid bacterium, *Lactobacillus plantarum* MF042018. *Scientific Reports, 10*, 314–321.
11. Angelova, V. R., Perifanova-Nemska, M., Uzunova, G., Ivanov, K., & Lee, H., (2016). Potential of sunflower (*Helianthus annuus* L.) for phytoremediation of soils contaminated with heavy metals. *World J. Sci. Eng. Technol, 10*, 1–11.
12. Atlas, R. M., (1984). *Microbiology: Fundamentals and Applications* (pp. 1–28). London: Macmillan Publishing Company.
13. Atlas, R. M., (1981). Microbial degradation of petroleum hydrocarbons: An environmental perspective. *Microbiol. Rev., 45*, 180–209.
14. Atlas, R. M., (1981). *Microbial Ecology: Fundamentals and Applications* (p. 640). Boston - USA: Addison-Wesley Publishing Company.
15. Atlas, R. M., & Bartha, R., (1993). *Microbial Ecology: Fundamentals and Applications* (3rd edn., p. 560). Boston - USA: Addison-Wesley Publishing Company.
16. Ayangbenro, A. S., & Babalola, O. O., (2017). A new strategy for heavy metal polluted environments: A review of microbial biosorbents. *Int. J. Environ. Res. Public Health, 14*, 94–99.
17. Baker, K. H., & Herson, D. S., (1994). Introduction and overview of bioremediation. In: Baker, K. H., & Herson, D. S., (eds.), *bioremediation* (pp. 64–79). New York: McGraw-Hill.
18. Blaylock, M. J., Salt, D. E., Dushenkov, S., Zakharova, O., Gussman, C., Kapulnik, Y., Ensley, B. D., & Raskin, I., (1997). Enhanced accumulation of Pb in Indian mustard by soil-applied chelating agents. *Environ. Sci. Technol, 31*, 860–865.
19. Boon, N., Goris, J., De Vos, P., Verstraete, W., & Top, E. M., (2000). Bioaugmentation of activated sludge by an indigenous 3-chloroaniline-degrading *Comamonas testosterone* strain, I2gfp. *Appl. Environ. Microbiol., 66*, 2906–2913.

20. Bosso, L., Lacatena, F., Cristinzio, G., Cea, M., Diez, M. C., & Rubilar, O., (2015). Biosorption of pentachlorophenol by *Anthracophyllum discolor* in the form of live fungal pellets. *New Biotechnol., 32*, 21–25.
21. Brandt, S., Zeng, A. P., & Deckwer, W. D., (1997). Adsorption and desorption of pentachlorophenol on cells of mycobacterium *Chloro phenolicum* PCP-1. *Biotechnol. Bioeng., 55*, 480–489.
22. Breedveld, G. D., & Briseid, T., (1994). In-situ bioremediation of creosote contaminated soil: Column experiments. In: Hinchee, R. E. (ed.), *Applied Techniques for Site Remediation* (pp. 124–136). Boca Raton, USA: CRC Press.
23. Calabrese, E. J., & Blain, R. B., (2009). Hormesis and plant biology. *Environ. Pollut., 157*, 42–48.
24. Chandra, K., Salman, A. S., Mohd, A., Sweety, R., & Ali, K. N., (2015). Protection against FCA induced oxidative stress induced DNA damage as a model of arthritis and in vitro anti-arthritic potential of *Costus speciosus* rhizome extract. *Int. J. Pharm. Phytopharmacol. Res., 7*, 383–389.
25. Chaudhary, D. K., (2016). Bioremediation: An eco-friendly approach for polluted agricultural soil. *Emer. Life Sci. Res., 2*(1), 73–75.
26. Chaudhary, D. K., & Kim, J., (2016). *Novosphingobium naphthae* sp. from oil-contaminated soil. *International Journal of Systematic and Evolutionary Microbiology, 66*(8), 3170–3176.
27. Chaudhary, D. K., Varma, A., & Tuteja, N., (2017). *Plant-Microbe Interaction: An Approach to Sustainable Agriculture* (p. 509). Singapore: Springer Nature.
28. Chibuike, G., & Obiora, S., (2014). Heavy metal polluted soils: Effect on plants and bioremediation methods. *Applied and Environmental Soil Science, 2014*, 12–20.
29. Day, S. M., (1992). Accessing bioremediation technologies via tech transfer from government industry. *Genetic Engineering News, 12*(10), 4–11.
30. Derz, K., Schmidt, B., Schwiening, S., & Schuphan, I., (2006). Comparison of microbial pyrene and benzo[a]pyrene mineralization in liquid medium, soil slurry and soil. *Journal of Environmental Science and Health B, 41*(5), 471–484. Online, doi: 10.1080/03601230600701619.
31. Diez, M. C., (2010). Biological aspects involved in the degradation of organic pollutants. *J. Soil. Plant. Nutr., 10*, 244–267.
32. Dixit, R., Wasiullah, Malaviya, D., Pandiyan, K., Singh, U., Sahu, A., Shukla, R., et al., (2015). Bioremediation of heavy metals from soil and aquatic environment: An overview of principles and criteria of fundamental processes. *Sustainability, 7*(2), 2189–2221. doi:10.3390/su7022189.
33. Dong, C., Bai, X., Sheng, H., Jiao, L., Zhou, H., & Shao, Z., (2015). Distribution of PAHs and the PAH degrading bacteria in the deep-sea sediments of the high-latitude arctic ocean. *Bio Geosciences, 12*(7), 2163–2177. doi: 10.5194/bg-12-2163-2015.
34. Dotaniya, M. L., Panwar, N. R., Meena, V. D., Dotaniya, C. K., Regar, K. L., Lata, M., & Saha, J. K., (2018). Bioremediation of metal contaminated soil for sustainable crop production. In: Meena, V. S., (ed.), *Role of Rhizospheric Microbes in Soil: Stress Management and Agricultural Sustainability* (Vol. 1, pp. 143–173). Cham: Springer Nature.
35. Ebbs, S. D., & Kochian, L. V., (1997). Toxicity of zinc and copper to *Brassica* species: Implications for phytoremediation. *J. Environ. Qual., 26*, 776–781.

36. Ekperusi, O., & Aigbodion, F., (2015). Bioremediation of petroleum hydrocarbons from crude oil-contaminated soil with the earthworm: Hyperiodrilus africanus. *3 Biotech, 5*, 957–965.
37. Ellis, B., (1994). Reclaiming contaminated land: *In-situ/ex-situ* remediation of creosote and petroleum-contaminated land. In: Flatham, P., et al., (eds.), *Bioremediation: Field Experience Reclaiming Contaminated Land* (pp. 107–128). Boca Raton, USA: Lewis Publishers.
38. El-sanhoty, R., Al-Turki, I., & Ramadan, M., (2016). Application of lactic acid bacteria in removing heavy metals and aflatoxin B1 from contaminated water. *Water Sci. Technol, 74*, 625–638.
39. Fernandez-Luqueno, F., Thalasso, F., Luna-Guido, M. L., Ceballos-Ramírez, J. M., Ordoñez-Ruiz, I. M., & Dendooven, L., (2009). Flocculantin wastewater affects dynamics of inorganic N and accelerates the removal of phenanthrene and anthracene in soil. *Journal of Environmental Management, 90*(8), 2813–2818. doi: 10.1016/j.jenvman.2009.03.010.
40. Freeman, D. J., & Cattell, F. C. R., (1990). Wood burning as a source of atmospheric polycyclic aromatic hydrocarbons. *Environmental Science and Technology, 24*, 1581–1585.
41. Fung, K. Y., Lee, C. M., Ng, K. M., Wibowo, C., Deng, Z., & Wei, C., (2012). Process development of treatment plants for dyeing wastewater. *AI Che. J., 58*, 2726–2742.
42. Gadd, G. M., (2000). Bioremedial potential of microbial mechanisms of metal mobilization and immobilization. *Curr. Opin. Biotechnol., 11*, 271–279.
43. Gadd, G. M., (2001). *Fungi in bioremediation* (p. 217). London: Cambridge University Press.
44. Gan, S., Lau, E. V., & Ng, H. K., (2009). Remediation of soils contaminated with polycyclic aromatic hydrocarbons (PAHs). *J. Hazard. Mater., 172*, 532–549.
45. Gentry, T. J., Christopher, R., & Ian, P., (2004). New approaches for bioaugmentation as a remediation technology. *Crit. Rev. Env. Sci. Tec., 34*, 447–494.
46. Ghaly, A. E., Rushton, D. G., & Martinel, L. K., (2007). Effect of mixing on the biological degradation of toluene under aerobic conditions. *Am. J. Environ. Sci., 3*, 246–257.
47. Ghaly, A. E., Yusran, A., & Dave, D., (2013). Effects of biostimulation and bioaugmentation on the degradation of pyrene in soil. *J. Bioremed. Biodeg., 5*, 1–13. doi:10.4172/2155-6199.S7-005.
48. Goblenz, A., Wolf, K., & Bauda, P., (1994). The role of glutathione biosynthesis in heavy metal resistance in the fission yeast *Schizosaccharomyces pombe*. *FEMS Microbiol. Rev., 14*, 303–308.
49. Halttunen, T., Salminen, S., & Tahvonen, R., (2007). Rapid removal of lead and cadmium from water by specific lactic acid bacteria. *Int. J. Food Microbiol., 114*, 30–35.
50. Hamer, G., (1993). Integrated circuit-based biofabrication with common biomaterials for probing cellular biomechanics. *Trends of Biotech., 34*, 171–186.
51. Haritash, A. K., & Kaushik, C. P., (2009). Biodegradation aspects of polycyclic aromatic hydrocarbons (PAHs): A review. *J. Hazard. Mater., 169*, 1–15.
52. Harms, H., Schlosser, D., & Wick, L. Y., (2011). Untapped potential: Exploiting fungi in bioremediation of hazardous chemicals. *Nature Rev. Microbiol., 9*, 177–192.
53. Herbert, B. E., Bertsch, P. M., & Novak, J. M., (1993). Pyrene sorption by water-soluble organic carbon. *Environmental Science and Technology, 27*, 398–403.

54. Herrero, E., Lopez-Gonzalvez, A., Ruiz, M., Lucas-Garcia, J., & Barbas, C., (2003). Uptake and distribution of zinc, cadmium, lead, and copper in *Brassica napus* var. *Oleifera* and *Helianthus annus* grown in contaminated soils. *Int. J. Phytoremed., 5*, 153–167.
55. Hockenreiner, M., & Neugebauer, H., (2015). Ex-situ bioremediation method for the treatment of groundwater contaminated with PAHs. *Int. J. Environ. Sci. Technol., 12*, 285–296. doi: 10.1007/s13762-013-0427-5.
56. Jabeen, R., Ahmad, A., & Iqbal, M., (2009). Phytoremediation of heavy metals: physiological and molecular mechanisms. *Bot. Rev., 75*, 339–364.
57. Jain, S., & Arnepalli, D., (2016). Biomineralization as a remediation technique: A critical review. *Oral Presentation at Indian Geotechnical Conference (IGC2016)*, 8. Chennai, India.
58. Jones, K. C., Stratford, J. A., Waterhouse, K. S., & Vogt, N. B., (1989). Organic contaminants in Welsh soil: Polynuclear aromatic hydrocarbons. *Environmental Science and Technology, 23*, 540–550.
59. Jutsz, A. M., & Gnida, A., (2015). Mechanisms of stress avoidance and tolerance by plants used in phytoremediation of heavy metals. *Arch. Environ. Prot., 41*, 104–114.
60. Kabata-Pendias, A., (2010). *Trace Elements in Soils and Plants* (4th edn., p. 548). Boca Raton - USA: CRC Press.
61. Kalantary, R. R., Mohseni-Bandpi, A., Esrafili, A., Nasseri, S., Ashmagh, F. R., Jorfi, S., & Jafari, M., (2014). Effectiveness of biostimulation through nutrient content on the bioremediation of phenanthrene contaminated soil. *Journal of Environmental Health Science and Engineering, 12*(1), 143–149.
62. Kannan, K., Tanabe, S., Willians, R. J., & Tatsukawa, R., (1994). Persistent organochlorine residues in foodstuffs from Australia, Papua new guinea and the Solomon islands: Contamination levels and dietary exposure. *Sci. Total Env., 153*, 29–49.
63. Lim, P. E., Mak, K. Y., Mohamed, N., & Noor, A. M., (2003). Removal and speciation of heavy metals along the treatment path of wastewater in subsurface-flow constructed wetlands. *Water Sci. Technol., 48*, 307–313.
64. Lin, C. C., & Lin, H. L., (2005). Remediation of soil contaminated with the heavy metal (Cd^{2+}). *J. Hazard. Mater., 122*, 7–15.
65. Lin, J., Jiang, W., & Liu, D., (2003). Accumulation of copper by roots, hypocotyls, cotyledons and leaves of sunflower (*Helianthus annuus* L.). *Bioresour. Technol., 86*, 151–155.
66. Linz, D. G., & Nakles, D. V., (1997). *Environmentally Acceptable Endpoints in Soil: Risk-Based Approach to Contaminated Site Management Based on Availability of Chemicals in Soil* (1st edn., p. 630). Indiana - USA: American Academy of Environmental Engineers.
67. Liu, H., & Amy, G., (1993). Modeling partitioning and transport interaction between natural organic matter and polynuclear aromatic hydrocarbons in groundwater. *Environmental Science and Technology, 27*, 1553–1562.
68. Llado, S., Covino, S., Solanas, A. M., Vinas, M., Petruccioli, M., & Dannibale, A., (2013). Comparative assessment of bioremediation approaches to highly recalcitrant PAH degradation in a real industrial contaminated soil. *Journal of Hazardous Materials, 248,* 407–414. doi: 10.1016/j.jhazmat.2013.01.020.

69. Madejon, P., Murillo, J. M., Maranon, T., Cabrera, F., & Soriano, M., (2003). Trace element and nutrient accumulation in sunflower plants two years after the Aznalcollar mine spill. *Sci. Total Environ., 307*, 239–257.
70. Malik, A., (2004). Metal bioremediation through growing cells. *Environ. Int., 30*, 261.
71. Mani, S., (2015). Production of reactive oxygen species and its implication in human diseases. In: Vibha, R., & Umesh, C. S. Y., (eds.), *Free Radicals in Human Health and Disease* (pp. 3–15). New Delhi: Springer.
72. Mao, J., & Guan, W., (2016). Fungal degradation of polycyclic aromatic hydrocarbons (PAHs) by *Scopulariopsis brevicaulis* and its application in bioremediation of PAH-contaminated soil. *Acta Agriculture Scandinavica Section B: Soil and Plant Science, 66*(5), 399–405. doi: 10.1080/09064710.2015.1137629.
73. Marchiol, L., Fellet, G., Perosa, D., & Zerbi, G., (2007). Removal of trace metals by *Sorghum bicolor* and *Helianthus annuus* in a site polluted by industrial wastes: A field experience. *Plant Physiol. Biochem., 45*, 379–387.
74. María, S. V. M., Oscar, J. H. H., Juan, A., Maya, C., Juan, C., Díaz, C., & Roblero, J. J., (2016). Approaches for removal of PAHs in soils: Bioaugmentation, biostimulation, and bioattenuation. *Soil Contamination-Current Consequences and Further Solutions* (pp. 329–342). Rijeka, Croatia: InTech Open, online.
75. Martins, M., Rodrigues-Lima, F., Dairou, J., Lamouri, A., Malagnac, F., Silar, P., & Dupret, J. M., (2009). An acetyltransferase conferring tolerance to toxic aromatic amine chemicals: Molecular and functional studies. *J. Biol. Chem., 284*(28), 18726–18733.
76. Mathialagan, T., & Viraraghavan, T., (2009). Biosorption of pentachlorophenol from aqueous solutions by a fungal biomass. *Bioresour. Technol., 100*, 549–558.
77. McGuiness, M., & Dowling, D., (2009). Plant-associated bacterial degradation of toxic organic compounds in soil. *Int. J. Environ. Res. Public Health., 6*, 2226–2247.
78. Michalska, J., Pinski, A., Zur, J., & Morzik, A., (2020). Selecting bacteria candidates for bioaugmentation of activated sludge to improve the aerobic treatment of landfill leachate. *Water, 12*, 140–149. doi:10.3390/w12010140.
79. Morel, J. L., Echevarria, G., & Goncharova, N., (2002). *Phytoremediation of Metal-Contaminated Soils* (pp. 1–11). Amsterdam: IOS Press and springer in conjunction with the NATO public diplomacy Division.
80. Mrozik, A., & Piotrowska-Seget, Z., (2010). Bioaugmentation as a strategy for cleaning up soils contaminated with aromatic compounds. *Microbiol. Res., 165*, 363–375.
81. Mrvcic, J., Stanzer, D., Solic, E., & Stehlik-Tomas, V., (2012). Interaction of lactic acid bacteria with metal ions: Opportunities for improving food safety and quality. *World J. Microbiol. Biotechnol., 28*, 2771–2782.
82. Muszynska, E., & Hanus-Fajerska, E., (2015). Why are heavy metal hyperaccumulating plants so amazing? *Biotechnol. J. Biotechnol. Comput. Biol. Bionanotechnol., 96*, 265–271.
83. Nematian, M. A., & Kazemeini, F., (2013). Accumulation of Pb, Zn, C and Fe in plants and hyperaccumulator choice in Galali iron mine area, Iran. *Int. J. Agric. Crop Sci., 5*, 426–438.
84. Nzila, A., Razzak, S. A., & Zhu, J., (2016). Bioaugmentation: An emerging strategy of industrial wastewater treatment for reuse and discharge. *International Journal of Environmental Research and Public Health, 13*, 846–856. doi: 10.3390/ijerph13090846.

85. Ojuederie, O. B., & Babalola, O. O., (2017). Microbial and plant-assisted bioremediation of heavy metal polluted environments: A review. *International Journal of Environmental Research and Public Health, 14*, 1504–1515. doi:10.3390/ijerph14121504.
86. Onwubuya, K., Cundy, A., Puschenreiter, M., Kumpiene, J., & Bone, B., (2009). Developing decision support tools for the selection of "gentle" remediation approaches. *Sci. Total Environ., 407*, 6132–6142.
87. Pritchard, P. H., (1991). Bioremediation as a technology: Experiences with the Exxon Valdez oil spill. *J. Hazardous Materials, 28*, 115–130.
88. Rascio, N., & Navari-Izzo, F., (2011). Heavy metal hyperaccumulating plants: How and why do they do it? And what makes them so interesting? *Plant Sci., 180*, 169–181.
89. Raymond, J., Rogers, T., Shonnard, D., & Kline, A., (2001). Review of structure-based biodegradation estimation methods. *J. Hazard. Mater., 84*, 189–215.
90. Salem, H. M., Eweida, E. A., & Farag, A., (2000). Heavy metals in drinking water and their environmental impact on human health. In: *ICEHM 2000* (pp. 542–556). Giza - Egypt: Cairo University.
91. Samuel, M., Abigail, M., & Chidambaram, R., (2015). Isotherm modeling, kinetic study and optimization of batch parameters using response surface methodology for effective removal of Cr (VI) using fungal biomass. *PLoS One, 10*, 12. Article ID: e0116884.
92. Samuel, M., Abigail, M., & Chidambaram, R., (2015). Biosorption of Cr(VI) by *Ceratocystis paradoxa* MSR2 using isotherm modeling, kinetic study and optimization of batch parameters using response surface methodology. *PLoS One, 10*, 9. Article ID: e0118999.
93. Santanu, M., (2018). *In Situ* bioremediation: An overview. *Research Journal of Life Sciences, Bioinformatics, Pharmaceutical and Chemical Sciences, 4*(6), 576–597. doi: 10.26479/2018.0406.45.
94. Satinder, K. B., Verma, M., Surampalli, R. Y., Misra, K., Tyagi, R. D., & Meunier, N., (2006). Bioremediation of hazardous wastes: A review. *Pract. Period. Hazard. Toxic. Radioact. Waste Manag., 10*, 59–72.
95. Sayara, T., Borras, E., Caminal, G., Sarra, M., & Sanchez, A., (2011). Bioremediation of PAHs-contaminated soil through composting: Influence of bioaugmentation and biostimulation on contaminant biodegradation. *International Biodeterioration and Biodegradation, 65*(6), 859–865. doi: 10.1016/j.ibiod.2011.05.006.
96. Schlautman, M. A., & Morgan, J. J., (1993). Effects of aqueous chemistry on the binding of polycyclic aromatic hydrocarbons by dissolved humic materials. *Environmental Science and Technology, 27*, 961–969.
97. Shuttleworth, K. L., & Cerniglia, C. E., (1995). Environmental aspects of PAH biodegradation. *Appl. Biochem. Biotechnol., 54*, 291–302.
98. Singer, A. C., Vander-Gast, C. J., & Thompson, I. P., (2005). Perspectives and vision for strain selection in bioaugmentation. *Trends Biotechnol., 23*, 74–77.
99. Sprocati, A. R., Alisi, C., Tasso, F., Marconi, P., Sciullo, A., Pinto, V., Chiavarini, S., et al., (2011). Effectiveness of a microbial formula, as a bioaugmentation agent, tailored for bioremediation of diesel oil and heavy metal co-contaminated soil. *Process Biochemistry, 47*(11), 1649–1655. doi: 10.1016/j.procbio.2011.10.001.
100. Strong, P. J., & Burgess, J. E., (2008). Treatment methods for wine related distillery wastewaters: A review. *Biorem. Jou., 12*, 7087–7085.
101. Tak, H. I., Ahmad, F., & Babalola, O. O., (2013). Advances in the application of plant growth-promoting rhizobacteria in phytoremediation of heavy metals. In: Whitacre,

D. M., (ed.), *Reviews of Environmental Contamination and Toxicology* (Vol. 223, pp. 33–52). New York: Springer.

102. Uqab, B., Mudasir, S., Sheikh, A. Q., & Nazir, R., (2016). Bioremediation: A management tool. *J. Bioremediat. Biodegrad., 7*, 331–339. doi: 10.4172/2155- 6199.1000331.
103. Validi, M., (2001). Bioremediation. An overview. *Pure Appl. Chem., 73*(7), 163–1172.
104. Verma, J. P., & Jaiswal, D. K., (2016). Book review: Advances in biodegradation and bioremediation of industrial waste. *Front. Microbiol., 6*, 1555–1559.
105. Verma, P., George, K., Singh, S., Juwarkar, A., & Singh, R., (2006). Modeling rhizofiltration: Heavy metal uptake by plant roots. *Environmental Modeling and Assessment, 11*, 387–394.
106. Wang, M., Xu, J., Wang, J., Wang, S., Feng, H., Shentu, J., & Shen, D., (2013). Differences between 4-fluoroaniline degradation and autoinducer release by *Acinetobacter* sp. TW: Implications for operating conditions in bacterial bioaugmentation. *Environ. Sci. Pollut. Res. Int., 20*, 6201–6209.
107. Wilson, S. C., & Jones, K. C., (1993). Bioremediation of soil contaminated with polynuclear aromatic hydrocarbons (PAHs): A review. *Environ. Pollut., 81*, 229–249.
108. Wistreich, G. A., & Lechtman, M. D., (1998). *Microbiology* (p. 220). New York: Macmillan Publishing Company.
109. Yu, K. S. H., Wong, A. H. Y., Yau, K. W. Y., Wong, Y. S., & Tam, N. F. Y., (2005). Natural attenuation, biostimulation and bioaugmentation on biodegradation of polycyclic aromatic hydrocarbons (PAHs) in mangrove sediments. *Marine Pollution Bulletin, 51*(8–12), 1071–1077. doi: 10.1016/j.marpolbul.2005.06.006.
110. Zeyaullah, M., Atif, M., Islam, B., Azza, S., Abdelkafe, Sultan, P., Mohamed, A., El-Saady, & Ali, A., (2009). Bioremediation: A tool for environmental cleaning. *African Journal of Microbiology Research, 3*(6), 310–314.

CHAPTER 10

ASSESSMENT OF REMEDIATION EFFICIENCY: MICROBIAL INDICATORS OF SOIL HEALTH

LAILA SHAHZAD, UMAIR RIAZ, and FAIZA SHARIF

ABSTRACT

The current chapter highlights the significance of soil microorganisms in marinating healthy soils. Soil health is an outcome of continuous efforts of conservation and protection to represent a vital living ecosystem. An exceptional balance is therefore needed between chemical-biological (microbial) and physical components of soil to perform well. The chapter explicitly discusses soil health and the risks and benefits of utilizing microbes as early indicators of changes in the setting. There has been a growing interest in different methods and techniques of bioremediation to improve the efficiency of soil. Different techniques use different approaches to deal with the diverse nature of contaminants. The future trend is discussed at the end of the chapter to lay light on the need for scientific and cross-disciplinary research in this growing technique.

10.1 INTRODUCTION

The soil as a living medium is instrumental in performing life support functions (LSF), which safeguards the planet earth. A healthy soil performs better functions resulting in a better ecosystem. The health of soils is indispensable for the veracity of terrestrial ecosystems to bear anthropogenic pressures and disturbances, such as drought, pest invasion, climatic changes, land degradation, and pollution [1, 3, 23]. Therefore, soil health becomes imperative to have ecological processes in place today and eco-sustainability for the future.

The soil microbiome act drivers of LSF by the provision of the prolific ground of food, fibers, feeds, etc., by protecting from erosion, by acting as a sink of atmospheric CO_2, and by filtering the pollutants to enter the groundwater reservoir [15, 34]. Thus, the role of soil in combating climatic changes by absorbing greenhouse gases (GHGs) is fundamentally significant [18, 27]. However, this multifunction nature of the soil is under significant threat due to land grazing, global climate changes, prolonged desertification and droughts, pollution, and mining. An additional burden is due to urbanization, where fertile soils are increasingly converted to livable lands [11, 21, 22]. Therefore, some robust indicators are needed to assess and monitor the settings of the soil for the health of ecosystems and sustainable development at large [13, 21]. In developing and developed countries today, with growing populations and greater demands of goods and services, lands are converted more rapidly, resulting in higher degradation of environmental resources.

The degradation produces physical, chemical, and biological perturbances to the soil in terms of contaminants that can be removed from soil through remediation methods. Remediation through a biological medium is a way of recycling the contaminants to be used by another organism. Bioremediation is a process of using microbes, plants, and their enzymes to detoxify the soil contaminants and the environment [4, 22, 35].

The bioremediation enhances the rate of degradation of contaminants in soil by supplementing the indigenous microorganisms with carbon sources, nutrients, or other electron donors (bio-stimulation). Bioremediation principally relies on microbes for the conversion of contaminants into food and energy. The effectiveness of this process needs appropriate temperature, light, and nutrients, which otherwise reduce the efficiency. Favorable conditions can be generated by adding amendments to soil [12, 13, 30].

Therefore, if we have to look for soil quality and soil health in the context of sustainable soil management (which is being degraded by the anthropogenic activities mostly), we have to consider not only physical and chemicals indicators, but also biological parameters, which are becoming increasingly significant [5, 18]. Soil health indicators are further described as quantitative surrogates for ecological processes, which systematically tell us if the soil functions normally.

This chapter explores the microbial indicators for the dynamics and quality of soil health to assess the efficiency of remediation techniques. Remediation can provide a sustainable solution by using microorganisms to enhance the life of soils and ecosystems.

10.2 SOIL HEALTH

The importance of soil cannot be denied. When defining the soil health, the central theme needs to be broad enough to comprehend the multifunction it serves. To describe the soil health, there need to be specific standards, which already exist for air and water [15, 26, 28]. The standards of these mediums (air, water) indicate specific concentrations of contaminants to be allowed, which causes no harm to humans and animals [23]. However, in case of soil health, there are now developed soil regulations around the world, which tell us about their significance for ecosystems and their health.

Soil health indicates the outcome of conservation efforts and degradation process, which depend on its biological, chemical, and physical components [16]. There are many definitions of soil health given in the last decade to justify this broad term; although the soil quality is considered to explain the status of the soil. In the mid-90s, yet the term of soil health was first presented, and a broadened definition appeared as it is the capability of the soil to perform as a dynamic system, which has biological productivity by maintaining the quality of its environments for a sustainable ecosystem [13].

The definition was appreciated widely, showing the dynamic nature of soil, its functions and significance. Soil health was afterwards explained in more depth of its functions, such as the capability of soils to maintain environmental worth and promote biological throughput. It is also defined as the ability of soil to support ecosystems (animal and plants) to function for human health and habitation, by maintaining its physical, biological, and chemical media [13].

Therefore, soil health indicates the potential of soil to act as a living system to sustain plant growth and animal production, maintaining air and water quality and the environment in general. The indicators of soil health are measured in terms of its properties providing comprehensive ecological and sustainable benefits. The chemical and physical soil indicators are pH, EC, nutrient cycling, etc.; whereas the biological indicators are generally of more interest as these are closer to proper land management [29, 33]. Healthy soils are accountable for providing a habitat for the microorganisms, which play a significant role in the biogeochemical cycling of nutrients [5, 6].

With advances in science and technology, the new emphasis is on sustainable agriculture, of which healthy soils are integral to better production. Quality of soil and presence of organic matter develops a soil of improved quality leading to sustainable production of goods and services and healthy ecosystems. Through growing food demands of human beings in the 21st century, there is an increased demand for organic food, which definitely cannot be handled until and unless soils are preserved from environmental pollutions. Hence, there is a dire need of global research on soil health for envisioning the sustainable agriculture.

10.2.1 PHYSICAL, CHEMICAL, AND BIOLOGICAL COMPONENTS OF SOIL

Soil particles vary in size, shape, arrangement, and composition. These physical properties provide the soil with different soil crust, porosity, texture, structure, color, and density. On the other hand, the chemical character of soil depends on the organic, inorganic, and colloidal contents. The more the organic matter, more fertile will be the soil, whereas inorganic content decides the type of soils. Another parameter is the solid phase of soil particles, which are always distributed with water and gases. There is a continuous exchange of molecules and ions among three forms of matter, which mediates a balance between physical, biological, and chemical components of soil [8, 13]. There is still limited research done on the third component of soil, which is biological in nature. Although in some cases, it is difficult to say, which characteristics of soil comes under which component of soil. Therefore, an in-depth analysis and description is needed.

The soil texture determines the type of soil on the basis of particle diameter size, such as sand, silt, and clay. The size of particles is important to absorb and release water and nutrients. Structure of soils tells us about particle aggregation: the way particles are bounded. The aggregation stability provides resistance to external forces thus maintains healthy soils [4, 21]. Organic matter is the key ingredient of the physical properties of soils, the constituents of which are humic and fulvic acid. The organic matter highly influences the soil aggregation. Another significant soil property is water content and water retention (field capacity and permanent wilting percentage), which determines the health of soils and can be a limiting factor under stressed environment [10, 33].

The chemical properties of soil are of equal prominence, such as soil acidity (soil pH) and cation exchange capacity (CEC). CEC is influenced by the mineralogical constituents of soil and mineral types. Soil acidification is responsible for most of the chemical reactions in the soil ecosystems; and this can be enhanced by plant, animals, and soil management techniques.

10.2.2 MICROORGANISMS AS INDICATORS OF SOIL HEALTH

Microorganisms are primarily concentrated in the topsoil, from a few to 30 cm in-depth and constitute as the biological medium of soil. Although these organisms make a tiny fraction in comparison to the total biological part of the soil, yet they have a marvelous role to perform. Microorganisms are significant in decomposing the organic matter and recycling of carbon, phosphorus, nitrogen, and sulfur [7, 10].

Microorganisms are also associated with the removal of waste contents and synthetic organic compounds [30]. Microorganisms produce extracellular polysaccharides, which maintain the soil structure by having a cementing property [14]. These microorganisms therefore also affect the infiltration rate, drainage capacity, erodibility, and water holding capacity of the soil. These organisms could provide an integrated degree of soil health, as they respond and adapt rapidly to the changing environment. Therefore, the change in microbial activities and their population will serve as an indicator of change in soil health [33]. Another significant change in microbial activities leads to discovering the change in physical and chemical properties, which further alarm about soil degradation [28].

As discussed earlier, the structure and functions of good soil depend on physical, biological, and chemical fertility. The soil microorganisms do not exist in isolation, as there is a strong interaction among activities of different microbes and these interactions are the strength of the soil health. These microorganisms work in groups to enhance the soil health and productivity [11, 31]. Each of these microbes has different characteristics [25]. The microorganisms which play a significant role in developing soil health are fungi, bacteria, algae, protozoa, actinomycetes, and nematodes.

10.2.3 FRAMEWORK FOR EVALUATING SOIL HEALTH

To monitor soil health, it is imperative to have some frameworks to analyze the external forces and resulting degradation. Soil health is subjected to have some properties, which will be changed in external pressures. As a result, its performance will be changed too. The positive change can lead to positive performance and vice versa. Figure 10.1 demonstrates a simple framework, which can be considered to evaluate the performance indicators of soil health.

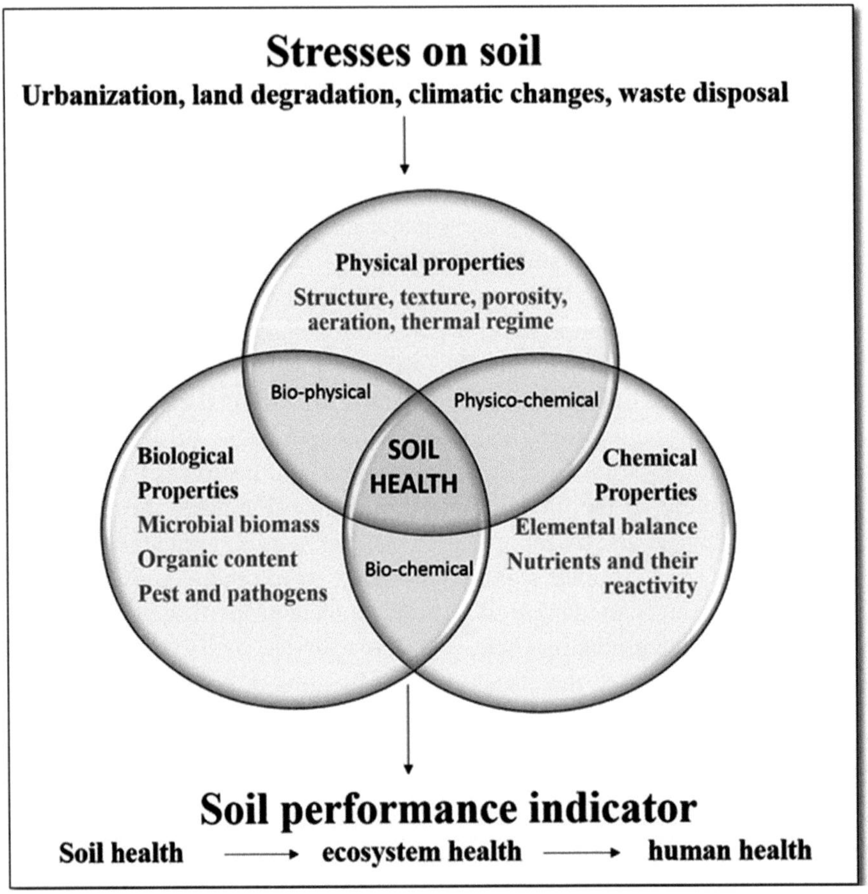

FIGURE 10.1 Framework of the soil health representative external pressures and leading to performance indicators.

10.3 SOIL ECOSYSTEM PARAMETERS AND MICROBIAL INDICATORS

A good indicator has characteristics of specificity, validity, relevance, and reliability. In terms of ecosystem parameters, the microbial indicators need to be the tools for analytical measures of soil health.

Literature has shown that there are some reliable parameters of soil ecosystem which provide desirable endpoints [17, 20]. The set of these indicators are the consistent ways to monitor the soil performance for a site, region, or at national level. There is a need for creating minimum criteria to answer the appropriateness of microbial indicators for the soil ecosystem health. Table 10.1 shows a broad category of soil ecosystem parameters and resulting performance targets. Literature has supported those microbes are complex organisms, and more research is required on the specificity of their gene markers to indicate their performance in soil medium. Therefore, soil LSF are the drivers to bring change and mostly depend upon soil microorganisms to have a framework for assessing the soil quality. Yet, there is no consensus for constituting the set of indicators for the soil quality assessment [24].

Microbial indicators provide a range of biological parameters owing to the remarkable characteristics of microbial populations within the soil environment.

TABLE 10.1 Parameters and Indicators to Attain Healthy Soils

Soil Ecosystem Parameter	Microbial Indicators	Performance Standing (Target Points)
Biodiversity	Genetic diversity	• Soil health.
	Functional diversity	• Ecosystem health.
Microbial biomass	Fungi	• Plant and animal health.
	Fungal bacterial ratio	• Soil microbial health. Atmospheric balance.
	Protozoa	
Microbial activity	Community growth physiology	
	Bacteria DNA/protein synthesis	
	Bacteriophages	
C-cycling	Decomposition of organic matter	
	Soil enzyme activity	
	Soil respiration	
N-cycling	Fixation	
	Denitrification	
	Nitrification	

10.3.1 INDICATORS OF BIODIVERSITY

Measurement of microbial diversity is suggested in soil quality monitoring systems and reflects measurements at the community level. A community's diversity is the biodiversity of the species and the relative significance of each population to the total number of organisms [6, 17]. In general, the medium to high variability is known to suggest healthy soil quality. Both current and future roles are dependent on the genomic resources available in the system [27]. A measure of genetic potential is the genetic richness of soil microorganisms. The most widely observed genetic diversity in bacteria is the variability of the 16S rDNA (ribosomal DNA) genes that exist in all bacteria and display the difference in the base composition between organisms [17, 27].

Protozoa and bacteria are common in soil and live in very complex and harsh conditions. They often emulate bacteria in that they are essential for soil health and productivity, they respond rapidly to changes in the climate, they are omnipresent, and they do not travel easily in soil [6, 13]. Protozoa are an important part of all soil systems and had been introduced as advance warning instruments. Measurement of protozoan diversity is typically achieved through taxonomic association to species, classes, or family's dependent on morphological characteristics. This approach involves experts and is challenged by the inadequate taxonomic classification of the protozoa. Protozoan abundance can instead be calculated by molecular markers [17, 27].

The soil's multiple functions are essential for the variety of functions within a microbial community. Microbial populations have shown their functional complexity to be very responsive to environmental changes. Whether the distribution of various enzymes (carbon usage patterns, extra-cellular enzyme rhythms) or the variety of organic molecules (mRNA, rRNA) inside cells will assess the functional complexity of microbial populations in soil, the later often representing the unique enzymatic processes functioning in the cells [6, 27].

Functional diversity metrics are often markers of microbial behavior and therefore reflect complexity and function. Soil enzymatic behavior is primarily microbial, originating from intracellular, cell-associated, or free metabolites [27]. In evaluating the variety of enzyme activities in soil samples, only the enzymatic action of ecto-enzymes and free enzymes are used. Segregation among free oxidative metabolism and cell-associated behavior can be accomplished by a quick filtration process to differentiate microbial cells in the soil [13, 17].

10.3.2 INDICATORS OF CARBON CYCLING

The decomposition of organic matter is an essential task of soil microorganisms. Soil microorganisms are typically heterotrophic, and are dependent on energy from outside the microorganisms of carbon resources. Biomass in soil is mainly extracted from plant species comprised of cellulose (15–60%), hemicellulose (10–30%) and lignin (5–30%) [5, 27].

The carbon cycling metrics reflect ecosystem-level calculations. In the C-cycle of all terrestrial environments, soil respiration, which seems to be the biochemical oxidation of organic compounds to CO_2 by aerobic species, especially microbes, plays a vital role [24]. This is the primary mechanism through which photosynthetically fixed carbon is put back into the atmosphere. Evaluating CO_2 production and/or absorption of O_2 will measure the physiological functions of soil microbes. Soil respiration assessment is one of the earliest and the most commonly used methods for quantifying biological populations in soil. Soil respiration correlates strongly with organic matter in the soil and biological productivity and microbial growth [5, 24, 27].

Enzymes are the main facilitators of mineral and organic soil elements for the cellular oxidation. Such catalysts hence provide a valid measurement of the reaction mechanism for critical soil processes. Enzyme function may be calculated as transition levels on *in situ* substrates or as distinct combinations if the emphasis is more analytical.

10.3.3 INDICATORS OF NITROGEN CYCLING

Soil organic nitrogen (N) mineralization by soil microorganisms by nitrate to gaseous N_2 is a very critical step of global N-cycling. This process involves nitrification, denitrification, and N_2-fixation [5]. Nitrogen cycling metrics reflect measures at the level of the environment. Organic N is mineralized by a large range of soil microbes to ammonium (NH_4^+), which represents the recycling of organic content in soil and the accessible endogenous N-pools to plants. Nitrification is a relevant indicator than the mineralization, since the method requires just a limited amount of microbes and the nitrifiers [6].

Nitrification observations reveal the nitrifiers population size, because ammonium is indeed an important substrate for these species. In fact, these quantities along with denitrification measurements can suggest ammonia accumulation in N-limited environments. Because denitrification is an anaerobic process, the extent of denitrification present in the soil is strongly reliant on abiotic stresses, such as precipitation, and compaction of soils [27]. Soil conservation activities therefore have an impact on the quantity of denitrification observed in farming activities. Denitrification quantities can suggest accumulation of ammonia in N-limited environments, along with nitrification readings [6, 27].

A simple pot test will establish the abundance and variety of rhizobium in soil, where a varied collection of legume seeds is sown in the test soil, and the amount of nodules produced after a given growth cycle are calculated. Additionally, the bacteria may also be measured using specific growth factors and biochemical and anatomical characteristics through direct extraction from the soil [6, 13, 27]. Cyanobacteria is primarily used as markers of pollution by heavy metals (HMs) (e.g., through application of waste sludge) in soils. The majority of studies revealed a negative association between the amount of cyanobacteria or nitrogenase production and the HM concentration [6, 24, 27].

10.3.4 INDICATORS OF SOIL BIOMASS

Biomass is important for soil reactions to take place, and therefore the measurement of microbial biomass is an ecosystem level indicator. Soil microbial biomass is a component of the soil essential for energy and nutrient storage and for controlling the transition of organic matter [13]. Several studies have reported strong association among the microbial biomass of the soil, rate of oxidation and N-mineralization [24]. It has also been shown that microbial biomass correlates favorably with the grain yield in organic but not in traditional farming. Lastly, microbial activity in the soil leads to soil composition and soil stabilization [13, 27]. Plant microbial biomass has indeed been proposed as soil organic matter marker.

10.3.5 INDICATORS OF MICROBIAL ACTIVITY

Microbial activity measures the soil reflect ecosystem-level assessments (e.g., pathways controlling the oxidation of agricultural residues and nutrient uptake, in particular nitrogen, sulfur, and phosphorus). Community-level tests include both bacterial DNA and protein synthesis. Bacteriophage occurrence is an indicator of population density [13, 27]. DNA synthesis is a necessary condition for the division of microbial species, so therefore an indicator of bacterial development. The synthesis of bacterial proteins is closely associated with bacterial behavior [24]. Protein synthesis tests would be more precise than DNA synthesis because of the comparatively higher protein content of cells. Through protein production, the RNA molecules, ribosomal RNA (rRNA) and messenger RNA (mRNA), perform important roles [13, 24]. Therefore, the extent of RNA in individual cells or in a population may be interpreted as a measure of protein synthesis, and hence microbial development.

10.4 FACTORS AFFECTING MICROBE ASSISTED REMEDIATION

The process of bioremediation involves bacteria, fungi, and plants to alter or remove the environmental pollutants that vary in chemical and physical nature. Microorganisms perform their activities by acting as biocatalysts, therefore they only perform well when they have access to materials, which help them to get energy and nutrients. The factors that affect the rate and efficiency of bioremediation are physicochemical nature of pollutants and the bioavailability to microorganisms. Nature of substrate, environmental conditions (EC) (EC, pH, moisture content of soil, temperature) and microbial biomass also make an immense alteration to the process of remediation [9].

10.5 FUTURE PROSPECTS OF BIOREMEDIATION FOR INCREASING SOIL HEALTH

In the current settings, industrialization (which has degraded the land, air soil and water bodies), urbanization (which has established a trend of urban growth and urban pollution) and a full scale of other anthropogenic activities are among the global concerns for the sustainable development of our planet and humans. Therefore, eradication of contaminants is a must.

Bioremediation is a natural, cost-effective, and green technology to restore the degraded soils through biological agents. This technique is emerging with new research by providing a range of different contaminants to be removed from soils and converting these into functioning conditions. Bio-surfactants are synthesized by living cells as surface-active substances; microbial surfactants are now thought to be ideal in considering their role for bioremediation. This is because of their eco-friendly behavior, selectivity, and diversity [32]. In aerobic bioremediation, oxygen acts as a rate-limiting factor. Use of oxygen releasing compound (ORC) can significantly enhance the rate of remediation. It was successfully tested with soils contaminated with kerosene [2].

Similarly, the role of tools and techniques of molecular biology and molecular genetics cannot be denied in the field of bioremediation. These can bring a revolution in the scientific arena. As most of the microbes are not cultivable in laboratory conditions, yet their particularity in soils cannot be denied. Therefore, the molecular techniques can be used to trace the genes and enzymes linked in bioremediation of targeted compounds. This can further identify the microbial floras tangled in the process.

Bioinformatics can identify and analyze metabolic pathways, gene, and protein functions, and other interactions. This technique uses computational tools to log various biodegrading abilities of microbes [19]. Nanotechnology involves the application of nanosize particles or by controlling the size to a nanosize to clean the environment from these contaminants. Nanotechnology is making its way in bioremediation (Nanoremediation) by advancing the research on characterization and designing of microbe-based treatments [12].

10.6 SUMMARY

In soil health management, soil microorganisms play a significant role in suitability and versatility. Soil microbiome has LSF in soil ecosystem health and sustainability. In an era of development and higher squalor, the health of soils is obligatory to have sustainable productivity and viable life going processes. This chapter has covered the microbial indicators of healthy soils to answer the quest of ecosystem survival. Further, a framework is presented, which shows external pressures on soil and resulting degradation. Also, it is described that the physical, chemical, and biological characteristics of soils makes a more significant difference. Much research is going on to attract policymakers and to develop a sustainable, eco-friendly solution for the diverse kinds of contaminants. Coupling of remediation techniques is essential to deal with the broad range of pollutants and environmental factors. However, fields like nanotechnology and bioinformatics are providing innovative ways to look at bioremediation to enhance the efficiency of soil microbiomes.

KEYWORDS

- **bioremediation**
- **contaminants**
- **ecosystem health**
- **microbial indicators**
- **soil health**
- **sustainable development**

REFERENCES

1. Abatenh, E., Gizaw, B., Tsegaye, Z., & Wassie, M., (2017). The role of microorganisms in bioremediation: A review. *Open Journal of Environmental Biology, 2*(1), 38–46.
2. Agarry, S. E., Owabor, C. N., & Yusuf, R. O., (2012). Enhanced bioremediation of soil artificially contaminated with kerosene: Optimization of biostimulation agents through statistical experimental design. *Journal of Petroleum and Environmental Biotechnology, 3*(3), 120–128. http://dx.doi.org/10.4172/2157-7463.1000120.
3. Antizar-Ladislao, B., Lopez-Real, J., & Beck, A. J., (2006). Investigation of organic matter dynamics during in-vessel composting of an aged coal-tar contaminated soil using fluorescence excitation-emission spectroscopy. *Chemosphere, 64*(5), 839–847.
4. Asha, A. J., Dixit, R. M., & Sharma, J., (2014). Recent trends in bioremediation. Chapter 5. In: Parmar, N., & Singh, A., (eds.), *Geomicrobiology and Biogeochemistry* (pp. 81–100). Berlin - Heidelberg: Springer.
5. Bhowmik, A., Fortuna, A. M., Cihacek, L., Bary, A., Carr, P. M., & Cogger, C. G., (2017). Potential carbon sequestration and nitrogen cycling in long term organic management systems. *Renewable Agriculture and Food Systems, 32*, 498–510.
6. Bhowmik, A., Fortuna, A. M., Cihacek, L., Bary, A., & Cogger, C. G., (2016). Use of biological indicators of soil health to estimate reactive nitrogen dynamics in long-term organic vegetable and pasture systems. *Soil Biology and Biochemistry, 103*, 308–319.
7. Bloem, J., De Ruiter, P., & Bouwman, L. A., (1997). Food webs and nutrient cycling in agro-ecosystems. In: Van, E. J. D., Trevors, J. T., & Wellington, E. M. H., (eds.), *Modern Soil Microbiology* (pp. 245–278). New York - USA: Marcel Dekker Inc.
8. Bonaventura, C., & Johnson, F. M., (1997). Healthy environments for healthy people: Bioremediation today and tomorrow. *Environmental Health Perspectives, 105*(Suppl. 1), 5–20.
9. Boopathy, R., (2000). Factors limiting bioremediation technologies. *Bioresource Technology, 74*, 63–67.
10. Brooks, R. R., (1998). Geobotany and hyperaccumulators. In: Brooks, R. R., (ed.), *Plants that Hyperaccumulate Heavy Metals: Their Role in Phytoremediation, Microbiology, Archaeology, Mineral Exploration, and Phytomining* (pp. 55–94). London: CAB International.
11. Chaudhry, T., Hayes, W., Khan, A., & Khoo, C., (1998). Phytoremediation: Focusing on accumulator plants that remediate metal-contaminated soils. *Australasian Journal of Ecotoxicology, 4*, 37–51.
12. Chauhan, R., Yadav, H. O. S., & Sehrawat, N., (2020). Nano-bioremediation: A new and a versatile tool for sustainable environmental cleanup - overview. *Journal of Materials and Environmental Sciences, 11*(4), 564–573.
13. Doran, J. W., & Safley, M., (1997). Defining and assessing soil health and sustainable productivity. In: Pankhurst, C., Doube, B. M., & Gupta, V. V. S. R., (eds.), *Biological Indicators of Soil Health* (pp. 1–28). Wallingford, Oxon, UK: CAB International.
14. Elliott, L. F., Lynch, J. M., & Papendick, R. I., (1996). The microbial component of soil quality. In: Stotzky, G., & Bollag, J. M., (eds.), *Soil Biochemistry* (pp. 1–21). New York: Marcel Dekker Inc.
15. Gan, S., Lau, E. V., & Ng, H. K., (2009). Remediation of soils contaminated with polycyclic aromatic hydrocarbons (PAHs). *Journal of Hazardous Materials, 172*(2, 3), 532–549. https://doi.org/10.1016/j.jhazmat.2009.07.118.

16. Halvorson, J. J., Smith, J. L., & Papendick, R. I., (1997). Issues of scale for evaluating soil quality. *Journal of Soil and Water Conservation, 52*, 26–30.
17. Huber, S., Syed, B., Freudenschuss, A., Ernstsen, V., & Loveland, P., (2001). *Proposal for a European Soil Monitoring and Assessment Framework* (p. 43). Technical Report No. 61, European Environment Agency, Copenhagen, Denmark.
18. Kalogerakis, N., (2011). *Ex-situ* bioremediation of contaminated soils: From biopiles to slurry phase bioreactors. Chapter 12. In: Koukkou, A. I., (ed.), *Microbial Bioremediation of Non-Metals: Current Research* (pp. 261–278). London: Caister Academic Press.
19. Khan, N. K., (2018). Integration of bioinformatics in bioremediation. *International Journal of Biomedical Data Mining, 7*(1), 9. Article ID: 1000130. http://dx.doi.org/10.4172/2090-4924.1000130.
20. Larson, W. E., & Pierce, F. J., (1994). The dynamics of soil quality as a measure of sustainable management. Chapter 3. In: Doran, J. W., Coleman, D. C., Bezdicek, D. F., & Stewart, B. A., (eds.), *Defining Soil Quality for a Sustainable Environment* (pp. 37–51). Madison, Wisconsin: Soil Science Society of America, Inc. https://doi.org/10.2136/sssaspecpub35.c3.
21. Malik, S., Andrade, S. A. L., Mirjalili, M. H., Arroo, R. R. J., Bonfill, M., & Mazzafera, P., (2016). Biotechnological approaches for bioremediation: *In vitro* hairy root culture. In: Jha, S. (ed.), *Transgenesis and Secondary Metabolism, Reference Series in Phytochemistry* (pp. 579–619). Champ, Switzerland: Springer. https://doi.org/10.1007/978-3-319-27490-4_28-1.
22. Miller, R. R., (1996). *Phytoremediation: Technology Overview* (pp. 1–26). Report TO-96-03; Pittsburgh, PA: Ground-Water Remediation, Technologies Analysis Center.
23. Parmar, N., & Singh, A., (2014). *Geomicrobiology and Biogeochemistry* (p. 81). Soil biology series 39, Verlag-Berlin, Heidelberg: Springer. https://doi.org/10.1007/978-3-642-41837-2_5.
24. Pereira-Silva, M. C., Semenov, A. V., Schmitt, H., van-Elsas, J. D., & Salles, J. F., (2012). Microbe-mediated processes as indicators to establish the normal operating range of soil functioning. *Soil Biology and Biochemistry, 57*, 995–1002.
25. Riaz, U., Abbas, Z., Zaman, Q., & Mubashir, M., (2018). Evaluation of groundwater quality for irrigation purposes and effect on crop yields: A GIS-based study of Bahawalpur. *Pakistan Journal of Agricultural Research, 31*(1), 29–36.
26. Sayara, T., Sarra, M., & Sanchez, A., (2010). Effect of compost stability and contaminant concentration on the bioremediation of PAHS contaminated soil through composting. *Journal of Hazardous Materials, 179*(1–3), 999–1006.
27. Schloter, M., Nannipieri, P., Sorensen, S. J., & Elsas, J. D., (2018). Microbial indicators for soil quality. *Biology and Fertility of Soils, 54*(1), 1–10. https://doi.org/10.1007/s00374-0 17-1248-3.
28. Sojka, R. E., & Upchurch, D. R., (1999). Reservations regarding the soil quality concept. *Soil Science Society of America Journal, 63*, 1039–1054.
29. Strong, P. J., & Burgess, J. E., (2008). Treatment methods for wine-related distillery wastewaters: A review. *Bioremediation Journal, 12*, 70–87.
30. Tomei, C. M., & Daugulis, A. J., (2013). *Ex-situ* bioremediation of contaminated soils: An overview of conventional and innovative technologies. *Critical Reviews in Environmental Science and Technology, 43*(20), 2107–2139. https://doi.org/10.1080/10643389.2012.672056.

31. Torstensson, L., Pell, M., & Stenberg, B., (1998). Need of a strategy for evaluation of arable soil quality. *AMBIO, 27*, 4–8.
32. Tribelli, P. M., Di Martino, C., Lopez, N. I., & RaigerIustman, L. J., (2012). Biofilm lifestyle enhances diesel bioremediation and biosurfactant production in the antarctic polyhydroxyalkanoate producer *Pseudomonas Extremaustralis. Biodegradation, 23*(5), 645–651.
33. Wall, D. H., & Virginia, R. A., (1999). Controls on soil biodiversity: Lessons from extreme environments. *Applied Soil Ecology, 13*, 127–150.
34. Whelan, M. J., Coulon, F., Hince, G., Rayner, J., McWatters, R., Spedding, T., & Snape, I., (2015). Fate and transport of petroleum hydrocarbons in engineered biopiles in polar regions. *Chemosphere, 131,* 232–240. http://dx.doi.org/10.1016/j.chemosphere.2014.10.088.
35. Williams, J., (2006). *Bioremediation of Contaminated Soils: A Comparison of In Situ and Ex Situ Techniques* (p. 12). Corpus ID: 9623160. http://home.eng.iastate.edu/~tge/ce421-521/jera.pdf (accessed on 3 August 2021).

CHAPTER 11

RECENT TRENDS IN TRANSGENIC PLANTS FOR ENHANCED PHYTOREMEDIATION

SHAHNAZ ANJUM, SHAYISTA YOUSUF, and URFEYA MIRZA

ABSTRACT

Plants can be genetically modified to overcome the immanent limitations of their detoxification abilities, following an approach of generation of transgenic crops capable of transforming the organic forms of contaminants into less toxic and volatile forms. The remediation efficiency of transgenic plants for various soil contaminants has also been reviewed in this chapter. Optimization of several plant processes and mechanisms is required for the proper engineering of phytoremediator plants.

11.1 INTRODUCTION

Phytoremediation began to mark its significance in the scientific literature since the 1990s, when the practical application of higher plants in detoxification was enabled through some related technologies developed in these years [16, 76, 77, 84]. Besides being able to manufacture their food from carbon dioxide (CO_2) using sunlight, plants bank on the root system to take nutrients, minerals, and water from the soil. Through diverse and radical detoxification mechanisms, plants also absorb various toxic compounds as a secondary effect [1, 24].

Plants also allow growth and maintenance of a microbial community by supplying enormous nutrients to the rhizosphere bacteria, thus, paving their way to further detoxification strategies [11, 64]. The natural defense mechanisms in plants against the range of allele-chemicals being produced by contending organisms, including microorganisms, and other plants have given rise to the pollutant degrading enzymes in these plants [1, 91]. This vantage point about plants presents them as the solar-powered and natural pump-and-treat systems for detoxifying the contaminated surroundings, guiding to the phytoremediation concept [1, 45, 71].

The phytoremediation implicates an extensive range of technologies being applied to rectify the pollutants through volatilization or degradation involving metabolism, accumulation, and stabilization. Because of its advantages as compared to conventional technologies, this method has drawn greater attention and has already been chosen as a research platform by a number of researchers [11, 17, 64]. Phytoremediation has taken the lead by making use of the naturally occurring and inimitable capabilities of the plant root systems to extract the toxic chemicals from soil, air, and other surroundings utilizing energy from sunlight, along with the bioaccumulation, translocation, and contaminant detoxification abilities of the whole plant body. Because of its execution *in situ* and minimal maintenance, phytoremediation is assessed to be tenfold cheaper when compared to the thermal, chemical, and physical remediation methods [42].

Plants play a dynamic role in diminishing the amount of contaminated soil, thus, acting as the soil stabilizers. Plants, as large biomass autotrophic structures, entail a modest amount of nutrient input and also thwart the contaminant dispersion through wind and water erosion [11, 64, 74]. Phytoremediation also turns out as the savior of the habitat, for example, the profusion of small mammals and birds harbored by poplar trees [65].

This chapter explores the scope of phytoremediation and rhizoremediation methods for cleaning up the contaminated environment using the plant genetic engineering.

11.2 CLASSES OF CONTAMINANTS

The pollutants can vary from highly toxic concentrations from the fortuitous spill to hardly visible concentrations that may prove harmful to human health after long-term exposure [2]. Phytoremediation has been used to treat a variety of contaminants, such as metals, polycyclic aromatic hydrocarbons (PAHs), solvents, explosives, petroleum, and other organic contaminants. There are two major classes of contaminants, such as organic and inorganic, based upon their physical and chemical attributes.

Organic contaminants comprise of compounds, such as PAHs, polychlorinated biphenyls (PCBs), chlorinated solvents, halogenated hydrocarbons, for example, trichloroethylene (TCE), and nitroaromatic (explosives: trinitrotoluene (TNT)). A study to estimate the capacity of the microbial community to degrade hydrocarbons and the effect of microbial remediation on soil mass showed an increase in the metabolic latency of the microbial community by phytoremediation treatments due to the alterations in its taxonomic structure [90].

An increase in the population of bacteria containing some necessary and specific genes has proved as the paramount remediation therapy to lessen the hydrocarbons in soil. Nevertheless, it is mandatory to first identify the suitable plant species responsible for alterations in microbial diversity for remediation of soil. It is the type of pollutant that helps us to estimate the probability of a plant being a phytoremediator. In a study, it was mentioned that the remediation of TNT by the use of fungi is a quite valid process; and the rhizoremediation by microbes colonizing the rhizosphere of plants will provide a competent and fast mechanism for the removal of this pollutant [26]. Some other types of contaminants also resist long in the soil, for example, persistent organic pollutants (POPs, e.g., dioxins), dichlorodiphenyltrichloroethane (DDT), and PCB.

Inorganic contaminants comprise of: radionuclides (such as chromium, strontium, uranium, cesium, tritium); heavy metals (HMs) (such as lead, mercury, cadmium); and the non-metallic compounds, e.g., arsenic. Some metals are critical for the normal life processes, but when present in greater amounts, prove extremely toxic and thus lead an individual to oxidative stress with enormous production of detrimental free radicals, which in turn are adversely harmful to the human body [12, 47]. Among these metals, the predominantly reactive ones hinder the proper functioning and structure of proteins in living systems and also prove as the major cause of substitution of various essential nutrients [30, 74, 98]. The sundry pollutants prefer the diverse technologies of phytoremediation according to their compatibility, for example, constructed wetlands have been practiced in the case of various organics, such as explosives and herbicides, and inorganics including metals, phosphates, cyanides, nitrates [43, 46, 86].

The phenomenon of hyperaccumulation has gained immense interest and might be used in the detoxification of heavy-metal contaminated soil [59, 79]. The sources of radionuclides are the atomic bomb explosions or the nuclear power plant accidents, such as in Fukushima, Japan. The rate of radioactive decay, environmental chemistry of each contaminant gives the inkling about the selection of suitable cleaning technology of these contaminated areas. The range of physicochemical methods available is ion exchange, soil washing, leaching by various chelating agents, osmosis-ultrafiltration (UF), and flocculation.

The use of biological methods in the removal of radionuclides has aroused great interest among the researchers [22]. The metal absorption by plants depends wholly on the number of metals (metal relative bioavailability) present in the contaminated area. A decrease in pH and certain other changes in the chemistry of the contaminated soil may lead to an increase in the accessibility of many metals for absorption.

11.3 FROM POLLUTED SOILS TO PLANT ADULTERATION

For the storage and translocation of both micro and macronutrients, plants have developed highly specific mechanisms, for example, maintenance of intracellular concentration of the metal ions in physiological range and transporter protein-mediated metal transport across the biological membranes [4, 5]. Generally, the uptake mechanism selectively depends on the metal transporters: their structure and properties [84, 85]. Phytoremediation has always proven to be a leading and effective clean-up technology when compared with other technologies of this function because of its less commotion of the environment, low maintenance and installation costs, and the positive side effects like production of biofuel and sundering out of carbon [19, 21]. Though phytoremediation is well adapted because of its advantages, yet some of its limitations are also been noticed, such as the potential increase in bioavailability, incomplete metabolism, and slow rate of removal of the toxic contaminants [24, 71].

Absence of substantial detoxification results in the accumulation of toxic metabolites in the plant tissues and ultimately their return to the soil or volatilization in the atmosphere. In the 1940s, the recognition of plants being capable of transforming xenobiotic compounds emerged when plants were revealed to metabolize pesticides [87]. Because of the absence of catabolic enzymes essential to achieve full mineralization of organic molecules in plants, there is an accumulation of toxic metabolites [24]; and thus, following a strategy similar to develop transgenic crops, the idea of plant biodegradation by genetic transformation was developed [19, 24].

11.4 DETOXIFICATION

Plants have developed various detoxification mechanisms in response to a diverse range of natural and synthetic toxic compounds absorbed by them [24]. The enzymes responsible for contaminant degradation in plants are originated from the natural defense systems, which exist because of the wide array of allelochemicals secreted by the competing organisms, pests, various microorganisms, and other plants as well [91]. This perspective regarding plants makes them a hero of environmental detoxification; and thus, leading to the concept of phytoremediation [71]. For the detoxification of a variety of organic contaminants (such as polyaromatic hydrocarbons, explosives, and chlorinated solvents) [71, 86].

The hyperaccumulators possess an efficient detoxification capacity by the proficient root-to-shoot transportation system, enhanced metal tolerance, and high bioconcentration factor [19, 62, 71, 86]. A study was conducted, in which transgenic poplars were developed with enhanced metabolism and uptake of volatile toxic contaminants [21]. The development of these transgenic varieties has elaborated the concept of phytoremediation.

11.5 CELLULAR MECHANISMS IN PHYTOREMEDIATION

In plants, a number of symptoms are caused as a result of the exposure to contaminants. The action of these contaminants due to possible damages of essential components can result in the inhibition of cellular activity or breakdown of cell structure [15]. In the detoxification of both organic and inorganic contaminants (such as explosives, HMs, solvents, and herbicides), plants show some latent cellular or molecular mechanisms, which can be related to the cell wall composition, plasma membrane integrity, complexation with ligands, enzymatic transformation, root environment and vacuolar compartmentalization [11, 35]. Plants can use the remediation systems based on the nature of the contaminant (whether organic or inorganic) [11, 15, 64].

11.5.1 CELL WALL AND CONTAMINATION

In contact with the plant roots, organic contaminants are occluded to the root structure. The likely uptake of these organic contaminants is also determined by their hydrophilic and hydrophobic nature. The hydrophobic organic pollutants are effectively bound by the lipid bilayer membranes and the hemicellulose present in the cell wall [71]. In the metal hyperaccumulators, cell walls are the main sources of detoxification in plants acting as metal (Ni and Zn/Cd). In the apoplast cell walls, almost 60–70% of Ni and/or Zn accumulated is distributed [50, 54, 108]; however, molecular concept of cell wall guided detoxification of metals is not quite well understood yet [108].

Various biochemical processes are used to metabolize the herbicides in higher plants and include three major phases and two enzyme systems (such as GSTs and P450s, which play key roles in herbicide degradation). Among the three phases, phase III includes the recognition and transfer of conjugates by ATP-dependent membrane pumps. These conjugates are settled in vacuoles and are carried to apoplast, or assimilated in cellulose, hemicellulose, pectin, and lignin fractions of cell-wall [58].

11.5.2 PLASMA MEMBRANE AND METAL TOLERANCE

In plants, metal transporters are involved in the uptake of metals and overall homeostasis and play a main role in the metal tolerance [109]. The uptake of metals athwart the plasma membranes is facilitated by the intracellular high affinity binding sites and transport proteins. For framing the effective strategies for the development of genetically engineered plants proficient in accumulating specific metals, there needs to be an inclusive understanding of the metal transport processes in plants [108]. The metal and alkali cation transport through the plasma membranes and organellar membranes in plants is crucial for proper growth and development, phytoremediation of the toxic metals, and signal transduction [11].

Some of the transporters of plasma membrane and tonoplast concerned with the increase of storage and metal uptake are HMA4, NRAMP (natural resistance-associated macrophage protein), YCF1 (yeast cadmium factor protein), CAX4, IRT1 (iron-regulated transporter 1), and aquaporin. In plants, numerous classes of proteins (such as ZIP (zinc-iron permease), Nramp, CPx-type (or HM) ATPases and CDF (cation diffusion facilitator) family of proteins [105]) are associated with the transportation of HMs [108, 109]. Out of these, the CPx-type HM ATPases are concerned with the transport of potentially toxic and essential metals (such as Cu, Pb, Zn, Cd) across the cell membranes [105, 108, 109]. These ATPases make use of ATP to drive a wide variety of charged molecules across the cell membranes and are notably responsible for the formation of charged intermediates during the reaction cycle [105, 108, 109]. Besides these efflux mechanisms, the maintenance of structural or physical integrity of a plasma membrane is necessary for the prevention of these HMs to enter the cell, in the presence of HMs [15, 35].

Numerous approaches have been developed to expedite volatilization in the plant tissues in case of Hg and Se; for example, to enable the increase in Hg-volatilization, specific genes from bacteria have been used for the reduction of ionic mercury [82], and Se-Cys methyltransferase and lyase have been used to increase the Se-volatilization [70].

11.5.3 ENZYMATIC CATALYSIS AND PHYTOTRANSFORMATION

In the process of enzymatic transformation in plants, there is an uptake of organic contaminants (chloroacetanilide, TNT, TCE, atrazine); and then, their transformation either fully into inorganic compounds (such as CO_2) and water or partly into stable compounds that are later stored in the plant [61]. The major classes of enzymes involved in phytodegradation of organic contaminants are peroxidases (POXs), phosphatases, peroxygenases, dehalogenases, lactases, and nitroreductases [106]. After uptake and translocation, these organic pollutants pass through three transformation stages [11, 68]:

- Chemical modification (oxidation, hydrolysis, reduction): P450 monooxygenases and carboxylesterases are the plant enzymes responsible for the catalysis in this stage of the reactions [9, 14].
- Conjugation (by glutathione (GSH), amino acids, glucose): this stage involves conjugation with GSH, amino acids or sugars, resulting in the formation of soluble and polar compounds [58]; for example, in the cytosol, a group of GSTs (glutathione S transferases) mediate the conjugation of organic pollutants to GSH [51, 66]. The way for the development of plants with enhanced tolerance to certain organic pollutants (such as phenanthrene, atrazine, and metolachlor) is paved by the overexpression of enzymes responsible for glutathione synthesis (ECS and GS) paves [28].
- Compartmentalization or sequestration (in which there is a conversion of conjugants to other ones and then their deposition in the plant vacuoles or binding to the cell wall and lignin) stage of plant metabolism also involves the soluble metabolite storage in the vacuoles or cell wall matrix.

The membrane pumps (ATP-dependent) help in the active passage of glutathione S conjugates to the apoplast or vacuole [60]. The engineering plants with higher enzyme activities involved in rate-limiting steps are important to increase the efficiency of phytoremediation of the organic contaminants.

11.6 CONTROL OF TRANSLOCATION FROM ROOTS TO SHOOTS

The metal absorption by plants depends on the relative metal bioavailability in the contaminated array. The availability of metals for absorption by the roots can be increased by certain changes in the soil chemistry, such as decrease in pH. Significant levels of metals can be taken by the plant roots under some soil conditions. In plants, the use of some chelates surges the HM translocation and it has also proved as a principal requirement for an efficient phytoremediation [8, 89]. In the soil, the plants and plant roots are capable of creating an environment able to change the availability of organic contaminants, rich in microbial activity, capable of increasing the degradation of some organic compounds, including hydrocarbons derived from petroleum. The first entry point for metal uptake in plants is the root cells, and this step is of great importance for the process of phytoextraction [1].

Efficient phytoextraction happens only if the soil metals are available to the root cells and then these absorbed metals are transported from roots to the shoots, giving rise to the term "translocation," which is predominantly controlled by two processes: leaf transpiration and root pressure. The majority of metals present in the soil are unavailable to the plant roots; therefore, the accumulation of certain chelating agents and subsequent development of the metal-chelate complexes averts the precipitation of metals in soil, thus, making these metals available for plant uptake [107].

11.7 GENETICALLY ENGINEERED PLANTS FOR PHYTOREMEDIATION

In plants, the understanding of the essential metal metabolism components can be enhanced by the suitable knowledge of molecular genetics model organisms. Regulatory control of the uptake and translocation processes in plants, a vital understanding of these activities in normal plants and metal hyperaccumulators, and usage of tissue-specific promoters have proposed great assurance that the use of molecular biology tools can provide the scientists with the skills to develop operative and commercial phytoremediation plants for the metals in soil [10, 29].

The capacity of plants to act as remediators can be improved by genetic manipulation and plant transformation technologies [72, 75, 100, 104]. For example, to overexpress the glutamylcysteine synthetase, certain plants (such as *Nicotiana tobaccum*, *Silene cucubalis* or *Populus angustifolia*) have been genetically modified to provide improved HM accumulation capability as compared to the analogous wild type plant [29]. The successful modeling of transgenic plants with enhanced phytoremediation capacity, the identification of particular genes from the hyperaccumulators, and their successive transfer to plant species showing faster growth, have proven to be advantageous [25, 53].

11.7.1 TRANSGENIC PLANTS FOR REMEDIATION OF METALS AND METALLOIDS

To enhance the efficiency of metal phytoremediation, the selected plant species are bred through genetic engineering technology. The introduction of genes from the other species is considered much faster than the conventional breeding. For the identification of plant genes encoding the metal ion transporters and extensive advancement have been made lately. It is consequently anticipated that genetic engineering in plants may offer a powerful tool to improve the remediation capacity to environmental pollutants [64, 109].

In a study conducted, *Escherichia coli (*gshl) gene was used for genetic engineering in *Brassica juncea* with a view to investigate the rate-limiting elements for phytochelatin and GSH production. The γ-ECS (γ-glutamylcysteine synthetase) transgenic saplings indicated a greater Cd tolerance and possessed higher amounts of phytochelatins, GSH, γ-GluCys, and overall nonprotein thiols as compared to the wild type seedlings [29, 69]. In plants, various genes with enhanced cadmium (Cd) resistance have been introduced so far, such as

- *AtNramps* [101];
- *AtPcrs* [93];
- *CAD1* [34] from *Arabidopsis thaliana;*
- Cadmium-induced cDNAs (complementary DNA) from *Datura innoxia* [55];
- *gshI*, *gshII* [110];
- PCS (phytochelatin synthase) cDNA clone [39] from *Brassica juncea.*

Transgenic plants bank on the overexpressing genes that are involved in the biosynthesis pathways of metal-binding peptides or proteins [5, 31, 63, 103]. For the enhanced metal uptake, three different engineering approaches have been seen [13]: (i) an increase in the number of uptake sites; (ii) modification of the specificity of uptake system to lessen the competition by unwanted cations; and (iii) increase in the intracellular binding sites. There are numerous examples of transgenic plants that are meant for metal tolerance/phytoremediation, such as

- *Arabidopsis thaliana* tolerant to Na, Cu, and Al with *Glutathione-S-transferase* gene from tobacco [27];
- Indian mustard tolerant to Se transformed with a bacterial GSH reductase in both cytoplasm and chloroplast [18];
- Tobacco and rice [32] with increased iron accumulation with gene *Ferretin* from soybean;
- Tobacco resistant to Hg and *A. thaliana* with the bacterial *merA* gene [7, 23, 80];
- Tobacco with lead accumulation and nickel tolerance with gene *Nt CBP4* from tobacco [3];
- Tobacco with the accumulation of Ca, Mn, and Cd transformed with *CAX-2* gene from *A. thaliana* [40];
- Transgenic *A. thaliana* plants expressing ArsC/SRSIp and γ-ECS/ACT 2p together showed high tolerance to As [3].

11.7.2 TRANSGENIC PLANTS FOR REMEDIATION OF HERBICIDES AND ORGANIC COMPOUNDS

Plants possess the property of removing organic contaminants (such as herbicides) by pulling them out from the environment through roots and leaves. These contaminants, once absorbed, are biochemically converted into non-toxic metabolites that later accumulate in the plant tissues, in a process known as phytoremediation [19, 21]. In the case of higher plants, several biochemical processes are used for detoxification of herbicides. Today, weeds have developed resistance against known classes of herbicides, such as enhanced detoxification, changes in the uptake, translocation, modification of the target site, and compartmentalization of the herbicide [44].

Constructed wetlands are used for several organic pollutants, such as herbicides and explosives and many inorganic pollutants (such as cyanides, nitrates, phosphates, metals) [43, 46, 88]. The metabolic capacity of the plant system began to be evaluated with the establishment of the pesticide industry lately. The majority of the modern herbicides are grounded on the discernment of crops due to the metabolic differences between various plant species, and thus forming the cornerstone of the highly profitable market of herbicides. The herbicide compounds are rapidly metabolized into the nontoxic ones by the desirable plants, whereas, the undesirable plants lack this capacity leading them to wilt or death.

The natural selection of plants gives rise to this overall mechanism, which hypothetically, proves to be quite exploitable in the remediation of contaminated soils. The majority of the toxic foreign metabolites (including herbicides) can be metabolized by plants meant for removal or degradation of toxic contaminants and rhizospheric microorganisms for enhancement of the accessibility of hydrophobic compounds. These plants act as the main reservoirs of genes involved in the metabolism of herbicides. Research study indicated that the absorption of herbicides by plants is drastically affected by the environmental conditions (EC), chemical properties of the compounds, and the other major characteristics of plant species [73].

11.7.3 TRANSGENIC PLANTS FOR RHIZOREMEDIATION OF POLLUTANTS

Till now, enormous experiments have shown the ability of transgenic plants to degrade the contaminants in the environment and that are either recalcitrant or poorly degraded by the native plant enzymatic systems. To know the alteration of the microbial diversity for soil remediation, it is necessary to identify the species of suitable plants. Several reports have proposed the use of plant-microbe strategies, particularly rhizoremediation [20, 83].

In a review article, the potential use of transgenic plants generated by the optimization of polychlorinated biphenyls (PCBs) degradation process and the expression of bacterial genes were investigated. In total, two main processes were shown in the degradation of PCB: (i) the gene expression of biphenyl 2, 3 dioxygenase (the first multi-component enzyme) in degradation pathway initiates the overall degradation process; and (ii) generation of transgenic plants with the expression of 2, 3-dihydroxybiphenyl dioxygenase would be concealing bphC and avoiding the inability of plants to metabolize the toxic dihydroxybiphenyls.

These newly generated transgenic plants were expected to show more resistance to PCBs than the wild-type ones, thus demonstrating the potential efficacy of these plants for effective rhizoremediation of PCBs [97]. According to a study, it was evaluated that oxygenated metabolism for aromatic compounds in TNT is not carried out by bacteria because of the biochemical properties of TNT-producing compounds, which are not metabolized by microorganisms [26]. Nevertheless, due to the lack of oxygen, anaerobic processes have advantages over the aerobic ones; therefore, considerable interest has been generated by the use of fungi for the bioremediation of TNT. It was later concluded that the rhizoremediation by microbes was able to colonize the plant rhizospheres to provide an efficient and competent mechanism for the removal of TNT; and the remediation of this pollutant by inorganic organisms is an effective process [26].

11.7.4 TRANSGENIC PLANTS FOR THE ENHANCED REMEDIATION OF EXPLOSIVES

Phytoremediation is proficient for the treatment of organic compounds, including polyaromatic hydrocarbons, chlorinated solvents, and explosives [1, 71, 86]. Mixed success was also achieved by the involvement of microorganisms in engineered bioremediation systems. The major explosives known for their properties, toxicity, and persistent environmental pollution are the nitro-substituted explosives, e.g., TNT, GTN (glycerol trinitrate), and RDX. Most of these explosives are toxic to almost all species of organisms, including plants, bacteria, algae, invertebrates, and mammals [41, 52, 78, 92, 96, 99].

The usage of plants has provided the best-known examples of phytodegradation for detoxification of halogenated hydrocarbons (like TCE) and explosives (like TNT) [37, 48, 67]. Large scale contamination has been due to the toxic wastes generated by decommissioning of ammunition stocks, testing, and firing on military ranges, and manufacture of explosives [38, 94]. The U.S. EPA has listed TNT and RDX as 'priority pollutants' and 'possible human carcinogens' [49, 52]. Before any treatment by incineration or landfilling, soil excavation is required for the traditional remediation of explosive-contaminated sites [36]. Plant or microorganism-mediated bioremediation represents a striking alternative for the refinement of explosive-contaminated sites.

11.8 CRITICAL FEATURES OF PLANT GENETIC ENGINEERING

Plant genetic engineering specifically represents a type of genetic variation involving an intended targeted change in the gene sequence of a plant or animal to affect an explicit result by making use of the rDNA technology. This technology is considered as a "shortcut" to obtain rapid results that can otherwise take a longer duration of time using the classical method of plant breeding. In the past to explicate the molecular and genetic mechanisms, genetic crosses between the complementary plant phenotypes have been fairly useful [57].

The sexual transference of the peculiar phytoremediation capabilities among the sister species has been observed. Since the 1980s, the probability of genetic engineering to hasten the bioremediation process has been confirmed with initial endeavors being focused on microorganisms [20, 81]. Genetic engineering has always been playing an essential role in disposing of the plants amassing HMs. The overexpression of some particular genes necessary for the uptake, transport, and metabolism makes the direct method to be applied for augmenting the efficiency of plants towards the eradication of toxic contaminants [1, 20, 56]. For the success of this venture, genes from plants, microbes, animals, and humans are being used. A critical issue in the case of heterologous genes introduced using this technology is its level of expression in the novel genetic and cellular context.

11.9 PERSPECTIVES

The environmental contamination by organic and inorganic pollutants holds great importance due to its bearings on human and animal health. Hence, the most effective and economical technologies to promote detoxification are essential in the retrieval of affected biomes. The main cause of adaptation of plants to the incidence of contaminants, including metal accumulation, metabolism, or degradation of the toxic compounds is genetic drift, mutation, migration, or natural selection. The genotypic and phenotypic plasticity can act as a cause for the increase in the adaptation and fitness of the majority of plants; and in this process, there is genome modification by various environmental factors.

There is a significant production of new gene variants that are meant for final selection, and simultaneously, the genome expression into phenotypes is being affected by various environmental factors [95]. To designate the genetic variability of natural populations using molecular markers directed towards candidate genes for accumulation and metal tolerance in the model plant-*Thlaspi caerulescens*, several attempts have been made [6]. When studying the response of a plant to environmental stress or stimuli, a surfeit of "epigenetic" phenomena is emerging [102]. In the identification of potential candidate genes for transgenic experiments, by confining the number of possible targets, proteomic experiments could prove quite helpful.

11.10 SUMMARY

Plants could be visualized as the solar-energized, natural pump and treat systems for detoxifying the contaminated environment, bringing out the idea of phytoremediation that takes benefit from the naturally occurring, selective, and unique uptake proficiencies of the plant root systems, along with bioaccumulation, translocation, and pollutant degradation or storage capabilities of the whole plant. For the enhancement of phytoremediation ability in plants, it is necessary to find new genes that could be introduced and used to develop novel transgenic plants to carry out phytoremediation through the technology of plant genetic engineering.

KEYWORDS

- **bioremediation**
- **detoxification**
- **persistent organic pollutants**
- **phytoremediation**
- **transgenic plants**
- **trichloroethylene**

REFERENCES

1. Aken, B. V., (2008). Transgenic plants for phytoremediation: Helping nature to clean up environmental pollution. *Trends in Biotechnology, 26*(5), 225–237.
2. Alexander, M., (1999). *Biodegradation and Bioremediation* (2nd edn., p. 453). Ithaca - NY, USA: Academic Press.

3. Arazi, T., Sunkar, R., Kaplan, B., & Fromm, H. A., (1999). Tobacco plasma membrane calmodulin-binding transporter confers Ni+ tolerance and Pb2+ hypersensitivity in transgenic plants. *Plant Journal, 20*, 171–182.
4. Begonia, G. B., Davis, C. D., Begonia, M. F. T., & Gray, C. N., (1998). Growth responses of Indian mustard (*Brassica juncea*) and its phytoextraction of lead from contaminated soil. *Bulletin in Environmental Contamination Toxicology, 61*, 38–43.
5. Bennett, L. E., Burkhead, J. L., Hale, K. L., Terry, N., Pilon, M., & Pilon-Smits, E. A. H., (2003). Analysis of transgenic Indian mustard plants for phytoremediation of metal-contaminated mine tailings. *Journal of Environmental Quality, 32*, 432–440.
6. Besnard, G., Basic, N., Christin, P. A., Savova-Bianchi, D., & Galland, N., (2009). *Thlaspi caerulescens* population genetics in western Switzerland: Is the genetic structure affected by natural variation of soil heavy metal concentrations? *New Phytol., 181*, 974–984.
7. Bizily, S. P., Rugh, C. L., & Meagher, R. B., (2000). Phytodetoxification of hazardous organomercurials by genetically engineered plants. *Nature Biotechnology, 18*, 213–217.
8. Blaylok, D. E., Salt, S., Dushenkov, O., Zakharova, C., Gussman, Y., Kapulnik, B. D., Ensley, Raskin, I., (1997). Enhanced accumulation of Pb in Indian mustard by soil-applied chelating agents. *Environmental Science Technology, 31*, 860–865.
9. Burken, J. G., (2003). Uptake and metabolism of organic compounds: Green-liver models. In: McCutcheon, S. C., & Schnoor, J. L., (eds.), *Phytoremediation: Transformation and Control of Contaminants* (pp. 59–84). New York: Wiley.
10. Chaney, R. L., Malik, M., Li, Y. M., Brown, S. L., Angle, J. S., & Baker, A. J. M., (1997). Phytoremediation of soil metals. *Current Opinion in Biotechnology, 8*, 279–284.
11. Cherian, S., & Oliveira, M. M., (2005). Transgenic plants in phytoremediation: Recent advances and new possibilities. *Environmental Science and Technology, 39*(24), 9377–9390.
12. Cho, U. H., & Park, J. O., (2000). Mercury induced oxidative stress in tomato seedlings. *Plant Sci., 156*, 1–9.
13. Clemens, S., Palmgren, M. G., & Krämer, U., (2002). A long way ahead: Understanding and engineering plant metal accumulation. *Trends in Plant Science,* 7, 309–314.
14. Coleman, J. O. D., Blake-Kalff, M. M. A., & Davies, T. G. E., (1997). Detoxification of xenobiotics by plants: Chemical modification and vacuolar compartmentation. *Trends in Plant Science, 2*, 144–151.
15. Coutinho, D. H., & Barbosa, A. R., (2007). *Fitorremediação: Considerações Gerais e Características de Utilização* (Phytoremediation: General considerations and usage Characteristics). *Silva Lusitana, 15*(1), 103–117.
16. Cunningham, S. D., & Berti, W. R., (1993). The remediation of contaminated soils with green plants: An overview. *In Vitro Cell Dev. Biol: Plant, 29*, 207–212.
17. Cunningham, S. D., Shann, J. R., Crowley, D. E., & Anderson, T. A., (1997). Phytoremediation of contaminated water and soil: Phytoremediation of soil and water contaminations. *ACS Symposium Series, 664*, 2–17.
18. D'Souza, M. P., Pilon-Smits, E. A. H., & Terry, N., (2000). The physiology and biochemistry of selenium volatilization by plants. In: Raskin, I., & Ensley, B. D., (eds.), *Phytoremediation of Toxic Metals: Using Plants to Clean up the Environment* (pp. 171–188). New York: Wiley.
19. Dietz, A. C., & Schnoor, J. L., (2001). Advances in phytoremediation. *Environ. Health Perspect., 109*, 163–168.

20. Doty, S. L., (2008). Enhancing phytoremediation through the use of transgenic plants and endophytes. *New Phytologist, 179*, 318–333.
21. Doty, S. L., Andrew, J. C., Moore, A. L., Vajzovic, A., Singleton, G. L., Ma, C. P., Khan, Z., et al., (2007). Enhanced phytoremediation of volatile environmental pollutants with transgenic trees. *Proc. Natl. Acad. Sci. U.S.A, 104*, 16816–16821.
22. Dushenkov, S., (2003). Trends in phytoremediation of radionuclides. *Plant and Soil, 249*, 167–175.
23. Eapen, S., & D'Souza, S. F., (2005). Prospects of genetic engineering of plants for phytoremediation of toxic metals. *Biotechnology Advances, 23*, 97–114.
24. Eapen, S., Singh, S., & D´Souza, S. F., (2007). Advances in the development of transgenic plants for remediation of xenobiotic pollutants. *Biotechnology Advances, 25*, 442–451.
25. Ellis, D. R., Sors, T. G., Brunk, D. G., Albrecht, C., Orser, C., Lahner, B., Wood, K. V., et al., (2004). Production of se-methyl-selenocysteine in transgenic plants expressing selenocysteine methyltransferase. *BMC Plant Biol., 4*, 1–10.
26. Estevez-Núnez, A., Caballero, A., & Ramos, J. L., (2001). Biological degradation of 2,4,6-trinitrotoluene. *Microbiology and Molecular Biology Reviews, 65*(3), 335–352.
27. Ezaki, B., Gardner, R. C., Ezaki, Y., & Matsumoto, H., (2000). Expression of aluminum induced genes in transgenic *Arabidopsis* plants can ameliorate aluminum stress and/or oxidative stress. *Plant Physiology, 122*, 657–665.
28. Flocco, C. G., Lindblom, S. D., & Smits, E. A., (2004). Overexpression of enzymes involved in glutathione synthesis enhances tolerance to organic pollutants in *Brassica juncea. Int. J. Phytochem., 6*(4), 289–304.
29. Fulekar, M. H., Singh, A., & Bhaduri, A. M., (2008). Genetic engineering strategies for enhancing phytoremediation of heavy metals. *African Journal of Biotechnology, 8*(4), 529–535.
30. Garbisu, C., Hernandez, A. J., Barrutia, O., Alkorta, J., & Bacerril, J. M., (2002). Phytoremediation: A technology using green plants to remove contaminants from polluted areas. *Reviews of Environmental Health, 17*(3), 173–188.
31. Gisbert, C., Ros, R., De Haro, A., Walker, D. J., Bernal, M. P., Serrano, R., & Navarro-Avino, J., (2003). A plant genetically modified that accumulates Pb is especially promising for phytoremediation. *Biochemical and Biophysical Research Communities, 303*, 440–445.
32. Goto, F., Yoshihara, T., & Saiki, H., (1998). Iron accumulation in tobacco plants expressing soybean ferritin gene. *Transgenic Research, 7*, 173–180.
33. Goto, F., Yoshihara, T., Shigemoto, N., Toki, S., & Takaiwa, F., (1999). Iron accumulation in rice seed by soybean ferritin gene. *Nature Biotechnology, 17*, 282–286.
34. Ha, S. B., Smith, A. P., Howden, R., Dietrich, W. M., Bugg, S., O'Connell, M. J., Goldsbrough, P. B., & Cobbett, C. S., (1999). Phytochelatin synthase genes from *Arabidopsis* and the yeast *Schizosac saccharomyces pombe. Plant Cell, 11*, 1153–1163.
35. Hall, J. L., (2002). Cellular mechanisms for heavy metal detoxification and tolerance. *Journal of Experimental Botany, 53*, 1–11.
36. Hannink, N. K., Rosser, S. J., & Bruce, N. C., (2002). Phytoremediation of explosives. *CRC Crit. Rev. Plant Sci., 21*; 511–538.
37. Hannink, N., Rosser, S. J., French, C. E., Basran, A., Murray, J. A. H., Nicklin, S., & Bruce, N. C., (2001). Phytoremediation of TNT by transgenic plants expressing a bacterial nitroreductase. *Nat. Biotechnol., 19*, 1168–1172.

38. Hawari, J., Beaudet, S., Halasz, A., Thiboutot, S., & Ampleman, G., (2000). Microbial degradation of explosives: Biotransformation versus mineralization. *Appl. Microbiol. Biotechnol., 54*, 605–618.
39. Heiss, S., Wachter, A., Bogs, J., Cobbett, C., & Rausch, T., (2003). Phytochelatin synthase (PCS) protein is induced in *Brassica Juncea* leaves after prolonged Cd exposure. *Journal of Experimental Botany, 54*, 1833–1839.
40. Hirschi, K. D., Korenkov, V. D., Wilganowski, N. L., & Wagner, G. J., (2000). Expression of *Arabidopsis* CAX2 in tobacco altered metal accumulation and increased manganese tolerance. *Plant Physiology, 124*, 125–133.
41. Honeycutt, M. E., Jarvis, A. S., & McFarland, V. A., (1996). Cytotoxicity and mutagenicity of 2,4,6-trinitrotoluene and its metabolites. *Ecotoxicol. Environ. Safety, 35*, 282–287.
42. Hooda, V., (2007). Phytoremediation of toxic metals from soil and wastewater. *Journal of Environmental Biology, 28*(2), 367–376.
43. Horne, A. J., (2000). Phytoremediation by constructed wetlands. In: Terry, N., & Bañuelos, G., (eds.), *Phytoremediation of Contaminated Soil and Water* (pp. 13–40). Boca Raton - FL: Lewis.
44. Hwang, J., Lee, S., & Kim, J., (2015). Interpretation and estimation for dynamic mobility of chlorpyrifos in soils containing different organic matters. *Environ. Geochem. Health, 37*(6), 1017–1027.
45. Interstate Technology and Regulatory Cooperation (ITRC), (2001). *Phytotechnology Technical and Regulatory Guidance Document.* Available from: https://connect.itrcweb.org/HigherLogic/System/DownloadDocumentFile.ashx?DocumentFileKey=9d763797-c32e-4f2b-8e07-ed81b323245c (accessed on 4 August 2021).
46. Jacobson, M. E., Chiang, S. Y., Gueriguian, L., Weshtholm, L. R., & Pierson, J., (2003). Transformation kinetics of trinitrotoluene conversion in aquatic plants. In: McCutcheon, S. C., & Schnoor, J. L., (eds.), *Phytoremediation: Transformation and Control of Contaminants* (pp. 409–427). New York: Wiley.
47. Jones, P., Kortenkamp, A., O'Brien, P., Wang, G., & Yang, G., (1991). Evidence for the generation of hydroxyl radicals from a chromium (V) intermediate isolated from the reaction of chromate with glutathione. *Biochim. Biophys. Acta, 286*, 652–655.
48. Kassel, A. G., Ghoshal, D., & Goyal, A., (2002). Phytoremediation of trichloroethylene using hybrid poplar. *Physiol. Mol. Plants, 8*(1), 3–10.
49. Keith, L. H., & Telliard, W. A., (1979). Priority pollutants: Perspective view. *Environ. Sci. Technol., 13*, 416–423.
50. Krämer, U., Pickering, I. J., Prince, R. C., Raskin, I., & Salt, D. E., (2000). Subcellular localization and speculation of nickel in hyperaccumulator and non-accumulator *Thlaspi* species. *Plant Physiology, 122*, 1343–1353.
51. Kreuz, K., Tommasini, R., & Martinoia, E., (1996). Old enzymes for a new job: Herbicide detoxification in plants. *Plant Physiology, 111*, 349–353.
52. Lachance, B., Robidoux, P. Y., Hawari, J., Ampleman, G., Thiboutot, S., & Sunahara, G. I., (1999). Cytotoxic and genotoxic effects of energetic compounds on bacterial and mammalian cells *in vitro*. *Mutat. Res., 444*, 25–39.
53. LeDuc, D. L., Tarun, A. S., Montes-Bayon, M., Meija, J., Malit, M. F., Wu, C. P., Abdel, S. M., et al., (2004). Overexpression of selenocysteine methyltransferase in *Arabidopsis*

and Indian mustard increases selenium tolerance and accumulation. *Plant Physiol., 135*(1), 377–383.
54. Li, T. Q., Yang, X. E., He, Z. L., Yang, J. Y., & Long, X. X., (2005). Zn accumulation and subcellular distribution in leaves of Zn hyperaccumulator species of *Sedum alfredii* hance. *Pedosphere, 16*(5), 616–623.
55. Louie, M., Kondor, N., & Witt, J. G., (2003). Gene expression in cadmium-tolerant *Datura Innoxia*: Detection and characterization of cDNAs induced in response to Cd2+. *Plant Molecular Biology, 52*, 81–89.
56. Macek, T., Kotrba, P., Svatos, A., Novakova, M., Demnerova, K., & Mackova, M., (2008). Novel roles for genetically modified plants in environmental protection. *Trends in Biotechnology, 26*, 146–152.
57. Maestri, E., Marmiroli, M., Visioli, G., & Marmiroli, N., (2010). metal tolerance and hyperaccumulation: Costs and trade-offs between traits and environment. *Environ. Exp. Bot., 68*, 1–13.
58. Marrs, K. A., (1996). The functions and regulation of glutathione S-transferase in plants. *Annual. Review of Plant Physiology and Plant Molecular Biology, 47*, 127–158.
59. Martínez, M., Bernal, P., Almela, C., Vélez, D., García-Agustín, P., Serrano, R., & Navarro- Aviñó, J., (2006). An engineered plant that accumulates higher levels of heavy metals than *Thlaspi caerulescens*, with yields of 100 times more biomass in mine soils. *Chemosphere, 64*, 478–485.
60. Martinoia, E., Grill, E., Tomasini, R., Kreuz, K., & Amrehin, N., (1993). ATP-dependent glutathione S-conjugate 'export' pump in the vacuolar membrane of plants. *Nature, 364*, 247–249.
61. McCutcheon, S. C., & Schnoor, J. L., (2003). Overview of phytotransformation and control of wastes. In: McCutcheon, S. C., & Schnoor, J. L., (eds.), *Phytoremediation: Transformation and Control of Contaminants* (pp. 3–58). New York: Wiley.
62. McGrath, S. P., & Zhao, F. J., (2003). Phytoextraction of metals and metalloids from contaminated soils. *Curr. Opin. Biotechnol., 14*, 277–282.
63. Mejäre, M., & Bülow, L., (2001). Metal-binding proteins and peptides in bioremediation and phytoremediation of heavy metals. *Trends in Biotechnology, 19*, 67–73.
64. Mello-Farias, P. C., & Chaves, A. L. S., (2008). Biochemical and molecular aspects of toxic metals phytoremediation using transgenic plants. In: Tiznado-Hernandez, M. E., Troncoso-Rojas, R., & Rivera, D. M. A., (eds.), *Transgenic Approach in Plant Biochemistry and Physiology* (pp. 253–266). Kerala - India: Research Signpost.
65. Moser, B. W., Pipas, M. J., Witmer, G. W., & Engeman, R. M., (2002). Small mammal use of hybrid poplar plantations relative to stand age. *Northwest Science, 76*, 158–165.
66. Neuefeind, T., Reinemer, P., & Bieseler, B., (1997). Plant glutathione S-transferases and herbicide detoxification. *Biological Chemistry, 378*, 199–205.
67. Newman, L. A., Strand, S. E., Choe, N., Duffy, J., Ekuan, G., Ruszaj, M., Shurtleff, B. B., et al., (1997). Uptake and biotransformation of trichloroethylene by hybrid poplars. *Environ. Sci. Technol., 31*, 1062–1067.
68. Ohkawa, H., Imaishi, H., & Shiota, N., (1999). In: Brooks, J. T., & Roberts, T. R., (eds.), *Pesticide Chemistry and Biosciences: The Food- Environment Challenge* (pp. 259–264). Cambridge - UK: Royal Society of Chemistry.
69. Ow, D. W., (1996). Heavy metal tolerance genes-prospective tools for bioremediation. *Resources, Conservation and Recycling, 18*, 135–149.

70. Pilon-Smits, E. A. H., & LeDuc, D. L., (2009). Phytoremediation of selenium using transgenic plants. *Current Opinion Biotechnol., 20*, 207–212.
71. Pilon-Smits, E., (2005). Phytoremediation. *Annu. Rev. Plant Biol., 56*, 15–39.
72. Pilon-Smits, E., & Pilon, M., (2002). Phytoremediation of metals using transgenic plants. *Crit. Rev. Plant Sci., 21*(5), 439–456.
73. Pires, F. R., Souza, C. M., Silva, A. A., Procópio, S. O., & Ferreira, L. R., (2003). Fitorremediação de solos contaminados com herbicidas (Phytoremediation of contaminated soils with herbicide). *Planta Daninha, 21*(2), 335–341.
74. Pulford, I. D., & Watson, C., (2003). Phytoremediation of heavy metal-contaminated land by trees: A review. *Environment International, 29*, 529–540.
75. Raskin, I., (1996). Plant genetic engineering may help with environmental cleanup. *Proc. Natl. Acad. Sci. U.S.A., 93*(8), 3164–3166.
76. Raskin, I., Kumar, P. B. A. N., Dushenkov, S., & Salt, D. E., (1994). Bioconcentration of heavy metals by plants. *Curr Opin Biotechnol., 5*, 285–290.
77. Reeves, R. D., Baker, A. J. M., Borhidi, A., & Berazain, R., (1999). Nickel hyperaccumulation in the serpentine flora of Cuba. *Ann. Bot., 83*, 29–38.
78. Robidoux, P. Y., Bardai, G., Paquet, L., Ampleman, G., Thiboutot, S., Hawari, J., & Sunahara, G. I., (2003). Phytotoxicity of 2,4,6-trinitrotoluene (TNT) and octahydro-1,3,5,7-tetranitro-1,3,5,7-tetrazocine (HMX) in spiked artificial and natural forest soils. *Arch. Environ. Contamin. Toxicol., 44*, 198–209.
79. Robinson, B. H., Brooks, R. R., Howes, A. W., Kirkman, J. H., & Gregg, P. E. H., (1997). The potential of the high biomass nickel hyperaccumulator *Berkheya coddii* for phytoremediation and phytomining. *Journal of Geochemical Exploration, 60*, 115–126.
80. Rugh, C. L., Bizily, S. P., & Meagher, R. B., (2000). Phytoremediation of environmental mercury pollution. In: Raskin, I., & Ensley, B. D., (eds.), *Phytoremediation of Toxic Metals Using Plants to Clean Up the Environment* (pp. 151–171). New York: Wiley and Sons.
81. Rugh, C. L., Seueoff, J. F., Meagher, R. B., & Merkle, S. A., (1998). Development of transgenic yellow poplar for mercury phytoremediation. *Nature Biotechnology, 16*, 925–928.
82. Ruiz, O. N., & Daniell, H., (2009). Genetic engineering to enhance mercury phytoremediation. *Curr. Opin. Biotechnol., 20*, 213–219.
83. Rylott, E. L., & Bruce, N. C., (2009). Plants and soil: Engineering plants for the phytoremediation of explosives. *Trends in Biotechnology, 27*, 73–81.
84. Salt, D. E., Blaylock, M., Kumar, P. B. A. N., Dushenkov, V., Ensley, B. D., Chet, I., & Raskin, I., (1995). Phytoremediation: A novel strategy for the removal of toxic metals from the environment using plants. *Biotechnology, 13*, 468–475.
85. Salt, D. E., Prince, R. C., Pickering, I. J., & Raskin, I., (1995). Mechanisms of cadmium mobility and accumulation in Indian mustard. *Plant Physiology, 109*, 1427–1433.
86. Salt, D. E., Smith, R. D., & Raskin, I., (1998). Phytoremediation. *Annual Review of Plant Physiology and Plant Molecular Biology, 49*, 643–668.
87. Sandermann, H., (1994). Higher plant metabolism of xenobiotics: The green liver concept. *Pharmacogenetics, 4*, 225–241.
88. Schnoor, J. L., Licht, L. A., McCutcheon, S. C., Wolf, N. L., & Carreira, L. H., (1995). Phytoremediation of organic and nutrient contaminants. *Environmental Science and Technology, 29*, 318–323.

89. Seth, C. S., Chaturvedi, P. K., & Mishra, V., (2008). The role of phytochelatins and antioxidants in tolerance to Cd accumulation in *Brassica juncea* L. *Ecotoxicology Environmental Safety, 71*, 76–85.
90. Siciliano, S. D., Germida, J. J., Banks, K., & Green, C. W., (2003). Changes in microbial community composition and during a polyaromatic hydrocarbon phytoremediation field trial. *Applied and Environmental Microbiology, 69*(1), 483–489.
91. Singer, A., (2006). The chemical ecology of pollutants biodegradation. In: Mackova, M., (ed.), *Phytoremediation and Rhizoremediation: Theoretical Background* (pp. 5–21). Cham: Springer.
92. Smock, L. A., Stoneburner, D. L., & Clark, J. R., (1976). The toxic effects of trinitrotoluene (TNT) and its primary degradation products on two species of algae and the fathead minnow. *Water Res., 10*, 537–543.
93. Song, W. Y., Martinoia, E., Lee, J., Kim, D., Kim, D. Y., Vogt, E., Shim, D., et al., (2004). A novel family of Cys-rich membrane proteins mediates cadmium resistance in *Arabidopsis. Plant Physiology, 135*, 1027–1039.
94. Spain, J. C., (1995). Biodegradation of nitroaromatic compounds. *Annu. Rev. Microbiol., 49*, 523–555.
95. Sultan, S. E., (2000). Phenotypic plasticity for plant development, function and life history. *Trends Plant Sci., 5*, 537–542.
96. Sunahara, G. I., Dodard, S., Sarrazin, M., Paquet, L., Hawari, J., Greer, C. W., Ampleman, G., et al., (1999). Ecotoxicological characterization of energetic substances using a soil extraction procedure. *Ecotoxicol. Environ. Safety, 43*, 138–148.
97. Sylvestre, M., Macek, T., & Mackova, M., (2009). Transgenic plants to improve rhizoremediation of polychlorinated biphenyls (PCBs). *Current Opinion in Biotechnology, 20*, 242–247.
98. Taiz, L., & Zeiger, E., (2002). *Plant Physiology* (3rd edn., p. 690). London - UK: Sinauer Associates.
99. Talmage, S. S., Opresko, D. M., Maxwell, C. J., Welsh, C. J., Cretella, F. M., Reno, P. H., & Daniel, F. B., (1999). Nitroaromatic munitions compounds: Environmental effects and screening values. *Rev. Environ. Contam. Toxicol., 161*, 1–156.
100. Terry, N., Sambukumar, S. V., & LeDuc, D. L., (2003). Biotechnological approaches for enhancing phytoremediation of heavy metals and metalloids. *Acta Biotechnol., 23*(2, 3), 281–288.
101. Thomine, S., Wang, R., Ward, J. M., Crawford, N. M., & Schroeder, J. I., (2000). Cadmium and iron transport by members of a plant metal transporter family in *Arabidopsis* with homology to *Nramp* genes. In: *Proceedings of the National Academy of Sciences USA* (Vol. 97, pp. 4991–4996).
102. Urano, K., Kurihara, Y., Seki, M., & Shinozaki, K., (2010). "Omics" analyses of regulatory networks in plant abiotic stress responses. *Curr. Opin. Plant Biol., 13*, 132–138.
103. Zhu, Y. L., Pilon-Smits, E. A. H., Tarun, A. S., Weber, S. U., Jouanin, L., & Terry, N., (1999). Cadmium tolerance and accumulation in Indian mustard is enhanced by overexpressing g-glutamyl cysteine synthetase. *Plant Physiology, 121*, 1169–1177.
104. Van, H. T., Terry, N., & Pilon-Smits, E. A., (2004). Exploring the selenium phytoremediation potential of transgenic indian mustard overexpressing ATP sulfurylase or cystathionine gamma-synthase. *Int. J. Phytorem., 6*(2), 111–118.

105. Williams, L. E., Pittman, J. K., & Hall, J. L., (2000). Emerging mechanisms for heavy metal transport in plants. *Biochimica et Biophysica Acta, 1465*, 104–126.
106. Wolfe, N. L., & Hoehamer, C. F., (2003). Enzymes are used by plants and microorganisms to detoxify organic compounds. In: McCutcheon, S. C., & Schnoor, J. L., (eds.), *Phytoremediation: Transformation and Control of Contaminants* (pp. 159–187). New York: Wiley.
107. Wu, L. H., Luo, Y. M., Xing, X. R., & Christie, P., (2004). EDTA-enhanced phytoremediation of heavy metal contaminated soil with indian mustard and associated potential leaching risk. *Agriculture Ecosystems and Environment, 102*, 307–318.
108. Yang, X., Feng, Y., He, Z., & Stoffella, P., (2005). Molecular mechanisms of heavy metal hyperaccumulation and phytoremediation. *Journal of Trace Elements in Medicine and Biology, 18*, 339–353.
109. Yang, X., Jin, X. F., Feng, Y., & Islam, E., (2005). Molecular mechanisms and genetic bases of heavy metal tolerance/ hyperaccumulation in plants. *Journal of Integrative Plant Biology, 47*(9), 1025–1035.
110. Zhu, Y. L., Pilon-Smits, E. A. H., Jouanin, L., & Terry, N., (1999). Overexpression of glutathione synthetase in indian mustard enhances cadmium accumulation and tolerance. *Plant Physiology, 119*, 73–79.

CHAPTER 12

NANOBIOREMEDIATION: AN INTEGRATED APPROACH FOR MANAGEMENT OF CONTAMINATED SITES

RAYEES AHMAD ZARGAR, KHURSHEED AHMAD WANI, and JUNAID AHMAD MALIK

ABSTRACT

Nanotechnological approaches seek to reduce contaminant concentrations to a level that is accepted by the soil ecosystems and may not cause deleterious effects to the subsequent food chains. Nanomaterials can eliminate the toxins from the soil ecosystem to a level that are considered to be safe. Further, nanoscience is a more effective, comparatively fast, and eco-friendly remediation technique as compared to other processes. This chapter offers a brief description of nanotechnology, polymer-based nanomaterials, iron oxide nanomaterials, carbon-based nanomaterials, and other nanomaterials for removing the pollutants from soil matrices.

12.1 INTRODUCTION

Soil is made up of organic and inorganic substances (such as oxides, silicate, phosphate, carbonate, sulfate rock minerals). Different microorganisms are also part of the soil ecosystem. The soil is contaminated by organic waste biomaterials and chemical substances, such as heavy metals (HMs), pesticides, chemical fertilizers, etc. [10, 28, 71]. HM toxins are discharged into the soil primarily by the anthropogenic activities, such as ore excavating, plating, chemical factories, agro-wastes, and domestic wastewater, etc.

Adulterated soil (due to HMs, fertilizers, and pesticides) has been reported to have direct effects as well as impairs the ecological balance and thereby impends the whole environment. Hence, the elimination of toxic substances from the contaminated soil has turn out to be a major environmental need [17, 44].

Nanotechnology and nanobioremediation can help to remove the contaminants owing to its minute particle size, high surface area to volume ratio, simple infusion at the spot location, suppleness for both *in situ* and *ex situ* presentations. In recent years, the advent of nanotechnology has been the prime focus of comprehensive research, interconnecting with numerous additional discourses of science and influencing all parts of life [49]. It is among the fastest-growing fields of logical science and innovation around the globe [81].

Nanotechnology provides unlimited benefits to the environment for its conservation and remediation, which are categorized into: treatment and remediation, observations, and detections, and avoidance of contaminants. The basic nanotechnologies in this chapter focus on the site remediation and treatment of soil contaminated with various pollutants.

Smaller molecule size takes into consideration the arrangement of smaller sensors; which can be used successfully in far-off areas. When food production demand increases for the expanding human population, the demand and consumption of land to satisfy these needs also increases. However, at the same time, soil contamination is becoming a limiting factor that needs to be addressed and nanoscience seems to be a viable option. For example, the nanoparticles (NPs) of manganese oxide (17β-estradiol MnO_2) are known to eliminate 88% of estrogens from the soil. It is a well-known fact that the reduction in the injection speed and increase in the concentration of NPs enhances the degradation of estrogen [31].

Soil remediation by nano-remediation may provide a sustainable solution for rejuvenating the depleted soil resources. Applications dependent on nanotechnology are cost-productive and simple to utilize, and propose increasingly successful strategies for treatment and remediation that can significantly reduce the soil contamination [30].

The combination of NPs utilizing microorganisms is a quickly developing field of examination in green nanotechnology around the world, with different organic elements being utilized in the amalgamation of NPs continually giving an imputable option to NPs customary compound and physical techniques [65, 68, 69]. Nanotechnologies are vectors of foundational applications in our economy. Recent developments in nanotechnology-based advancements have consolidated numerous innovations together in a single system. Nanotechnology could decontaminate the site in an eco-friendly and time efficient manner [3, 11, 23, 54, 65, 68–70]. Sustainable remediation is aimed to diminish the grouping of contaminants to the precise levels on the basis of the pollutant type and its hazard.

Nano-bioremediation is a coordinated innovation that fuses nanotechnology and bioremediation to accomplish a progressively productive, less tedious, and ecologically well-disposed remediation than the individual procedures that are in opposition to ordinary bioremediation. Palladium NPs are known to decontaminate the soil polluted with pentachlorobiphenyl. Palladium NPs combined with supercritical liquid CO_2 are fit for separating entire PCBs from the soil at 200 atm as well as within a broad range of temperatures. Cheng et al. [14] found that graphene oxide silver nanoparticles (rGO-Ag) could display photocatalytic elimination of such toxic substances.

This chapter focuses on nanotechnology and nanoremediation-based decontamination of the adulterated soils. It also discusses the use of polymer-based nanomaterials, iron oxide nanomaterials, carbon-based nanomaterials and other nanomaterials for removing pollutants from soil matrices.

12.2 TOXIC EFFECTS OF HEAVY METALS (HMS) IN POLLUTED SOILS

HMs (such as As, Cd, Cu, Ni, Zn, Pb, Hg, etc.), are very poisonous and cancer-causing elements and belong to the group of trace elements (metals and metalloids) with elemental density >5 g/cm^3 [3, 24]. The natural resources of toxic metal ions enter into soil/wastewater from the volcanic activities, soil erosion, and weathering of rocks and minerals and anthropogenic activities (such as fuel combustion, mineral processing, landfills, urban run-offs, mining industrial effluents, agricultural activities, metal finishing and plating, printed board manufacturing, textile dyes, semiconductor manufacturing, etc.). Owing to their high stability, migration rate and solubility in water, these are known to induce severe environmental changes and health hazards, such as

- Arsenic is introduced into water and soil by both natural and man-made sources; and it causes lung and bladder cancer, and excessive consumption may lead to death [24, 25].

- Cadmium acute inhalation for short interval of time can result in flu like symptoms and damages lungs, but its chronic exposure causes kidney, bones, and lung diseases.
- Excessive zinc develops Nausea, vomiting, loss of appetite, stomach pain, headache, diarrhea.
- Lead damages kidney, liver, reproductive system, basic cellular processes, brain functions, and so forth.
- Mercury vapor effects the nervous, digestive, and immune systems, lungs, and kidneys, and may be fatal if exposed for long duration [3, 25]. The inorganic salts of mercury are corrosive to the skin, eyes, and gastrointestinal tract, and induce kidney toxicity if ingested.
- The persistent long-term exposure of copper causes irritation in eyes, nose, and mouth, headache, stomach ache, dizziness, vomiting, and diarrhea and high intake may cause liver and kidney problems and even death [3, 8].
- Though Ni and Zn as micronutrients are required by human beings yet their excessive concentrations are however harmful. Nickel/compound dust, when inhaled by the workers doing jobs in refineries or nickel processing plants causes chronic bronchitis, nasal sinus, lung problem, and cancer.

12.3 NANOTECHNOLOGY AND SOIL REMEDIATION

The different types of nanomaterials synthesized have been found to be effective in remediating various pollutants from soil. These types of nanomaterials can enhance *in situ* treatment with high capacity to remove or degrade the pollutants. The main classes of nanomaterials in soil remediation are (i) polymer-based NPs; (ii) nanoscale metal oxides; and (iii) carbon-based nanomaterials.

Other NPs (such as MnO_2, and calcium peroxide (CP) NPs, etc.), are used for the removal of Chromium and for degradation HM contaminants. These methods are novel in approach; however, further studies are recommended for their wider applicability [55].

12.3.1 POLYMER-BASED NANOMATERIALS

Despite the fact that the high area-to-volume proportion of nanomaterials subsidizes more to the reactivity with concomitant enhanced productivity, because of lack of practicality, aggregation rate, non-applicability; and low strength can restrict the utilization of these nanotechnological methods. An option, in contrast to improving the dependability of nano-scale materials, is the utilization of a host material, the reason for which is polymers are generally utilized for the recognition and disposal of pollutants (e.g., manganese, nitrate, iron, asbestos, and so on.), gasses (e.g., CO, SO_2, NO_x), compound toxins (e.g., aliphatic, and aromatic hydrocarbons, pharmaceuticals, or VOCs) and a wide assortment of biologicals (e.g., organisms, growths, infections, and so on.).

Polymeric hosts (for instance, surfactants, emulsifiers, stabilizers) are also used to improve security, stability, and solve the disadvantages of pristine NPs, and to impart some of the desirable properties, such as, improved mechanical force, warm soundness, versatility, and recyclability of the material concerned.

Amphiphilic polyurethane (APU) NPs have been created to remediate polynuclear aromatic hydrocarbons (PAHs) from soil, approving the speculation that natural NPs with desired properties can be designed [79]. The hydrophilic covering of the NPs rejuvenates its portability in the soil, whereas the inner hydrophobic side of the substance offers affinity to the natural hydrophobic pollutants.

Phenanthrene was expelled from polluted aquifer sand by APU NPs (e.g., 80% recovery). The affinity of APU NPs for phenanthrene was expanded, when the size of the hydrophobic backbone expanded correspondingly. An increase in the quantity of ionic groups on the precursor chain also prompted a decrease in the aggregation of APU particles in presence of multipurpose cations [79]. While the utilization of such materials in nature could be helpful for the remediation of toxins, there is no report on the biodegradability of these materials, which adds the doubts about their purpose.

For wastewater remediation, poly(amidoamine) or dendrimers (PAMAM) are utilized for water tests adulterated with metal particles, for example, Cu^{2+}. Such dendritic nanopolymers encompass functional groups (for example, essential amines, carboxylates, and hydroxamates) that can epitomize an enormous assortment of water solutes, and cations (e.g., Cu^{2+}, Ag^{+}, Au^{+}, $Fe^{2+,}$ Fe^{3+}, Ni^{2+}, Zn^{2+}, and U^{6+}) [19]. They are utilized as chelating agents and ultrafilters to bind with metal particles and along these lines, reassure water cleansing. They have also been utilized as antibacterial/antivirus specialists [7].

The primary status of the dendritic nanopolymers is their low affinity to pass through the apertures of the ultrafiltration (UF) films than straight polymers of proportional atomic weight polymers because of their lower polydispersity and spherical form. Consequently, they were utilized to create procedures of UF and microfiltration (MF) to extract dissolved ions from aqueous solutions. Next, the reaction of functionalized dendritic nanopolymers is linked with polluted water, which then shifts to UF or MF units and the attached nanopolymer-contaminants to obtain the pure water. They can be disengaged by changing the system acidity (pH), and afterwards the recuperated concentrated solution is utilized for eradication, and again the nanopolymers can be reclaimed [4].

Guerra et al. [30]; and Campbel et al. [9] identified the use of functionalized biologically degradable and non-toxic polymeric nanomaterials to capture VOCs in a target-specific way through a different approach and concluded that biodegradability is an essential and desirable feature. Non-toxicity is one more significant issue, which ought to be addressed while planning a novel ecological remediation apparatus.

Polymer nanocomposites (PNCs) are composed of materials, which use a polymer as a key component on which NPs are scattered or speckled. This substance incorporates the required characteristics of the polymer host (i.e., excellent mechanical resistance) as well as the NPs (i.e., improved reactivity from its wide surface area). In the manufacture of membranes to integrate metal and metal oxide NPs for ecological decontamination applications, specific polymers have also been used [40, 59, 78, 84].

Polymers are commonly used as source materials in polymeric nanocomposites, such as NPs are responsible for the remediation of pollutants [91]. Nevertheless, the functionalized polymeric nanomaterials were also recognized as key agents accountable for the process of remediation. To build target-specific remediation technologies, the polymeric nanomaterials have been utilized for the compounds extracted from gaseous mixtures to avoid off-target entangling that otherwise may contribute to a decline in material output [9].

Polymer/inorganic mixed nanomaterials have also been researched for different applications in environmental management and studied in particular for the adsorptive elimination of numerous harmful pollutants in the wastewater; as they depict high stability to chemical and thermal properties [59, 79, 91]. However, these hybrid materials also display great capability for selective HM sorption from aqueous media. These were synthesized by using the modified techniques, such as sol-gel processes, self-assembly techniques, and assembly of nanobuilding parts, collecting the characteristics of both materials [79].

Khare et al. [39] produced chitosan-based carbon nanofibers (CNFs), which were combined with iron oxide NPs along with polyvinyl alcohol nanocomposite films. This unusual combination of materials revealed an effective water adsorption ability of Cr^{6+} and demonstrated a high metal absorption potential (~80 mg/g of composite chitosan/iron-CNF). Mittal et al. [52] reported that the adsorption of SiO_2 NPs in acrylamide gel increases the monolayer adsorption ability of the acrylamide hydrogel to remove 1408.67 mg/g of cationic dyes (MB) (efficacy of 96%) with a 0.2 g/L dose of hydrogel nanocomposite.

Apart from that absorption method has helped to remove methylene blue (MB) from aqueous solution by the help of SiO_2-NPs and gum karaya grafted with poly (acrylic acid co-acrylamide) nanocomposite containing hydrogel. Such results revealed that the implanted copolymerized gel karaya hydrogel could also be used to remove MB dye from industrial wastewater as an ecologically stable and effective adsorbent.

Biocompatibility and biodegradability are two main fundamentals for the utilization of polymeric nanocomposites. Sun et al. [76], for example, developed a green hybrid nanomaterial adsorbent composed of wheat xylan/poly(acrylic acid) NP hydrogel that encapsulates Fe_3O_4 NPs. The magnetic hydrogel content achieved 90% removal of MB dye with this formulation. Similarly, by formulating a nanogel composed of sodium alginate loaded with tetrasodium thiacalixarene tetrasulfonate and Fe_3O_4 [43, 76], high adsorption potential of Ni^{2+}, Co^{2+}, Pb^{2+}, Cd^{2+}, Cu^{2+}, and Cr^{2+} ions from aqueous solution were realized.

Ge et al. [24] prepared another formulation of Fe_3O_4 magnetic NP, modified by a mixture of 3-aminopropyltriethoxysilane and (acrylic/crotonic acid) copolymers. Then, the resulting hybrid NPs (diameter of 15–20 nm) were applied to extract Cu^{2+}, Cd^{2+}, Pb^{2+}, and Zn^{2+} from polluted metal water. The antibacterial activity has been observed when AgNPs was mixed with cellulose acetate fibers [67].

Antimicrobial properties against *Escherichia coli, Staphylococcus aureus, Aspergillus niger,* and *Salmonella enterica* have also been demonstrated, when AgNPs and Ag^+ were blended with polymethoxybenzyl and poly(L-lactic acid) co-poly(3-caprolactone) nanofibers by the same community [20, 36, 57, 63]. Furthermore, dispersing AgNPs (1–70 nm) into polysulfone membranes did not change the structure of the membrane. The impregnation to the NCs membrane of just 0.9 wt.% of AgNPs caused a 99% decrease in the *E. coli* bacteria, which could grow on the surface of the membrane. In addition, these AgNPs have been able to lower an *E coli* by 94% bacteria suspension on the surface of the submerged membrane [89]. Therefore, both polymer NPs and polymer NCs play a critical role in environmental remediation to effectively remove several pollutants through different mechanisms.

Currently the polymeric NPs have been used for the detection and removal of pesticides and HMs from the wastewater and groundwater only. However, recent advances and perspectives advocate those polymeric nanomaterials are also expected to play a significant role in the remediation of soil contaminated with varied toxins.

12.3.2 *IRON OXIDE NANOMATERIALS*

Iron is non-toxic in nature, highly absorbent, eco-friendly, and cost effective. Based on these characteristics, the development of iron-based remediation technologies has rapidly gained attention for environmental remediation [62]. Because of bonding of Fe ions to As, the Fe-based NPs can restore As-polluted soils [75]. The nano-scale zero-valent iron (nZVI) materials are now widely used as compared to traditional conventional iron powder to increase the efficiency of adsorption capability due to greater surface area [24, 25].

The ZVI NPs concentration has a direct effect on the metal removal process. The basic mechanisms involved in removal of HM (Hg^{2+}, Cd^{2+}, Ni^{2+}, Cr^{6+} and Pb^{2+}) are adsorption and reduction because of controllable particle size, through which HMs are transformed into less toxic species, immobilized in soil [8, 17, 44, 45]. On the other side, HMs (Cr, As, Cu) with huge positive E^0 value in comparison to ZVI NPs can be reduced easily in the soil.

The nZVI, when added with a metal catalyst (mainly Pd, Ni), has been a useful candidate for degradation of chlorinated solvents (such as PCBs, DCE, TCE) owing to its robust reducing ability and eco-friendly nature [80, 86]. Following equations show the reaction mechanism of ZVI corrosion and reduction of aliphatic chlorinated organics via ZVI [86]:

$$Fe^0 \rightarrow Fe^{2+} + 2e^- \quad (E^0 = 0.44 \text{ V}) \tag{1}$$

$$RCl + 2e^- + H^+ \rightarrow RH + Cl^- \quad (E^0 = 0.5 - 1.5 \text{ V, pH} = 7) \tag{2}$$

Usually, Pd is one of the cost effective and widely used components for the reduction of aliphatic chlorinated organic compounds [33]. Figure 12.1 depicts the mechanisms involved for reductive dechlorination process by nZVI with and without catalyst Pd. The Fe-Pd bimetal NPs can provide much quicker reaction rates along with better removal efficiency with no accumulation of waste as compared to monometallic nZVI [46].

The significant reactions comprise H_2 production, adsorption of H_2 on the metals, reactive hydrogen species formation and dechlorination reactions. Pd repressed in the iron oxide shell has been less reactive than on the iron coat [85]. It should be noted that nZVI's removal mechanism for a variety of HMs may vary by standard potential (E^0) values of HMs according to Huang et al. [34] and is useful for identification of reduction and adsorption process.

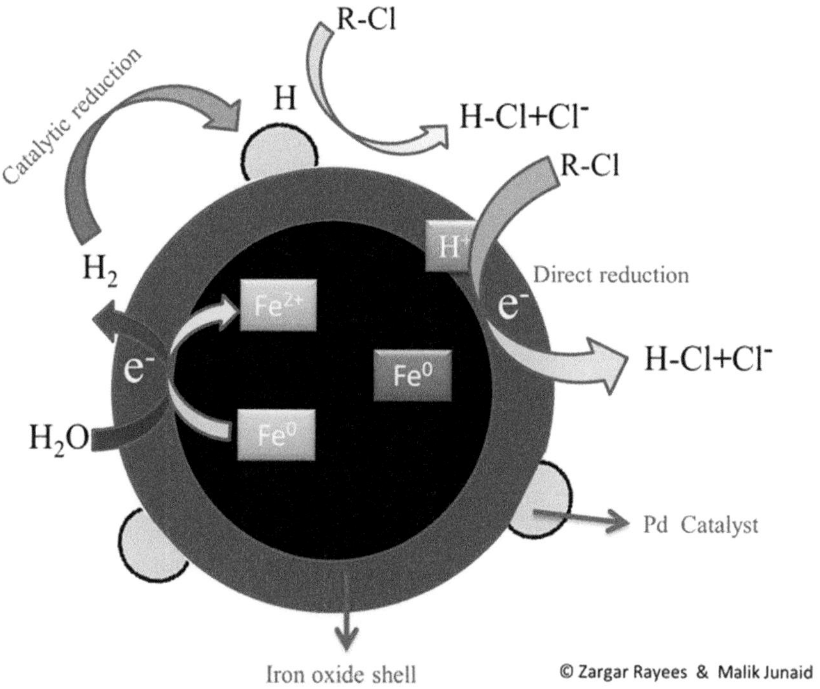

FIGURE 12.1 Pathways of reductive dechlorination through nZVI (with and without Pd catalyst).

Source: Modified from Ref. [92].

12.3.3 CARBON-BASED NANOMATERIALS

Carbon nanotubes (CNTs), which because of their high specific surface area, can attract a large number of organic wastes into its hydrophobic and micro-porous characters [61]. The mean diameter of pores is close to that of single atom molecule. Generally, CNT's are classified into two groups: single-walled carbon nanotubes (SWCNTs) and multi-walled carbon nanotubes (MWCNTs). SWCNTs have a diameter ranging from 0.5 to 1.5 nm compared to >100 nm for MWCNTs. Both can remove organic and inorganic pollutants from the soil [51]. Oxygen can affect the capturing capacity of CNTs. The synthesis mechanism also affects the efficiency of CNTs apart from the purification phenomenon. The structural changes bring about a new class of materials, such as changeable states of hybridization are grapheme, fullerene C_{60}, single-walled nanotubes, multi-walled nanotubes, and fullerene C_{540} [35].

The microbiota has a key function in altering the nature of different pollutants in the soil, and CNTs may aid in this process [41, 47, 60]. It is advocated that CNTs may vigorously fluctuate the chemical composition and microscopic functioning of the soil ecosystems, which will help to remediate the soils polluted with various pollutants [73]. However, at the same time, the toxicity of CNTs have been reported by various researchers for growing concerns of health [5, 56]. Further, low removal efficiency, restricted selectivity, and complexity [53] and low absorption capacities are the other issues with CNTs.

In this context, activated carbon (AC) is the ideal carbon material for removal of metallic chemical elements from the soil; besides (AC) is cheap, highly efficient, eco-friendly, and can be derived from non-usable materials or by-products [58]. Because of its excellent properties, AC is recommended to reduce the phytotoxicity of several residues of herbicides and many harmful chemicals from agricultural soils [74]. AC has also been used to clean up contaminated soil ecosystems with of primary pollutants, such as hydrocarbons, chlorinated solvents, polychlorinated biphenyls (PCB), and pentachlorophenol.

Figure 12.2 depicts the simple procedure how the biowaste is physically and chemically activated and converted into the carbon nanomaterial, which can be directly used in the remediation of contaminated soil. Cheng et al. [13, 14] reported that surface-modified AC is used for soil polluted with Cu and Zn through adsorption and complexion. They oxidized AC with KOH and nitric acid to introduce functional groups, which in turn increased surface cation exchange and complexion character of carbon. They concluded that the quantity of copper and zinc was amassed in the stem and root of ryegrass and it was considerably diminished with the escalation in AC added to the adulterated soil.

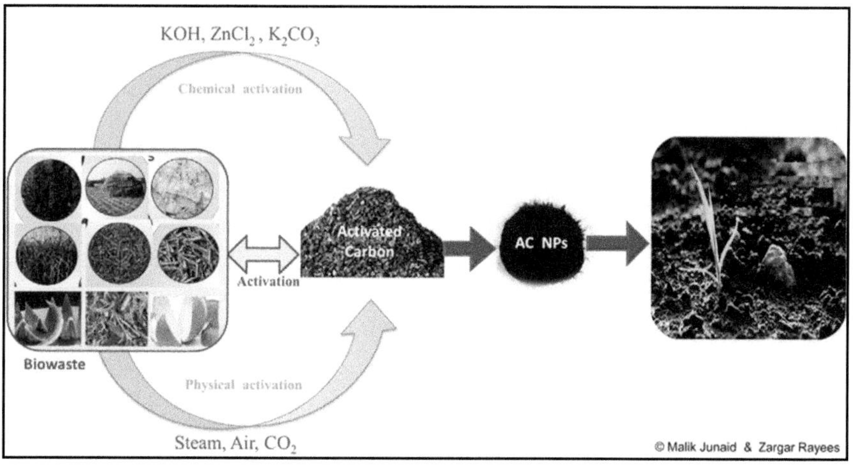

FIGURE 12.2 Derived activated carbon (AC) as a nano-fertilizer.

12.3.4 OTHER NANOMATERIALS

The use of soil amendments and improvements has been progressively been employed in soils. The nanotechnology delivers some sustainable remediation options to reduce the availability of harmful metals [15]. Chromium (Cr) is a HM, which is easily mobile in all types of soils [64]. Chromium (Cr) is essentially a highly toxic element and is a major threat for soil ecosystem [16]. Cr concentration in soils considerably may vary based on the composition of rocks and sediments, and the level of Cr increases due to atmospheric deposition or by solid wastes [21]. Generally, Chromium (III and IV) has been found in most agricultural soils, and its concentration varies up to 350 mg/kg, which is harmful to humans [6, 15, 16, 64].

The naturally occurring oxidizing agents of Cr is MnO_2, because of greater affinity for a variety of cations. Han et al. [32] synthesized MnO_2 NPs using carboxy-methyl-cellulose (CMC) as a stabilized and proved alternative for chemical degradation of polluted soil. Yuhuan et al. [88] also suggested the higher proportion of oxidized chromium in the soil. After leaching with Mn, amount of content of the oxidized Cr showed a significant decrease, and thus it may be used for remediation of Cr-contaminated soil.

Recently, CP-based NPs have been used as pollution eliminating agents [18]. Calcium peroxide (CaO_2: CP) has found its use in inorganic synthesis [50]. According to the literature survey, the dissolved oxygen (DO) concentration relatively shows an alarming increase if and when CP is added directly, which may demand oxygenation [22]. The reactive nature in CP changes due to changes in the concentrations with other materials. The CP in the modified iron can be added externally as Fe(II) (more common oxide of iron) or Fe(III) (also known as mineral hematite) salts or as telluric iron-containing minerals (e.g., goethite, pyrite, and magnetite) [29]. The remedial decomposition of CP producing free radicals is achieved by using natural iron as catalyst [26, 27].

One more important reason for using the multi-ferrite combination is that it provides strength to these nanostructures [82, 87]. Photo-catalysis is also an invincible feature of oxide NPs as they too show characteristics of semi-conductor materials; and this photo-catalysis phenomenon helps to degrade the dyes, which contaminate the soil [42].

The probe compounds and EPR (electrochemical potentiokinetic reactivation) tests identified the formation of hydroxide anion and O_2 from CP and their (HO and O_2) assistances to target pollutants degradation were assessed with free radical scavengers. PCE and TCE oxidation by HO resulted in complete transformation to HCl, CO_2 and H_2O. Microbial degradation of hydrocarbons takes place in the range of pH of 6.5–7.5 [48].

The soil conditions can be properly maintained and improved with the help of nano-fertilizers [12], where only a small percentage of the fertilizer is used compared with the conventional fertilizers. This technique has helped to retain the nutrient uptake in the rhizosphere, increase the uptake of beneficial molecules, and repair damaged soils [37]. Particles smaller than 2 micrometers belong to colloidal fractions, and there is the possibility to recognize the particles within a range of 1 to 100 nanometers as well that can be promoted for the remediation of soil. The nano-electron acceptors and donors, such as fulvic and humic acids can play a beneficial role in the biodegradation of contaminants in soil [83].

The 50% of minerals in the aerosol category come from world deserts, which include hydroxides, metal oxides, and oxyhydroxide-based metallic elements (such as Fe, Mn, and Al), from which different types of NPs have been synthesized [2, 72].

Gibbsite and Boehmite in soil are the aluminum nanostructures accomplished by geological processes and the iron hydroxides are natural NPs in soil that have a role in the process of nutrients absorption without causing any internal damage to soil [77]. The amalgamation of nanoscience with biohydrometallurgical processes facilitated the degradation of minerals. Other interesting processes are bioaccumulation, bioflotation, bioprecipitation, and biomineralization and are now thought to be eco-friendly and more efficient alternative way for polluted soil bioremediation [66].

12.4 DISADVANTAGES OF NANOBIOREMEDIATION

The high degree of aggregation that usually restricts movement of these NPs has been observed, due to their coating with suitable polymers. Although different behavior and different course of action of these NPs at different scales has been observed by Klaine et al. [41], yet specific surface and different chemical properties are embedded in these NPs for their application (which are not found in natural particles). Irrespective of this, NPs display certain limitations, such as pulmonary toxicity, oxidative damage due to generation of Fe(IV) [1, 38]. The adverse health effects from exposure to final uptake is certainly a major issue that must be addressed.

12.5 FUTURE PROSPECTS

As part of a scientific approach, the biodegradable materials provide very cumbersome and sophisticated applications for the treatment of pollutants. Also, for consumer acceptance and non-waste production, this is by far the more efficient technology. Nanotechnology, in combination with physical and chemical changes of surface of materials is focused to obtain suitable mixture materials for contaminant remediation. Nanomaterials are relatively target specific, cost effective, non-toxic, recyclable when developed under various favorable conditions, which are sometimes very difficult to achieve under normal environmental factors. There are also some operations needed to prevent agglomeration, enhance mono-dispersity and increased stability. Production of toxic NPs as by-products from remediation sites also limits their use. Hence, a more prominent nanotechnological bioremediation method is required, which should be easily implemented, with efficient performance and little or no production of toxic by-products.

12.6 SUMMARY

The integration of nanoremediation with the help of polymer-based nanomaterials, iron oxide nanomaterials, carbon-based nanomaterials, and other nanomaterials to remove different types of pollutants is a feasible alternative to conventional remediation technologies either sequentially or simultaneously. There is still a lack of information on the synergetic impact of NPs and polymer science; iron oxide nanomaterials, carbon-based NPs during a cycle of nanobioremediation and how these are combined to react to pollutants in a diverse nature. Present nanotechnologies may be used to decontaminate polluted soil environment, but more cost-effective manufacturing methods of different polymers need to be conceived. Regulatory structure is an important concern on the use of nanomaterials that need to be taken into consideration.

ACKNOWLEDGMENTS

The author R. A. Zargar expresses sincere thanks and acknowledges the financial support provided by the BGSB University under TEQIP-III research grant.

KEYWORDS

- **amphiphilic polyurethane**
- **carbon nanofibers**
- **carbon nanotubes**
- **nanoremediation**
- **pollutant**
- **polymers**

REFERENCES

1. Auffan, M., Achouak, W., Rose, J., Roncato, M. A., Chaneac, C., Waite, D. T., & Bottero, J. Y., (2008). Relation between the redox state of iron-based nanoparticles and their cytotoxicity toward *Escherichia coli. Environmental Science & Technology, 42*(17), 6730–6735.
2. Bakshi, S., He, Z. L., & Harris, W. G., (2015). Natural nanoparticles: Implications for environment and human health. *Critical Reviews in Environmental Science and Technology, 45*(8), 861–904.
3. Bharagava, R. N., Saxena, G., & Chowdhary, P., (2017). Constructed wetlands: An emerging phytotechnology for degradation and detoxification of industrial wastewaters.

In: Bharagava, R. N., (ed.), *Environmental Pollutants and their Bioremediation Approaches* (pp. 397–426). Boca Raton - FL: CRC Press, Taylor & Francis Group.
4. Biricova, V., & Laznickova, A., (2009). Dendrimers: Analytical characterization and applications. *Bioorg. Chem., 37*, 185–192.
5. Bottini, M., Bruckner, S., Nika, K., Bottini, N., Bellucci, S., Magrini, A., Bergamaschi, A., & Mustelin, T., (2006). Multi-walled carbon nanotubes induce T-lymphocyte apoptosis. *Toxicol. Lett., 160*(2), 121–126.
6. Branca, M., Dessi, A., Kozlowski, H., Micera, G., & Swiatek, J., (1990). Reduction of chromate ions by glutathione tripeptide in the presence of sugar ligands. *Journal of Inorganic Biochemistry, 39*, 217–226.
7. Brinkmann, N., Giebel, D., Lohmer, G., Reetz, M. T., & Kragl, U., (1999). Allylic substitution with dendritic palladium catalysts in a continuously operating membrane reactor. *J. Catal., 183*, 163–168.
8. Caliman, F. A., Robu, B. M., Smaranda, C., Pavel, V. L., & Gavrilescu, M., (2011). Soil and groundwater cleanup: Benefits and limits of emerging technologies. *Clean Techn. Environ. Policy, 13*, 241–268.
9. Campbell, M. L., Guerra, F. D., Dhulekar, J., Alexis, F., & Whitehead, D. C., (2015). Target-specific capture of environmentally relevant gaseous aldehydes and carboxylic acids with functional nanoparticles. *Chem. A Eur. J., 21*, 14834–14842.
10. Cetin, M., & Sevik, H., (2016). Measuring the impact of selected plants on indoor CO2 concentrations. *Pol. J. Environ. Stud., 25*(3), 973–980.
11. Chandra, R., Saxena, G., & Kumar, V., (2015). Phytoremediation of environmental pollutants: An eco-sustainable green technology to environmental management. In: Chandra R., (ed.), *Advances in Biodegradation and Bioremediation of Industrial Waste* (1st edn., pp. 1–30). Boca Raton - FL: CRC Press, Taylor & Francis Group.
12. Chen, H., & Yada, R., (2011). Nanotechnologies in agriculture: New tools for sustainable development. *Trends in Food Science & Technology, 22*(11), 585–594.
13. Cheng, J. M., Liu, Y. Z., & Wang, H. W., (2014). Effects of surface-modified nano-scale carbon black on Cu and Zn fractionations in contaminated soil. *Int. J. Phytoremediation, 16*, 86–94.
14. Cheng, J., Fan, M., Wang, P., & Su, X. O., (2020). The twice-oxidized graphene oxide/gold nanoparticles composite SERS substrate for sensitive detection of clenbuterol residues in animal-origin food samples. *Food Analytical Methods, 13*(4), 1–9.
15. Clare, M. M., Neil, D. G., Janette, Tourney, J., Davenport, R. J., Wade, M., Finlay, N., & Johnson, K. L., (2015). Remediation of a historically Pb contaminated soil using a model natural Mn oxide waste. *Chemosphere, 138*, 211–217.
16. Cohen, M. D., Kargacin, B., Klein, C. B., & Costa, M., (1993). Mechanisms of chromium carcinogenicity and toxicity. *Crit. Rev. Toxicol., 23*, 255–281.
17. Cundy, A. B., Hopkinson, L., & Whitby, R. L. D., (2008). Use of iron-based technologies in contaminated land and groundwater remediation: A review. *Sci. Total Environ., 400*, 42–51.
18. Dhal, B., Thatoi, H. N., Das, N. N., & Pandey, B. D., (2013). Chemical and microbial remediation of hexavalent chromium from contaminated soil and mining/metallurgical solid waste: A review. *J. Hazard. Mater., 250*, 272–291.
19. Diallo, M. S., Christie, S., Swaminathan, P., Johnson, J. H., & Goddard, W. A., (2005). Dendrimer enhanced ultrafiltration, part I: Recovery of Cu(II) from aqueous solutions

using PAMAM dendrimers with ethylene diamine core and terminal NH2 groups. *Environ. Sci. Technol., 39*, 1366–1377.

20. Elechiguerra, J. L., Burt, J. L., Morones, J. R., Camacho-Bragado, A., Gao, X., Lara, H. H., & Yacaman, M. J., (2005). Interaction of silver nanoparticles with HIV-1. *J. Nanobiotechnol., 3*, 6–13.
21. Fendorf, S. E., & Zasoski, R. J., (1992). Chromium (III) oxidation by delta-manganese oxide (MnO2)1 characterization. *Environmental Science & Technology, 26*(1), 79–85.
22. Fox, R., & Tuchman, M., (1996). The assessment and remediation of contaminated sediments (ARCS) program. *J. Great Lakes Res., 22*, 493–494.
23. Gautam, S., Kaithwas, G., Bharagava, R. N., & Saxena, G., (2017). Pollutants in tannery wastewater, pharmacological effects and bioremediation approaches for human health protection and environmental safety. In: Bharagava, R. N., (ed.), *Environmental Pollutants and their Bioremediation Approaches* (pp. 369–396). Boca Raton -FL: CRC Press, Taylor & Francis Group.
24. Ge, F., Li, M. M., Ye, H., & Zhao, B. X., (2012). Effective removal of heavy metal ions Cd2+, Zn2+, Pb2+, Cu2+ from aqueous solution by polymer-modified magnetic nanoparticles. *J. Hazard. Mater., 211*, 366–372.
25. Gil-Díaz, M., Alonso, J., Rodríguez-Valdés, E., Pinilla, P., & Lobo, M. C., (2014). Reducing the mobility of arsenic in brownfield soil using stabilized zero-valent iron nanoparticles. *J. Environ. Sci. Heal., 49*, 1361–1369.
26. Goi, A., & Trapido, M., (2010). Chlorophenols contaminated soil remediation by peroxidation. *Journal of Advanced Oxidation Technologies, 13*(1), 50–58.
27. Goi, M., Viisimaa, O., & Karpenko, O., (2012). DDT-contaminated soil treatment with persulfate and hydrogen peroxide utilizing different activation aids and the chemicals combination with biosurfactant. *J. Adv. Oxid. Technol., 15*(1), 41–52.
28. Gouda, S., Kerryr, G., Das, G., Paramitiotis, S., Shin, H. S., & Patra, J. K., (2018). Revitalization of plant growth promoting rhizobacteria for sustainable development in agriculture. *Microbiol. Res., 206*, 131–138.
29. Gryzenia, J., Cassidy, D., & Hampton, D., (2009). Production and accumulation of surfactants during the chemical oxidation of PAH in soil. *Chemosphere, 77*, 540–545.
30. Guerra, F. D., Campbell, M. L., Whitehead, D. C., & Alexis, F., (2017). Tunable properties of functional nanoparticles for efficient capture of VOCs. *Chemistry Select., 2*, 9889–9894.
31. Han, B., Zhang, M., & Zhao, D., (2017). In-situ degradation of soil-sorbed 17β-estradiol using carboxymethyl cellulose stabilized manganese oxide nanoparticles: Column studies. *Environmental Pollution, 223*, 238–246.
32. Han, B., Zhang, M., Zhao, D., & Feng, Y., (2015). Degradation of aqueous and soil-sorbed estradiol using a new class of stabilized manganese oxide nanoparticles. *Water Research, 70*, 288–299.
33. He, F., & Zhao, D., (2008). Hydrodechlorination of trichloroethene using stabilized Fe-Pd nanoparticles: Reaction mechanism and effects of stabilizers, catalysts and reaction conditions. *Appl. Catal. B: Environ., 84*(3), 533–540.
34. Huang, P., Ye, Z., Xie, W., Chen, Q., Li, J., Xu, Z., & Yao, M., (2013). Rapid magnetic removal of aqueous heavy metals and their relevant mechanisms using nanoscale zero-valent iron (nZVI) particles. *Water Res., 47*, 4050–4058.
35. Iijima, S., (1991). Helical microtubules of graphitic carbon. *Nature, 354*, 56–58.

36. Jin, G., Prabhakaran, M. P., Nadappuram, B. P., Singh, G., Kai, D., & Ramakrishna, S., (2012). Electrospun poly (l-lactic acid)-co-poly (ϵ-caprolactone) nanofibers containing silver nanoparticles for skin-tissue engineering. *J. Biomater. Sci. Polym., 23*, 2337–2352.
37. Johnston, C. T., (2010). Probing the nanoscale architecture of clay minerals. *Clay Minerals, 45*(3), 245–279.
38. Keenan, C. R., & Sedlak, D. L., (2008). Factors affecting the yield of oxidants from the reaction of nanoparticulate zero-valent iron and oxygen. *Environ. Sci. Technol., 42*(4), 1262–1267.
39. Khare, P., Yadav, A., Ramkumar, J., & Verma, N., (2016). Microchannel-embedded metal-carbon-polymer nanocomposite as a novel support for chitosan for efficient removal of hexavalent chromium from water under dynamic conditions. *Chem. Eng. J., 293*, 44–54.
40. Kim, J., & Der - Bruggen, B. V., (2010). The use of nanoparticles in polymeric and ceramic membrane structures: Review of manufacturing procedures and performance improvement for water treatment. *Environ. Pollut., 158*, 2335–2349.
41. Klaine, S. J., Alvarez, P. J., Batley, G. E., Fernandes, T. F., Handy, R. D., Lyon, D. Y., & Lead, J. R., (2008). Nanomaterials in the environment: Behavior, fate, bioavailability, and effects. *Environmental Toxicology and Chemistry: An International Journal, 27*(9), 1825–1851.
42. Korsvik, C., Patil, S., Seal, S., & Self, W. T., (2007). Superoxide dismutase mimetic properties exhibited by vacancy engineered ceria nanoparticles. *Chemical Communications, 10*, 1056–1058.
43. Lakouraj, M. M., Mojerlou, F., & Zare, E. N., (2014). Nanogel and superparamagnetic nanocomposite based on sodium alginate for sorption of heavy metal ions. *Carbohydr. Polym., 106*, 34–41.
44. Li, L., Fan, M., Brown, R. C., Leeuwen, J. H. V., Wang, J., Wang, W., Song, Y., & Zhang, P., (2006). Synthesis, properties, and environmental applications of nanoscale iron-based materials: A review. *Crit. Rev. Environ. Sci. Technol., 36*, 405–431.
45. Li, X. Q., Elliott, D. W., & Zhang, W. X., (2006). Zero-Valent iron nanoparticles for abatement of environmental pollutants: materials and engineering aspects. *Critical Reviews in Solid State and Materials Sciences, 31*(4), 111–122.
46. Lowry, G. V., & Reinhard, M., (1999). Hydrodehalogenation of 1- to 3-carbon halogenated organic compounds in water using a palladium catalyst and hydrogen gas. *Environ. Sci. Tech., 33*(11), 1905–1910.
47. Madsen, E. L., (2011). Microorganisms and their roles in fundamental biogeochemical cycles. *Curr. Opin. Biotechnol., 22*, 456–464.
48. Małachowska-Jutsz, A., & Niesler, M., (2015). The effect of calcium peroxide on the phenoloxidase and acid phosphatase activity and removal of fluoranthene from soil. *Water, Air, & Soil Pollution, 226*(11), 365–372.
49. Masciangoli, T., & Zhang, W., (2003). Environmental technologies. *Environ. Sci. Technol., 37*, 102–108.
50. Massalimov, I. A., Shayakhmetov, A. U., & Mustafin, A. G., (2010). Specific features of thermal decomposition of mechanically activated calcium peroxide. *Russ. J. Appl. Chem., 83*, 1794–1798.
51. Matos, M. P., Correia, A. A. S., & Rasteiro, M. G., (2017). Application of carbon nanotubes to immobilize heavy metals in contaminated soils. *Journal of Nanoparticle Research, 19*(4), 126–134.

52. Mittal, H., Maity, A., & Ray, S. S., (2015). Synthesis of co-polymer-grafted gum karaya and silica hybrid organic-inorganic hydrogel nanocomposite for the highly effective removal of methylene blue. *Chem. Eng. J., 279*, 166–179.
53. Moghaddam, H. K., & Pakizeh, M., (2015). Experimental study on mercury ions removal from aqueous solution by MnO2/CNTs nanocomposite adsorbent. *J. Ind. Eng. Chem., 21*, 221–229.
54. Mosa, K. A., Saadoun, I., Kumar, K., Helmy, M., & Dhankher, O. P., (2016). Potential biotechnological strategies for the cleanup of heavy metals and metalloids. *Front Plant Sci., 7*, 303–311.
55. Mueller, N. C., & Nowack, B., (2010). Nanoparticles for remediation: Solving big problem with little particles. *Elements, 6*, 395–400.
56. Muller, J., Huaux, F., & Lison, D., (2006). Respiratory toxicity of carbon nanotubes: How worried should we be? *Carbon, 44*(6), 1048–1056.
57. Nair, A. S., Binoy, N. P., Ramakrishna, S., Kurup, T. R. R., Chan, L. W., Goh, C. H., Islam, M. R., et al., (2009). Organic-soluble antimicrobial silver nanoparticle-polymer composites in gram scale by one-pot synthesis. *ACS Appl. Mater. Interfaces., 1*, 2413–2419.
58. Nethaji, S., Sivasamy, A., & Mandal, A. B., (2013). Preparation and characterization of corn cob activated carbon coated with nano-sized magnetite particles for the removal of Cr(VI). *Bioresour. Technol., 134*, 94–100.
59. Ng, L. Y., Mohammad, A. W., Leo, C. P., & Hilal, N., (2013). Polymeric membranes incorporated with metal/metal oxide nanoparticles: A comprehensive review. *Desalination, 308*, 15–33.
60. Nowack, B., & Bucheli, T. D., (2007). Occurrence, behavior and effects of nanoparticles in the environment. *Environ. Pollut., 150*, 5–22.
61. O'Brien, G., (1992). Estimation of the removal of organic priority pollutants by the powdered activated carbon treatment process. *Water Environ. Res., 64*, 877–883.
62. O'Carroll, D., Sleep, B., Krol, M., Boparai, H., & Kocur, C., (2013). Nanoscale zero-valent iron and bimetallic particles for contaminated site remediation. *Adv. Water Resour, 51*, 104–122.
63. Pal, S., Tak, Y. K., & Song, J. M., (2007). Does the antibacterial activity of silver nanoparticles depend on the shape of the nanoparticle? A study of the gram-negative bacterium *Escherichia coli. Appl. Environ. Microbiol., 73*, 1712–1720.
64. Palmer, C. D., & Wittbrodt, P. R., (1991). Processes affecting the remediation of chromium-contaminated sites. *Environ. Health Perspect., 92*, 25–40.
65. Perelo, L. W., (2010). *In situ* and bioremediation of organic pollutants in aquatic sediments. *J Hazard Mater., 177*(1–3), 81–89.
66. Pollmann, K., Kutschke, S., Matys, S., Kostudis, S., Hopfe, S., & Raff, J., (2016). Novel biotechnological approaches for the recovery of metals from primary and secondary resources. *Minerals, 6*(2), 54.
67. Rieger, K. A., Cho, H. J., Yeung, H. F., Fan, W., & Schiffman, J. D., (2016). Antimicrobial activity of silver ions released from zeolites immobilized on cellulose nanofiber mats. *ACS Appl. Mater. Interfaces., 8*, 3032–3040.
68. Saxena, G., & Bharagava, R. N., (2017). Organic and inorganic pollutants in industrial wastes, their ecotoxicological effects, health hazards, and bioremediation approaches. In: Bharagava, R. N., (ed.), *Environmental Pollutants and their Bioremediation Approaches* (1st edn., pp. 23–56). Boca Raton - FL: CRC Press, Taylor & Francis Group.

69. Saxena, G., Chandra, R., & Bharagava, R. N., (2016). Environmental pollution, toxicity profile and treatment approaches for tannery wastewater and its chemical pollutants. *Rev. Environ. Contam. Toxicol., 240*, 31–69.
70. Saxena, G., Purchase, D., Mulla, S. I., Saratale, G. D., & Bharagava, R. N., (2019). Phytoremediation of heavy metal-contaminated sites: Eco-environmental concerns, field studies, sustainability issues, and future prospects. In: Pim, V., (ed.), *Reviews of Environmental Contamination and Toxicology* (Vol. 249, pp. 71–131). Cham: Springer.
71. Sevik, H., Cetin, M., & Kapucu, O., (2016). Effect of light on young structures of Turkish fir (*Abies nordmanniana* subsp. *Born mulleriana*). *Oxid. Comm., 39*(I–II), 485–492.
72. Shi, Z., Shao, L., Jones, T. P., & Lu, S., (2005). Microscopy and mineralogy of airborne particles collected during severe dust storm episodes in Beijing, China. *Journal of Geophysical Research: Atmospheres, 110*, 8. Article ID: D01303.
73. Shrestha, B., Acosta-Martinez, V., Cox, S. B., Green, M. J., Li, S., & Canas-Carrell, J. E., (2013). An evaluation of the impact of multiwalled carbon nanotubes on soil microbial community structure and functioning. *J. Hazard. Mater., 261*, 188–197.
74. Strek, H. J., Weber, J. B., Shea, P. J., Mrozek, J., & Overcash, E., (1981). Reduction of polychlorinated biphenyl toxicity and uptake of carbon-14 activity by plants through the use of activated carbon. *J. Agr. Food Chem., 29*, 288–293.
75. Su, B., Lin, J., Owens, G., & Chen, Z., (2020). Impact of green synthesized iron oxide nanoparticles on the distribution and transformation of as species in contaminated soil. *Environ. Pollut., 258*, 9. Article ID: 113668.
76. Sun, X. F., Liu, B., Jing, Z., & Wang, H., (2015). Preparation and adsorption property of xylan/poly (acrylic acid) magnetic nanocomposite hydrogel adsorbent. *Carbohydr. Polym., 118*, 16–23.
77. Theng, B. K., & Yuan, G., (2008). Nanoparticles in the soil environment. *Elements, 4*(6), 395–399.
78. Trujillo-Reyes, J., Peralta-Videa, J. R., & Gardea-Torresdey, J. L., (2014). Supported and unsupported nanomaterials for water and soil remediation: Are they a useful solution for worldwide pollution? *J. Hazard. Mater., 280*, 487–503.
79. Tungittiplakorn, W., Lion, L. W., Cohen, C., & Kim, J. Y., (2004). Engineered polymeric nanoparticles for soil remediation. *Environ. Sci. Technol., 38*, 1605–1610.
80. Urbano, F. J., & Marinas, J. M., (2001). Hydrogenolysis of organohalogen compounds over palladium supported catalysts. *J. Mole. Catal. A-Chemica., 173*(1, 2), 329–345.
81. Vaseashta, A., Vaclavikova, M., Vaseashta, S., Gallios, G., Roy, P., & Pummakarnchana, O., (2007). Nanostructures in environmental pollution detection, monitoring, and remediation. *Sci. Technol. Adv. Mater., 8*, 47–59.
82. Vivekananthan, V., Selvapriya, A., Janani, D., & Narendhar, C., (2014). Synthesis of mixed oxides of cerium-iron nanostructures for effective removal of heavy metals from wastewater. *Research Journal of Recent Sciences, 3*, 212–217.
83. Wu, C. Y., Zhuang, L., Zhou, S. G., Yuan, Y., Yuan, T., & Li, F. B., (2013). Humic substance-mediated reduction of iron (III) oxides and degradation of 2, 4-D by an alkaliphilic bacterium, *Corynebacterium humireducens* MFC-5. *Microb. Biotechnol., 6*, 141–149.
84. Xu, J., & Bhattacharyya, D., (2005). Membrane-based bimetallic nanoparticles for environmental remediation: Synthesis and reactive properties. *Environ. Prog., 24*, 358–366.

85. Yan, W., Herzing, A. A., Kiely, C. J., & Zhang, W. X., (2010). Nanoscale zero-valent iron (nZVI): Aspects of the core-shell structure and reactions with inorganic species in water. *J. Contam. Hydrol., 118*(3), 96–104.
86. Yan, W., Lien, H. L., Koel, B. E., & Zhang, W. X., (2013). Iron nanoparticles for environmental clean-up: Recent developments and future outlook. *Environ. Sci. Proc. & Imp., 15*(1), 63–77.
87. Yi, P., Pignatello, J. J., Uchimiya, M., & White, J. C., (2015). Hetero-aggregation of cerium oxide nanoparticles and nanoparticles of pyrolyzed biomass. *Environmental Science & Technology, 49*(22), 13294–13303.
88. Yuhuan, S., Feng, G., Weiwei, Y., & Fayuan, W., (2019). Removal of chromium from a contaminated soil using oxalic acid, citric acid, and hydrochloric acid: Dynamics, mechanisms, and concomitant removal of non-targeted metal. *Int. J. Environ. Res. Public Health, 16*, 11. Article ID: 2771.
89. Zaman, J., & Chakma, A., (1994). Inorganic membrane reactors. *J. Memb. Sci., 92*, 1–28.
90. Zhang, W. X., (2003). Nanoscale iron particles for environmental remediation: An overview. *J. Nanoparticle Res., 5*, 323–332.
91. Zhao, X., Lv, L., Pan, B., Zhang, W., Zhang, S., & Zhang, Q., (2011). Polymer-supported nanocomposites for environmental application: A review. *Chem. Eng. J., 170*, 381–394.
92. Zhao, X., Liu, W., Cai, Z., Han, B., Qian, T., and Zhao, D. An overview of preparation and applications of stabilized zero-valent iron nanoparticles for soil and groundwater remediation, Water Research, *100*, 2016, 245–266.

INDEX

D

E

F

Q

R